PROCESSING FOODS

Quality Optimization and Process Assessment

FOOD ENGINEERING AND MANUFACTURING SERIES

R. Paul Singh, Series Co-Editor
University of California, Davis

Dennis R. Heldman, Series Co-Editor
Weinberg Consulting Group, Inc.
Washington, D. C.

Published Titles
Advances in Food Engineering
R. Paul Singh and M. A. Wirakartakusumah

Transport Phenomena of Foods and Biological Materials
Vassilis Gekas

*Minimal Processing of Foods and Process Optimization:
An Interface*
R. Paul Singh

Forthcoming Titles
Food Engineering: Principles and Applications
B. O. Balaban

Caking Phenomena in Powered Foods
Prabir K. Chandra and Jose Cal-Vidal

PROCESSING FOODS

Quality Optimization and Process Assessment

Edited by

Fernanda A.R. Oliveira, Ph.D.
*Instituto Inter-Universitário de Macau
and Universidade Católica Portuguesa
Porto, Portugal*

Jorge C. Oliveira, Ph.D.
*Instituto Inter-Universitário de Macau
Macau*

With
Marc E. Hendrickx, Ph.D.
*Katholieke Universiteit Leuven
Heverlee, Belgium*

Deitrich Korr, Ph.D.
*Berlin University of Technology
Berlin, Germany*

Leon G.M. Gorris, Ph.D.
*Unilever Research Laboratory
Vlaardingen, The Netherlands*

CRC Press
Boca Raton London New York Washington, D.C.

Contact Editor:	Lourdes Franco
Project Editor:	Ibrey Woodall
Marketing Managers:	Barbara Glunn, Jane Lewis, Arline Massey, Jane Stark
Cover design:	Dawn Boyd

Library of Congress Cataloging-in-Publication Data

Oliveira, Fernanda A. R.
   Processing Foods: quality optimization and process assessment / edited by Fernanda A.R. Oliveira, Jorge C. Oliveira, with Marc E. Hendrickx, Dietrich Knorr, Leon G.M. Gorris.
       p.   cm.
   Includes bibliographical references and index.
   ISBN 0-8493-7905-9 (alk. paper)
   1. Food industry and trade. I. Oliveira, Fernanda A. R. II. Oliveira, Jorge C.
TP370.P64 1999
664—dc21   98-46086
                                                                                                          CIP

No claim to original U.S. Government works
International Standard Book Number 0-8493-7905-9
Library of Congress Card Number 98-46086
Printed in the United States of America 1 2 3 4 5 6 7 8 9 0
Printed on acid-free paper

Preface

In 1993, a workshop was organized in Porto, Portugal, bringing together researchers from Europe and the U.S.A. to discuss the subject of process optimization and minimal processing of foods. The resulting communications were published in a book coedited by R.P. Singh and F.A.R. Oliveira and published by CRC Press in 1994. The basic idea behind the initiative was to foster discussion and cooperation between scientists of different disciplines pertaining to food processing (microbiology, engineering, chemistry, etc.) targeted at the development of food products of higher quality, where the concept of quality is related to the best possible preservation of the original characteristics of the product ("just fresh like" or "just cooked like"). This issue combines the development of novel processing technologies, of combined methods, and of process optimization.

Following this workshop, a concerted action project was promoted in Europe, involving researchers from European Union and Central and Eastern European countries, within the framework of the Copernicus program of the European Commission. The project organized three workshops and other activities from 1995 to 1997 and strengthened the purpose of dissemination of the scientific information generated by many research projects. A detailed presentation of the Copernicus project can be found in Chapter 1 of this book and at www.esb.ucp.pt/Copernicus.

This book combines contributions specifically selected and prepared by the project participants with the purpose of providing a large body of updated information in a style that can assist researchers and industrialists to make use of the new concepts, technologies, and approaches that are at the heart of the concern of modern food research. It is our expectation that it can be a useful tool in the interweaving of scientific and technological information that the increasing multidisciplinarity of food processing and preservation requires. Thus, the reader will largely find reviews of different subjects, as well as some specific studies that clarify the issues and their potential applications. A description of the book's content can be found in Chapter 1.

This book is therefore the result of much effort produced by many researchers, and the committed involvement of all participants in the Copernicus project is gratefully appreciated and acknowledged. Not only the contributors to this book, but many others who participated actively in the project and in its workshops have given invaluable contributions, and a sincere word of gratitude is due to their generosity in exchanging knowledge and skills.

The editors are obviously indebted in particular to the many contributors to this book, to whom the quality of this final work is entirely due.

A very special word must be given to the project area leaders of Thermal Processing (Prof. Marc Hendrickx), High Pressure (Prof. Dietrich Knorr), and Minimal and Combined Processes (Dr. Leon Gorris), whose expert guidance, selection of contributions, and coediting of the texts permitted us to achieve this final result. We would also like to acknowledge the area leaders of Freezing (Prof. Lásló Mészáros) and Drying (Prof. Andrzej Lenart) for the organization of their areas of the project and contribution to the selection of materials for this book.

We would also like to acknowledge Prof. Augusto Medina for his enthusiastic and inspiring networking that were at the origin of all these activities.

For the preparation of this book, management of the project and of its dissemination material, a lot of effort has been kindly provided by Mrs. Isabel Lino, Mr. Eduardo Luís Cardoso, Mrs. Manuela Pascoal, Ms. Assunção Teles, and Mr. Kai Sprecher, and their commitment is gratefully acknowledged.

Finally, we wish to leave a special word of appreciation to the European Commission officials of DG XII, who relentlessly pursue the difficult task of supporting high quality scientific food research in Europe, particularly those of the Copernicus program who accompanied this project, and the Agro-Industrial Research division. An overview of the chapters of this book is sufficient to see how much has been achieved thanks to their professional organization of challenging research programs, namely FLAIR, AIR, and FAIR, which over the past decade have truly changed the way that science can be done in Europe today in an environment of international collaboration. In the person of the Head of Division, Dr. Liam Breslin, we would like to give our sincere gratitude to the officials of this Division.

Dedication

To Catarina, Pedro, Patrícia, Filipe, hoping they will believe that all those times that mummy and daddy are out traveling the world they are actually working.

Editors

Fernanda A. R. Oliveira has a degree in Chemical Engineering from the University of Porto, Portugal (Faculty of Engineering) and a Ph.D. from the University of Leeds, U.K. (Department of Food Science). She is currently Vice President of the Inter-University Institute of Macau and Associate Professor of the College of Biotechnology of the Catholic University of Portugal. Her areas of research are in the field of food engineering, particularly heat and mass transfer applications, experimental design and data analysis methods, postharvest technology, and fresh food chain management. She has coedited one book, coauthored about 50 publications in international peer-reviewed journals, and presented over 100 communications in international congresses. She has also been active in international scientific and academic networks.

Jorge C. Oliveira has a degree in Chemical Engineering from the University of Porto, Portugal (Faculty of Engineering) and a Ph.D. from the University of Leeds, U.K. (Department of Food Science). He is Associate Professor at the Inter-University Institute of Macau and at the College of Biotechnology of the Catholic University of Portugal. His areas of research are in the field of food engineering, particularly enzyme thermal inactivation, process modeling, and application of rheology to food processes and properties. He has coauthored about 35 publications in international peer-reviewed journals and presented over 70 communications in international congresses. He has also been active in international scientific and academic networks.

Coeditors

Marc E. Hendrickx graduated in Food Science and Technology at the Katholieke Universiteit Leuven, where he later obtained a Ph.D. in the same field. He is currently Professor at the Laboratory of Food Technology in the K. U. Leuven Department of Food and Microbial Technology, and his current research unit consists of over 30 people (postdoctoral, doctoral coworkers, technicians, and master students) with a strong focus on high-pressure research. He has concentrated his research in the field of food preservation technology and engineering, with the long-term research objective of developing scientifically based systems, both novel and improved, to evaluate quantitatively the impact of physical preservation methods on food safety and quality in terms of microbiologic, sensorial, nutritional, (bio)chemical, and physical attributes of foods. He has been actively involved in European funded research, coordinating different shared-cost projects (AIR, FAIR), and fellowships (institutional fellowships HCM), and has published extensively in peer-reviewed international journals. In 1995, he received the IFT Samuel Cate Prescott Award for excellence in research and the Octaaf Callebaut Award from the "Koninklijke Academie van Wetenschappen en Schone Kunsten van Belgie" (The Belgian Royal Academy of Sciences and Arts).

Dietrich Knorr has obtained a Dipl. Eng. degree and later a Ph.D. degree (*cum laude*) in Food and Fermentation Technology at the Agricultural University of Vienna, Austria. He is currently Professor and Head at the Food Technology Department, Berlin University of Technology, Germany, Research Professor (Food Processing and Biotechnology) at the Department of Animal Science and Food Science, University of Delaware, U.S.A., and Director of the Institute of Food Technology, Berlin University of Technology. His areas of research span Food Process Engineering (inactivation kinetics of microorganisms and enzymes, process/product development, equipment design, high-pressure freezing/thawing, and thermal processing of foods) and Food Biotechnology (plant cell and tissue culture model systems, secondary metabolite biosynthesis, bioactive substances, and biopolymers). He has published extensively in the scientific literature (over 250 publications) and has several editorial tasks in major journals in the food science area, being the Editor of *Food Biotechnology*.

Leon G. M. Gorris graduated in Biology at the Catholic University of Nijmegen, where he later gained a Ph.D. degree in Microbiology. In 1990, he joined the Agrotechnological Research Institute (ATO-DLO) in Wageningen (the Netherlands), where he established the Department of Food Safety and Applied Microbiology. The main research areas in this department were mild food preservation systems, novel preservation methods, combined processing ("Hurdle Technology"), green chemicals, biological crop protection, detection of microorganisms, and health aspects of bioactive natural compounds. At ATO-DLO, he coordinated four shared-cost projects funded by the European Commission (AIR, FAIR, and INCO-A1 programs) and was a member of eight other shared-cost and concerted actions in these EU programs and in the COST program. In 1998, he moved to Unilever Research Laboratory, Vlaardingen (the Netherlands), where he heads the Microbiology and Preservation unit which operates mainly in the food preservation arena.

Contributors

Günter van Almsick
Intitut für Pharmazeutische
 Technologie und Biopharmazie
Gruppe Physikalische Chemie
Universität Heidelberg
Heidelberg, Germany

Éva Andrássy
Dept. of Refrigeration and Livestock
 Products Technology
University of Horticulture and
 Food Industry
Budapest, Hungary

Ana Andrés
Dept. Tecnología de Alimentos
Universidad Politecnica de Valencia
Valencia, Spain

Jacek Arabas
High Pressure Research Center
Polish Academy of Sciences
Warsaw, Poland

Ioannis Arvanitoyannis
Lab. of Food Chemistry and
 Biochemistry
Dept. of Food Science and
 Technology
School of Agriculture
Aristotle University of Thessaloniki
Thessaloniki, Greece

Isabel M.L.B. Ávila
Escola Superior de Biotecnologia
Universidade Católica Portuguesa
Porto, Portugal

José M. Barat
Dept. Tecnología de Alimentos
Universidad Politecnica de Valencia
Valencia, Spain

Jeffrey K. Brecht
Dept. of Horticultural Sciences
University of Florida
Gainesville, FL

Pilar M. Cano
Dept. of Plant Foods Science and
 Technology
Instituto del Frío
Madrid, Spain

Khe V. Chau
Dept. of Agricultural and Biological
 Engineering
University of Florida
Gainesville, FL

Amparo Chiralt
Dept. de Tecnología de Alimentos
Universidad Politécnica de Valencia
Valencia, Spain

Isabel B. Cruz
Escola Superior de Biotecnologia
Universidade Católica Portuguesa
Porto, Portugal

Begoña De Ancos
Dept. of Plant Foods Science and
  Technology
Instituto del Frío
Madrid, Spain

Josse De Baerdemaeker
Dept. of Agro-Engineering and
  Economics
Katholieke Universiteit Leuven
Heverlee, Belgium

Stefan Ditchev
Dept. of Food Engineering
  Equipment
Higher Institute of Food and Flavour
  Industries
Plovdiv, Bulgaria

József Farkas
Dept. of Refrigeration and Livestock
  Products Technology
University of Horticulture and Food
  Industry
Budapest, Hungary

Pablo S. Fernández
Universidad Miguel Hernández
Alicante, Spain

Pedro Fito
Dept. Tecnología de Alimentos
Universidad Politecnica de Valencia
Valencia, Spain

Monika Fonberg-Broczek
High Pressure Research Center
Polish Academy of Sciences
Warsaw, Poland

Susana C. Fonseca
Escola Superior de Biotecnologia
Universidade Católica Portuguesa
Porto, Portugal

Jesus M. Frías
Escola Superior de Biotecnologia
Universidade Católica Portuguesa
Porto, Portugal

Vassilis Gekas
Dept. of Food Engineering
University of Lund
Lund, Sweden
(currently at: Lab. of Transport
  Phenomena and Unit
  Operations
Dept. of Environmental
  Engineering
Technical University of Crete
Chania, Greece)

T. Ronan Gormley
Teagasc
The National Food Centre
Dunsinea
Castleknock
Dublin, Ireland

Leon G.M. Gorris
Agrotechnological Research Institute
  (ATO-DLO)
Bornsesteeg Wageningen
The Netherlands (currently at: Unit
  Microbiology and Preservation
Unilever Research Laboratory
Vlaardingen, The Netherlands)

Tamara Haentjens
Dept. of Food and Microbial
  Technology
Katholieke Universiteit Leuven
Heverlee, Belgium

Marc E. Hendrickx
Dept. of Food and Microbial
  Technology
Katholieke Universiteit Leuven
Heverlee, Belgium

Karel Heremans
Dept. of Chemistry
Katholieke Universiteit Leuven
Leuven, Belgium

Almudena Hernández
Dept. of Plant Foods Science and
  Technology
Instituto del Frío
Madrid, Spain

Krisztina Horti
Dept. of Refrigeration and Livestock
  Products Technology
University of Horticulture and
  Food Industry
Budapest, Hungary

Leonard Ilincanu
Faculty of Food Science
Aquaculture and Fishing
University "Dunãrea de Jos"
Galati, Romania

Dietrich Knorr
Dept. of Food Biotechnology and
  Food Process Engineering
Berlin University of Technology
Berlin, Germany

Ágota Koncz
Dept. of Refrigeration and Livestock
  Products Technology
University of Horticulture and
  Food Industry
Budapest, Hungary

E. Kostrzewa
Institute of Agricultural and
  Food Biotechnology
Warsaw, Poland

Harris N. Lazarides
Dept. of Food Science and
  Technology
School of Agriculture
Aristotelian University of
  Thessaloniki
Thessaloniki, Greece

Andrzej Lenart
Dept. of Food Engineering
Faculty of Food Technology
Warsaw Agricultural University
Warszawa, Poland

Isabel B.M. Lino
Escola Superior de Biotecnologia
Universidade Católica Portuguesa
Porto, Portugal

Horst Ludwig
Intitut für Pharmazeutische
  Technologie und Biopharmazie
Gruppe Physikalische Chemie
Universität Heidelberg
Heidelberg, Germany

William M. MacInnes
Nestlé Research Centre
Lausanne, Switzerland

Antonio Martínez
Instituto de Agroquímica y
  Tecnología de Alimentos
Valencia, Spain

Javier Martínez-Monzó
Dept. Tecnología de Alimentos
Universidad Politecnica de Valencia
Valencia, Spain

Nuria Martínez-Navarrete
Dept. Tecnología de Alimentos
Universidad Politecnica de Valencia
Valencia, Spain

F. Meersman
Dept. of Chemistry
Katholieke Universiteit Leuven
Leuven, Belgium

László Mészáros
Dept. of Refrigeration and Livestock
  Products Technology
University of Horticulture and
  Food Industry
Budapest, Hungary

Csilla Mohácsi-Farkas
Dept. of Microbiology and
  Biotechnology
University of Horticulture and
  Food Industry
Budapest, Hungary

Bart M. Nicolaï
Dept. of Agro-Engineering and
  Economics
Katholieke Universiteit Leuven
Heverlee, Belgium

María J. Ocio
Instituto de Agroquímica y
  Tecnología de Alimentos
Valencia, Spain

Thomas Ohlsson
The Swedish Institute for Food
  and Biotechnology
Göteborg, Sweden

Fernanda A.R. Oliveira
Escola Superior de Biotecnologia
Universidade Católica Portuguesa
Porto, Portugal
Currently at: SPI — Sociedale
  Portuguesa de Inovação
Porto, Portugal

Jorge C. Oliveira
Instituto Inter-Universitário de
  Macau
Macau
Currently at: SPI — Sociedale
  Portuguesa de Inovação
Porto, Portugal

Pedro M. Pereira
Escola Superior de Biotecnologia
Universidade Católica Portuguesa
Porto, Portugal

Dariusz Piotrowski
Dept. of Food Engineering
Faculty of Food Technology
Warsaw Agricultural University
Warszawa, Poland

Sylwester Porowski
High Pressure Research Center
Polish Academy of Sciences
Warsaw, Poland

Arnold Reps
Institute of Food Biotechnology
Olsztyn University of Agriculture
  and Technology
Olsztyn, Poland

Philip Richardson
Dept. of Food Process
  Engineering
Campden & Chorleywood Food
  Research Association
United Kingdom

Carmen Rodrigo
Dept. de Medicina Preventiva y
  Salud Pública
Bromatología
Toxicología y Medicina Legal
Facultad de Farmacia
Universitat
Valencia, Spain

Francisco Rodrigo
Instituto de Agroquímica y
  Tecnología de Alimentos
Burjassot
Valencia, Spain

Miguel Rodrigo
Instituto de Agroquímica y
  Tecnología de Alimentos
Valencia, Spain

P. Rubens
Dept. of Chemistry
Katholieke Universiteit Leuven
Leuven, Belgium

Tamás Sáray
Dept. of Refrigeration and Livestock
  Products Technology
University of Horticulture and Food
  Industry
Budapest, Hungary

Nico Scheerlinck
Dept. of Agro-Engineering and
  Economics
Katholieke Universiteit Leuven
Heverlee, Belgium

Christian Schreck
Intitut für Pharmazeutische
  Technologie und Biopharmazie
Gruppe Physikalische Chemie
Universität Heidelberg
Heidelberg, Germany

Cristina L.M. Silva
Escola Superior de Biotecnologia
Universidade Católica Portuguesa
Porto, Portugal

L. Smeller
Semmelweiss Medical University
Institute of Biophysics
Budapest, Hungary

Chantal Smout
Dept. of Food and Microbial
  Technology
Katholieke Universiteit Leuven
Heverlee, Belgium

J. Snauwaert
Dept. of Chemistry
Katholieke Universiteit Leuven
Leuven, Belgium

J. Szczawinski
Dept. of Food Hygiene
Faculty of Veterinary Medicine
Warsaw Agricultural University
Warsaw, Poland

Janusz Szczepek
High Pressure Research Center
Polish Academy of Sciences
Warsaw, Poland

Bernhard Tauscher
Institute of Chemistry and
  Biology
Federal Research Centre for
  Nutrition
Karlsruhe, Germany

Ann Van Loey
Dept. of Food and Microbial
  Technology
Katholieke Universiteit Leuven
Heverlee, Belgium

Pieter Verboven
Dept. of Agro-Engineering and
  Economics
Katholieke Universiteit Leuven
Heverlee, Belgium

G. Vermeulen
Dept. of Chemistry
Katholieke Universiteit Leuven
Leuven, Belgium

Bozena Windyga
National Institute of Hygiene
Warsaw, Poland

Erna Zackel
Dept. of Refrigeration and Livestock
  Products Technology
University of Horticulture and
  Food Industry
Budapest, Hungary

Contents

Chapter 1   Process optimization and minimal processing of foods:
an international collaborative approach in the framework of the
EU Copernicus program...1
Fernanda A.R. Oliveira, Isabel B.M. Lino, and Jorge C. Oliveira

Chapter 2   Enzymic time–temperature integrators for the
quantification of thermal processes in terms of food safety 13
*Ann Van Loey, Tamara Haentjens, Chantal Smout,
and Marc E. Hendrickx*

Chapter 3   Possibilities and limitations of computational fluid
dynamics for thermal food process optimization 41
*Pieter Verboven, Nico Scheerlinck, Josse De Baerdemaeker,
and Bart M. Nicolaï*

Chapter 4   Methodologies to optimize thermal processing
conditions: an overview 67
Isabel M.L.B. Ávila and Cristina L.M. Silva

Chapter 5   Microbial and quality assessment in thermal
processing...83
*António Martínez, Carmen Rodrigo, Pablo S. Fernández,
Maria J. Ocio, Francisco Rodrigo, and Miguel Rodrigo*

Chapter 6   Minimal processing of foods with electric
heating methods..97
Thomas Ohlsson

Chapter 7   Application of the concepts of biomaterials science
to the quality optimization of frozen foods 107
*Jorge C. Oliveira, Pedro M. Pereira, Jesús M. Frías, Isabel B. Cruz,
and William M. MacInnes*

Chapter 8   The influence of freezing and frozen storage time
on structural and other changes in plant tissue..........................131
Tamás Sáray, Krisztina Horti, Erna Zackel and Ágota Koncz

Chapter 9   Intensification of freezing...145
Stefan Ditchev and Philip Richardson

Chapter 10   Developments in fish freezing in Europe with
emphasis on cryoprotectants..163
T. Ronan Gormley

Chapter 11   Advances in osmotic dehydration...........................175
*Harris N. Lazarides, Pedro Fito, Amparo Chiralt, Vassilis Gekas,
and Andrzej Lenart*

Chapter 12   Rehydration of dried plant tissues: basic concepts
and mathematical modeling...201
Fernanda A.R. Oliveira and Leonard Ilincanu

Chapter 13   Recent advances in the drying of apples under
variable process conditions...229
Dariusz Piotrowski and Andrzej Lenart

Chapter 14   Process assessment of high-pressure processing
of foods: an overview 249
Dietrich Knorr

Chapter 15   A comparison between pressure and temperature
effects on food constituents 269
*Karel Heremans, F. Meersman, P. Rubens, L. Smeller, J. Snauwaert,
and G. Vermeulen*

Chapter 16   High-pressure treatment of fruit, meat, and
cheese products — equipment, methods, and results....................281
*Monika Fonberg-Broczek, Jacek Arabas, Ewa Kostrzewa,
Arnold Reps, Jacek Szczawinski, Janusz Szczepek,
Bozena Windyga, and Sylwester Porowski*

Chapter 17   Combined high-pressure/temperature treatments
for quality improvement of fruit-derived products.....................301
Pilar M. Cano, Almudena Hernández, and Begoña De Ancos

Chapter 18   Influence of culturing conditions on the pressure
sensitivity of *Escherichia coli* 313
Christian Schreck, Günter van Almsick, and Horst Ludwig

Chapter 19   Quality and safety aspects of novel minimal
processing technologies 325
Leon Gorris and Bernhard Tauscher

Chapter 20   Vacuum impregnation: a tool in minimally
processing of foods..341
*Amparo Chiralt, Pedro Fito, Ana Andrés, José M. Barat,
Javier Martínez-Monzó, and Nuria Martínez-Navarrete*

Chapter 21   Edible and biodegradable polymeric materials
for food packaging or coating 357
Ioannis Arvanitoyannis and Leon Gorris

Chapter 22   The role of ionizing radiation in minimal processing
of precut vegetables with particular reference to the control
of *Listeria monocytogenes*...373
*József Farkas, László Mészáros, Csilla Mohácsi-Farkas,
Tamás Sáray, and Éva Andrássy*

Chapter 23   Development of perforation-mediated modified
atmosphere packaging for fresh-cut vegetables 389
*Susana C. Fonseca, Fernanda A.R. Oliveira, Jeffrey K. Brecht,
and Khe V. Chau*

Index 405

chapter one

Process optimization and minimal processing of foods: an international collaborative approach in the framework of the EU Copernicus program

Fernanda A.R. Oliveira, Isabel B. M. Lino, and Jorge C. Oliveira

Contents

Summary 1
1 Introduction 2
2 The participants and management 3
3 The plugged-in research projects 6
   3.1 General overview 6
   3.2 Topical issues in the project and in this book 8
4 The activities 10
   4.1 The workshops 11
   4.2 The program of short stays 11
   4.3 The dissemination 11

Summary

Copernicus is the acronym for European Union (EU) Cooperation in Science & Technology with Central and Eastern European countries. One of the projects developed in the framework of this program studied Process Optimization and Minimal Processing of Foods, embracing 41 institutions from

0-8493-7905-9/99/$0.00+$.50
© 1999 by CRC Press LLC

17 European countries (10 EU member states, namely Belgium, France, Germany, Greece, Ireland, the Netherlands, Portugal, Spain, Sweden, and United Kingdom, and Bulgaria, Czech Republic, Hungary, Lithuania, Poland, Romania, and Slovenia). The project ran from January 1995 to June 1998 and operated using the concept of plugged-in research projects, i.e., participants included their ongoing research projects and the results were shared and discussed at meetings with a workshop format (one per year) and were also disseminated via personal contacts and through the Flair-Flow Europe dissemination project. Seventy four plugged-in projects were subdivided in five areas, namely, thermal processing, freezing, drying, high pressure, and minimal and combined processes. The workshops were held in 1995 in Porto (Portugal), in 1996 in Warsaw (Poland), and in 1997 in Leuven (Belgium). To increase interaction and concertation between partners, a program of short stays of participants from EU institutions in Eastern Institutions or vice-versa was also set up. The proceedings of the yearly meetings were published and the abstracts are available through the Internet (www.esb.ucp.pt/copernicus/). This book includes a number of selected papers resulting from this project and specifically written to foster wider dissemination, some of them involving collaboration between researchers of different institutions.

1. Introduction

Food processing involves a wide variety of industrial processes with a correspondingly large variety of products. One of the main functions served by food processing is the preservation of foods, which involves two main aspects: preserving fresh products and preserving cooked products. Optimum quality, as perceived by the consumer, relates most usually to the maintenance of product attributes in relation to the "original" — in the former case one can refer to "fresh-like"characteristics and in the latter to "just cooked-like" attributes. These include textural sensations, color, and aroma, but other lesser noticeable properties such as wholesomeness and nutritional quality are also of increasing importance, in addition to safety concerns.

The need for improved quality of processed foods continues to grow with consumers' demands for increased convenience and improved quality at an affordable cost. The food industry is challenged to produce new products with longer shelf life that also possess quality attributes that are more reminescent of the fresh or native state of a given food. These requirements may be met by (1) minimizing the detrimental influence of processing treatments on foods and (2) developing new technologies.

One of the topics where the general drive for improved quality has led to more important changes is the so-called minimally (or lightly) processed foods. Using new technological solutions, such as modified atmosphere packaging, new packaging materials and technologies, or combinations of different (mild) treatments, a variety of products much closer to fresh and wholesome products is increasingly reaching the market. High-pressure treatment is another very promising technology, of a more severe nature, but that has proven successful in preserving quality attributes and for that reason is sometimes included in this category, with "mild" referring to processing impact.

However, it must be borne in mind that the development of such new technologies is being made while facing a very conventional industry sector, where less technological ability means that optimization of conventional technologies will have greater potential impact, in the short term. Such lower technological innovation capacity is particularly found in peripheral regions (including Eastern Europe) and SME's (which in Southern European countries and Ireland are a backbone of industrial production and employment in the food sector).

The project that eventually gave rise to this book was set up with the view of bringing together a large number of possible solutions and pertinent information for the task of minimizing the impact of food handling and processing on product quality characteristics, either focusing on the production of fresh-like products or on the production of processed products with minimal deleterious changes. The most relevant work carried out by more than 40 institutions was concerted in this single action, joining updated scientifical and technological information on how to preserve foods by physical processes (thermal processing, freezing, drying, high pressure, and minimal and combined processes) with optimum quality characteristics. The overall objectives of this concertation were: (1) coordination of research efforts, (2) improvement of information flow, (3) dissemination to Central and Eastern European countries of the most advanced research efforts financed by the European Commission and (4) promotion of collaborative research projects.

2 The participants and management

The project was coordinated by the College of Biotechnology (ESB) of the Catholic University of Portugal and involved 45 other institutions from 17 European countries (see Table 1, a and b). It was divided into five areas, as above referred, with each coordinated by an area leader:

- Thermal processing: Marc E. Hendrickx, Catholic University of Leuven (Belgium)
- Freezing: László Mészáros, University of Horticulture and Food Industry (Hungary)
- Drying: Andrezj Lenart, Agricultural University of Warsaw (Poland)
- High pressure: Dietrick Knorr, Technical University of Berlin (Germany)
- Minimal and combined processes: Leon Gorris, ATO-DLO (the Netherlands)

The area leaders together with the project coordinator, Fernanda A.R. Oliveira from ESB, constituted the Steering Committee.

Table 1 Partners of the Copernicus Project

Country	Institution	Department	City
a. EU countries			
Belgium	Alma University Restaurant VZW	Sous Vide Competence Center	Leuven
	University of Ghent	Food Technology & Nutrition	Ghent
	Catholic University of Leuven	Agro-Engineering & Economics	Leuven
		Chemistry	Leuven
		Food & Microbial Technology	Leuven
France	Danone Group	Direction Scientifique	Paris
	Nat. Higher Ed. Sch. Ag.-Fd. Ind.	(ENSIA), Dpt. Food Industry	Massy
	Nat. Inst. Agronomic Res. (INRA)	Lab. Food Proc. Eng. & Tech.	Lille
	University of Montpellier II	Food Biochem. & Technology	Montpellier
Germany	Bundesforsch. fur Ernahrung (BFE)	Inst. Chemistry & Biology	Karlsruhe
	Technical University of Berlin	Food Technology	Berlin
	University of Heidelberg	Inst. Pharm. Tech. & Bioph.	Heidelberg
Greece	Aristotelean Univ. of Thessaloniki	Food Science & Technology	Thessaloniki
	National Agricultural Foundation	Inst. Tech. Ag. Products	Athens
	National Tech. University of Athens	Food Chemistry & Tech.	Athens
Ireland	National Food Center (TEAGASC)		Dublin
	University College Cork	Food Chemistry	Cork
Netherlands	Dienst Landbouwkundig Ond. (DLO)	Agro-Tech. Res. Inst. (ATO)	Wageningen
Portugal	Catholic University of Portugal	College of Biotechnology (ESB)	Porto
	Nat. Inst. Ind. Eng. & Tech. (INETI)	Inst. Biotech., Chem. & Fd. Tech.	Lisbon
	Port. Inst. Marine Res. (IPIMAR)	Valorization of Marine Resources	Lisbon

Country	Organization	Department	City
Spain	High Council of Sci. Res. (CSIC)	Cent. App. Edaf. & Safety (CEBAS)	Murcia
		Inst. Ag-Chem. & Fd. Tech. (IATA)	Valencia
		Inst. Refrigeration (Instituto del Frio)	Madrid
	Polytechnic University of Valencia	Food Technology	Valencia
Sweden	Swed. Inst. Fd. & Biotech. (SIK)	Processing Systems Division	Goteborg
	Tetra-Pak	Food Engineering	Lund
	University of Lund		Lund
U.K.	Campden & Chorleywood Fd. R. A.	Food Processing Engineering	Campden
	Unilever Research	Colworth Laboratory	Bedford
	University of Reading	Food Science & Technology	Reading

b. Central and Eastern European countries

Country	Organization	Department	City
Bulgaria	Higher Inst. of Food and Flavour Ind.	Food Engineering Equipment	Plovdiv
		Food Preservation	Plovdiv
Czech R.	Institute of Chemical Technology	Food Preserv. & Meat Tech.	Prague
Hungary	Univ. of Horticulture & Food Ind.	Refrig. & Livestock Prod. Tech.	Budapest
Lithuania	Technical University of Kaunas	Food Technology	Kaunas
Poland	Agricultural University of Warsaw	Food Engineering	Warsaw
	Polish Academy of Sciences	High Pressure Res. Center	Warsaw
	Technical University of Gdansk	Food Engineering	Gdansk
Romania	University of Galati	Fd. Ind., Aquac. & Sea Prod.	Galati
Slovenia	University of Ljubljana	Food Science & Technology	Ljubljana

3 The plugged-in research projects

3.1 General overview

During the lifetime of the project, over 70 collaborative or individually developed research projects of greater interest to the research groups formed the basis of the oral and poster presentations at the meetings (see Tables 2a through e).

Table 2   Individual Research Projects

Subjects

a. thermal processing

Heat distribution and heat penetration of modern overpressure rotary retorts
Comparison of thermobacteriological models
Contamination of foods by packaging materials during heat treatment
Pasteurization of tropical fruit juice
Enzyme and vitamin based product history indicators
Test methods for assessment of food processing equipment cleanability
Tubular heat exchangers for processing non-Newtonian fluids with solid particles
Experimental validation of optimum thermal processing conditions
High-temperature thermal resistance studies of microorganisms
Thermal degradation of anthocyanins during in-pack sterilization
Thermal degradation kinetics of carotenoids
Microwave sterilization and consumer interactions
Continuous thermal treatment of viscous products with solid particles
Process validation of microwave sterilization
Protein-mineral interactions and raw milk quality in aseptic processing
Thermal processing control in case of arbitrary temperature fluctuations
Thermal resistance of spore-forming bacteria
Development of new time–temperature integrators
Thermobacteriology in acidified mushrooms
Uniformity and realiability of the heat transfer to food products in ovens
Using anthocyanins, ascorbic acid, and other indicators of the degree of sterility

b. freezing

Freezing suitability of Spanish berry cultivars
Cryoprotectants and frozen fish mince
Process chains of osmo-chilling/freezing of fruits and vegetables
Evaluation of shelf life of frozen (thin flank) pork
Influence of freezing and frozen storage in the color, flavor, and nutritive value of fruits
Integrated freezing-thawing study
Methods and technologies for freezing intensification
Modeling and quality of frozen foods
Nonsensory techniques to monitor the quality of fish during frozen storage
Physical treatments, ultrasounds, ion radiation, and high-frequency power in food freezing
Physicochemical properties of starch hydrolyzis products

Table 2 (continued) Individual Research Projects

Subjects

c. drying

A novel approach to preserve the intrinsic quality of fruits and vegetables
The role of some drying procedures on aroma retention
Agglomeration and coating of powders to improve and change physicochemical
 properties
Air-suspension microencapsulation of food ingredients and additives
Analysis of mass transfer and structural changes during rehydration of dried apples
Controlled instantaneous flashing (Dic process) for the puffing of instant dry vegetables
Design and optimization of sublimation for continuous freeze-drying processes
Dispersion and immobilization of essential oils and extracts
Mass transfer analyis in frying of potatoes
New drying technologies
New methods of microencapsulation with modified biopolymers
Osmotic dehydration of fruits and vegetables
Solute diffusivity and conductivity in cellular structures
Effect of agglomeration on the physical properties of milk powder
Influence of drying on the flavor and antioxidant substances in spices and herbs
Influence of step changes in the medium temperature and velocity on drying kinetics

d. high pressure

Changes in color and myoglobin of minced beef meat due to high-pressure processing
Chemical changes in food under the conditions of ultrahigh-pressure processing
Effect of high-pressure treatment on gelled fruit products and sliced meat products
Development of a high-pressure laboratory and industrial food processing unit
Food preservation by ultrahigh pressure
Process assessment of high-pressure processing: kinetics of enzymic systems in UHP
High-pressure effects on food related biomacromolecules
Storage behavior of high-pressure-treated fruits and vegetables

e. minimal and combined processes

Perforation mediated modified atmosphere packaging for fruits and vegetables
Application of low-dose gamma irradiation to vacuum-packed chilled minced meat
Modified atmosphere technology to obtain a new ready-to-use vegetable product
Cavitation treatment of liquid foods
Changes in the internal atmosphere during storage of packed meat products
Combination of gamma irradiation and MAP to improve the shelf life of fresh pork
Development of minimally processed products by vacuum impregnation
Enzymatic ripening of pelagic fish species
Harmonization of safety criteria for minimally processed foods
Improvement of the safety and quality of refrigerated ready-to-eat foods
Enzymatic maceration for fruit juice production
Optimization of controlled atmosphere storage of fruits and vegetables
Preservation of fresh fish and vegetables by application of VP/MAP
Process options for intermediate moisture level products
Key-quality (bio)chemical indicators in minimal processed fruits and vegetables
Changes of fruit-vegetable-raw materials during storage aand processing
The microbial safety and quality of foods processed by the "sous-vide" system
The optimal solution of modified atmospheres

The purpose of this concertation was to promote the integration of knowledge drawn from different disciplines required to design quality-efficient food processes for the modern food industry and consumer markets. This integrated perspective allowed focus on two major concerns: careful, wide-ranging evaluation of processing impact, especially of new technologies, on food quality and safety; and effective dissemination to improve the knowledge base available to the food industry and the potential application of the novel approaches and technologies.

Detailed information on these projects can be found in the project booklets or obtained from the Internet site previously mentioned.

3.2   Topical issues in the project and in this book

An integrated perspective of the issues that raised wider interest can be obtained from analyzing the contents of this book.

Thermal processing is one of the more traditional forms of preservation, but it affects the quality of food products significantly. Minimization of the detrimental effects of heating while maintaining the lethality target is possible, as reviewed in Chapter 4. An accurate, safe, and reliable process design depends on the ability to assess the treatment delivered in terms of the impact on both quality and safety factors, as discussed in Chapters 2 and 5. Novel assessment methods are particularly important in many new technologies and processes, where conventional physical-mathematical methods cannot be used due to equipment restrictions. Quality optimization greatly depends on the ability to process foods in high-temperature short-time regimes, which is currently widely done for liquid foods and potentially to liquids with particles. Questions related to heat exchangers and to their efficiency in delivering safe, high-quality products were approached in some communications of the project concerned with equipment design and also with maintenance (fouling), though these are not being covered in the book.

A similarly great impact on product quality for large solid products that basically heat by conduction depends on the ability to develop suitable industrial systems based on new heating technologies, generally called volumetric heating (e.g., microwave, radio-frequency, ohmic), reviewed in Chapter 6. A thorough analysis of the thermal treatment delivered is particularly important in milder thermal treatments, and the analysis of the variability of the conditions imposed throughout the equipment is essential for ensuring safety without gross overprocessing. A mathematical approach using computational fluid dynamics (CFD) is discussed in Chapter 3. Greater understanding of the response of microbial cells to thermal stress is greatly needed, and it would be advantageous to escape from the oversimplification of current design strategies based on first-order death rates, worst scenarios, and, in general, deterministic principles, although a lot of work would still be required to build up a sufficient knowledge base for alternative approaches.

The use of milder thermal treatments associated with high pressure or other preservation hurdles are included in some communications of high pressure and minimal and combined processes, respectively.

In freezing, the concern for product quality is associated both with obtaining high freezing rates and preserving the high quality achieved in quality-efficient freezing technologies throughout storage and distribution, as discussed in Chapter 7. In relation to freezing rates, the intensification of air freezing following a new concept is discussed in Chapter 9, where the need for integrating storage in the final product quality assessment is also clearly presented. The dynamic dispersion medium concept is an inexpensive way to improve freezing rates in air blast systems. The effect of freezing on foods at cellular level is discussed in Chapter 8, for vegetable products, where the impact on texture can be more detrimental for sensory quality. The importance of the matrix characteristics on keeping quality is clarified in Chapters 7 and 10.

Many approaches to improving the quality of dried products were discussed in the project, with the greater concern being put on the preservation of aroma. There are many ongoing research areas, and in this book two major aspects are covered. Osmotic dehydration is gaining increasing interest due to its role as a pretreatment that can be part of effective integrated quality strategies for improving the characteristics of the matrix (as for freezing/thawing) or as one hurdle in a combined processes approach. The concepts and applications are discussed in Chapter 11. For many products, rehydration is just as important as drying in terms of the final consumer assessment of the product quality. Fundamentals and kinetics of rehydration are reviewed in Chapter 12. Chapter 13 shows the impact of process variables on product history in air drying.

High pressure is one of the more topical issues on food research today, not only due to its applications in providing high-quality products, but also as a different processing concept altogether, requiring more understanding of the underlying phenomena and responses, as they may enclose innovative routes to obtain truly novel food products. Assessment of high-pressure processing is reviewed in Chapter 14. The effect of high pressure on food composition is discussed in Chapter 15, comparing it to the better known thermal treatment effects. The impact of high pressure in a number of food products is examplified in Chapters 16 and 17. More details of microbial response under high-pressure stress are analyzed in Chapter 18, for *Escherichia coli*, where intrinsic variability factors that yield different responses are studied. The general concerns in relation to high pressure are the improvement of the lethality effect and the effectiveness of enzyme inactivation, as patent throughout these chapters.

A wide perspective on combined processes and the application of the hurdle concept is given in Chapter 19. The combination of different mild treatments is a promising way to obtain safe products of high quality and convenience. Many hurdles were discussed in the project, and in this book, three are discussed in more detail. The use of low-dose irradiation is discussed in Chapter 22 from a specific study concerned with *Listeria monocytogenes* in precut vegetables. Packaging and storage conditions play a major role in a combined processing strategy, as minimally processed foods are generally not stable. The potential of edible coatings is discussed in Chapter 21 while Chapter 23 discusses perforated-mediated packaging, a new form of promoting modified atmosphere that is applicable to large containers, using rigid impermeable materials with plug implements to control gas exchange.

It can be seen that in general the great concern of the researchers involved in the project is in general to obtain food products of higher perceived quality, improving the factors that are more important for consumer assessment and hence market success. This project, and this book as an image of it, basically put together engineering/technology approaches with scientific/fundamental understanding of microorganisms and foods with a view to design industrially applicable solutions able to generate such products. The communications presented in the project as well as in this book show that this concern for higher perceived quality goes well outside the factory floor. Food producers are increasingly unable to detach market success from the product history between factory and consumer. This adds a technological aspect to what already is a key economical issue: the need to improve the complementarity and cooperation between production and distribution as a basis for fulfilling efficiently the consumer demands.

Worldwide, the food industry is facing relatively stagnant markets in terms of quantity — the most important markets for most food companies still are in the part of the world where population is not growing. Economic success is greatly linked to the ability to pay off investments at attractive rates and is basically associated with growth scenarios. Therefore, the interest in market segments involving higher added-value products and with great growth potential is evident. The changes in lifestyles, consumption habits and patterns, and in the diet-health perception are the basic contours of such opportunities. Hence, the focus of researchers and industrialists on technologies and processing approaches and methodologies that can efficiently lead to adequate responses to the newer market demands. While it is likely that the bulk of food production may differ slightly from what it is today for many years, the greater interest for development lies in paths that may lead us to a completely new way of understanding and producing food products, even if market niches are at the moment comparatively small.

4   The activities

The main project activities were meetings (workshops), short stays, and dissemination.

4.1 The workshops

The main objective of these meetings was to provide a forum for open presentation and discussion of the research performed. The flow of information from experts to nonexperts was a basis of the exchange intended and therefore the meetings were organized around a series of plenary lectures, which were supplemented by specific short oral communications and poster presentations. The meetings had a workshop-type format and were divided into five sections, each devoted to a different project area. The meetings had 80 to 100 participants, with a total of 241 communications (59 oral communicatons, 182 posters). The abstracts of all comunications can be freely downloaded from the Internet site indicated in the Summary.

4.2 The program of short stays

This program was set up with the objective of promoting further interaction between East-West research groups by bringing together partners with similar interests. Thirty-four visits of 1 to 2 weeks took place along the project duration, involving more than twenty scientists. This program was very well received by the Central-Eastern European institutions in particular, with visits to EU partners accounting for 74% of the total number of visits.

4.3 The dissemination

The information pertaining to the different research subprojects was compiled in a booklet that includes the name and affiliation of the contact person, an overview of the project status, work performed and achievements to date, a description of the links to international collaborative projects, and future perpectives and publications. Information on the different institutions was also compiled in a booklet containing the name of the contact person and full address, a short presentation of the institution, a list of other researchers involved in the project, and a description of current interests in collaboration. Books of abstracts and proceedings were also published for every meeting. This information was widely distributed, and it is also currently available on the Internet, from where it can be freely downloaded (site address mentioned in the Summary).

In addition, at the end of the project an open seminar was organized, where selected projects were presented, together with other collaborative EU research projects in the areas covered by the Copernicus project. A book of abstracts was also prepared for this seminar, giving an overview of the current collaborative research ongoing in the EU.

chapter two

Enzymic time–temperature integrators for the quantification of thermal processes in terms of food safety

Ann Van Loey, Tamara Haentjens, Chantal Smout,
and Marc E. Hendrickx

Contents

Summary 14
1   Methods for process impact assessment: the need for TTIs 14
2   General aspects on time–temperature integrators 15
    2.1   Definition of a time–temperature integrator 15
    2.2   Criteria for a time–temperature integrator 15
    2.3   Classification of time–temperature integrators 16
3   Feasibility of thermostable enzymes for TTI development 17
4   Research objective 18
    4.1   Pasteurization processess 18
    4.2   Sterilization processes 18
5   Development of enzymic isolated extrinsic time–temperature
    integrators 19
    5.1   Isothermal calibration of enzymic time–temperature
          integrators 19
          5.1.1   Materials and methods 19
          5.1.2   Results and discussion 21
    5.2   Evaluation of integrating properties under variable
          temperature conditions 25
          5.2.1   Materials and methods 25
          5.2.2   Results and discussion 25

6   Application of enzymic time–temperature integrators for thermal
    process evaluation...29
    6.1   Determination of the coldest point in a particulate model
          food system...30
          6.1.1   Materials and methods 30
          6.1.2   Results and discussion..31
    6.2   Determination of the coldest zone in a retort...................................32
          6.2.1   Materials and methods 32
          6.2.2   Results and discussion..33
7   Conclusions...35
Acknowledgments...36
References...36

Summary

With the further development of new heating technologies, process assessment and process optimization in thermal processing of foods are limited
by the applicability of currently used process evaluation methodologies. This
text reviews the development of enzymic time–temperature integrators
(TTIs) that allow fast, easy, and correct quantification of the thermal process
impact in terms of food safety without the need for detailed knowledge of
the actual temperature history of the product. The experimental work
involves extensive kinetic studies of α-amylase at different environmental
conditions under steady state and nonsteady state conditions. After careful
isothermal calibration and validation under variable temperature conditions,
several TTIs are successfully applied as wireless devices to (1) determine the
influence of process and/or product parameters on the spatial distribution
of process-lethalities in a particulate model food system and to (2) evaluate
the lethality distribution and hence determine the coldest zone in a retort.

1   Methods for process impact assessment: the need for TTIs

Thermal processing, including blanching, pasteurization and sterilization,
has been and still is one of the most widely used physical methods of food
preservation. In the context of food preservation, the quantitative measurement of the impact of a thermal process in terms of food safety is of utmost
importance in process design, optimization, evaluation, and control. The *in
situ* method and the physical–mathematical approach are commonly used
process evaluation techniques. In the *in situ* method the level of the food
safety attribute of interest is evaluated before and after thermal processing
to provide direct and accurate information on the process impact, whereas
in the physical mathematical approach, based on the temperature history of
the product (either recorded or simulated) combined with knowledge on the

heat inactivation kinetics of the safety attribute, the impact of the thermal treatment on the parameter of interest is calculated.

Consumer demands for higher-quality convenience products and the striving of food companies for energy savings and better process control have resulted in the optimization of existing and the development and application of new heating technologies, such as continuous processing in rotary retorts, aseptic processing, ohmic and microwave heating, and combined processes. Because the *in situ* and the physical–mathematical evaluation methods have serious limitations with regard to their applicability in these technologies, considerable effort has been and will continue to be put into the development of TTIs.

2 General aspects on time–temperature integrators

2.1 Definition of a time–temperature integrator

A TTI can be defined as "a small measuring device that shows a time–temperature dependent, easily, accurately and precisely measurable irreversible change that mimics the change of a target attribute undergoing the same variable temperature exposure" (Taoukis and Labuza, 1989a, b; Weng et al., 1991a, b). The target attribute can be any safety or quality attribute of interest such as microorganism (spore) inactivation, loss of a specific vitamin, texture, or color. Of major interest are important hidden quality attributes like microbiologic safety.

The major advantage of TTIs is the ability to quantify the integrated time–temperature impact on a target attribute without the need for information on the actual temperature history of the product. All TTIs are by definition *post factum* indicators of the impact of a thermal process, because the calculation of this process impact is based on the change in status of the TTI after thermal treatment, as compared with its initial status. The time after thermal processing needed to calculate the process impact by readout of a TTI depends upon the nature of the monitoring system: e.g., microbiologic assays are time-consuming (several days), whereas in general the evaluation of enzymic activity or quantification of a chemical compound or a physical change is much faster (up to minutes).

It should be stressed that if temperature is not the only rate determining factor, using only a TTI to monitor food safety or quality loss would result in error, as other factors that change with time can be critical.

2.2 Criteria for a time–temperature integrator

From the definition of a TTI given above, several criteria that such a measuring device should meet can be formulated: (1) for convenience, the TTI has to be inexpensive, quickly and easily prepared, easy to recover, and give an accurate and user-friendly read-out; (2) the TTI should be incorporated into the food without disturbing heat transfer within the food, and it should experience the same time–temperature profile as the parameter under investigation; (3) the TTI should quantify the impact of the process on a target attribute. This results in specific kinetic requirements. The temperature dependency of the rate constants of TTI and target attribute should be described by the same law (e.g., Arrhenius model, TDT model). Because the aim of using a TTI is to calculate the processing value F relying solely on the TTI response, it is necessary that the TTI response kinetics obey a rate equation that allows separation of the variables (Hendrickx et al., 1995a). Furthermore, it can be easily shown that the temperature sensitivity of the rate constants (E_a-value or z-value) of the TTI and the target attribute should be equal to assure equality in process-values (Maesmans, 1993; Hendrickx et al., 1995a).* Contrary to the requirement that the z-value of the TTI and of the target attribute be equal, the reaction rate constant at reference temperature and reaction order of heat inactivation of TTI and target attribute may differ. It is important, however, that the reaction rate constant k (or decimal reduction time D) of the TTI should be sufficiently low (high) in the relevant temperature domain to induce a detectable response to the temperature history.

2.3   Classification of time–temperature integrators

TTIs can be classified in terms of working principle, type of response, origin, application and location in the food as shown in Figure 1 (Hendrickx et al., 1993; Hendrickx et al., 1995b, c, d; Van Loey et al., 1995, 1996a). Depending on the response property, TTIs can be subdivided into biological (microbiologic and enzymic), chemical and physical systems. The use of a TTI consisting of one component, characterized by a single activation energy (z-value) equal to the one of the target attributes, is straightforward and hence desirable. In case no such TTI is available, the use of multicomponent TTIs has been suggested to predict the change in status of a quality attribute from the reading of the process impact on a set of individual temperature-sensitive components, each characterized by its activation energy but deviating from the one of the target attribute. With respect to the origin of the TTI, extrinsic and intrinsic TTIs can be distinguished. An extrinsic TTI is incorporated into the food, whereas intrinsic TTIs are intrinsically present in the food and represent the behavior of another food aspect. With regard to the application of the TTI in the food product, three approaches can be identified: dispersed, permeable, or isolated. In dispersed systems, the TTI (extrinsic or intrinsic) is homogeneously distributed throughout the food, allowing evaluation of the volume-average impact of a process. Besides dispersion of an extrinsic TTI in the food, extrinsic TTIs may be permeable or isolated. All three approaches can be the basis for single-point evaluations of the process impact at specific locations within the food.

* On practical grounds, the allowed difference in z-value between TTI and target attribute to ascertain a pre-set accuracy in process impact determination has been studied theoretically (Van Loey *et al.*, 1995).

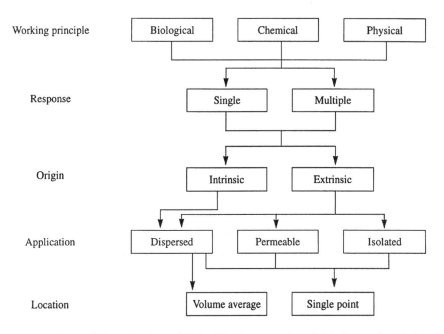

Figure 1 General classification of TTIs (Van Loey et al., 1996a). Printed with kind permission from Elsevier Science Ltd.

3 Feasibility of thermostable enzymes for TTI development

It follows from the foregoing criteria that it is not straightforward to find a system meeting all the conditions listed above. A possible approach is to identify systems that have some of the required properties and to engineer the remaining conditions. From this point of view, several features of thermostable enzymes are favorable for their use as a TTI: they are small in size, relatively low-priced, and easy to prepare. The relative easiness of read-out and handling of enzymic systems gives them a significant advantage over microbiologic TTIs. In enzyme-based monitoring systems, the amount of enzymic activity remaining after the heat treatment is often assayed to determine the thermal impact, although other properties, such as the heat of enzyme deterioration, can be determined instead. Both properties, enzymic activity and denaturation enthalpy, can be measured rapidly and accurately. Heat inactivation kinetics of enzymes can be manipulated in various ways: to a limited extent, by changing the conditions of an enzyme (e.g., through enzyme immobilization or protein engineering) and/or its environment (e.g., by "solvent engineering": changing the ion concentration, pH, moisture content, additives,...) the thermal sensitivity of the enzyme can be altered to match the kinetic behavior of a target parameter. In the context of enzyme-based TTIs, much work has been reported on α-amylase of *Bacillus* species (De Cordt et al., 1992, 1993, 1994; Van Loey et al., 1996a, 1996b, 1997a, 1997b, 1997c).

4　Research objective

It is the objective of the present work to develop enzymic TTIs to monitor the safety of thermal treatments, including pasteurization and sterilization processes.

In the context of safety assessment by use of a TTI, a first question that is to be answered is which microorganism(s) is (are) representative for the safety of thermally processed food. Second, knowledge of the heat inactivation kinetics of the target microorganism (i.e., its z-value) is essential to select an appropriate TTI as the main kinetic requirement for proper TTI functioning is equality in z-value between TTI and target.

4.1　Pasteurization processes

In pasteurization processes, a mild heat treatment is mostly combined with other inhibitory factors (acidification, refrigerated storage, low water content, preservatives,…) to ensure that any surviving organism is inhibited during the intended shelf life of the product. Therefore, other microorganisms might be the main cause of food poisoning and of particular concern depending on these secondary preservative barriers (Brown, 1992). Several target microorganisms, having z-values in the range from 5 to 12°C, have been suggested in literature depending on the type of product (pH, a_w, etc.) and conservation process applied (sous-vide, in pack pasteurization, smoking, roasting, frying, etc.); however, without arriving at an agreement. A second aspect of thermal process evaluation is related to the extent of the heat treatment. A pasteurization process should be designed to achieve a reduction of the number of the most heat-resistant pathogenic/spoilage microorganism relevant to the type of product under consideration, by a preset level, to ensure that the product formulation and the storage conditions applied inhibit the growth of any surviving cells during the intended shelf life of the product. Recommended heat treatments for different types of products vary from a $^zF_{80°C}$-value of 0.002 min to 400 min (Betts, 1992; Brown, 1992; Gaze, 1992; Lund and Notermans, 1993). It can be stated that at the moment the design of a pasteurization process should be based on a detailed and thorough knowledge of the product and associated microbial flora, i.e., their heat resistance and potential growth under the defined product formulation and intended storage conditions.

4.2　Sterilization processes

In sterilization of low-acid canned foods (LACF, 7.0>pH>4.6) there are three specific types of final product spoilage that are of concern to food microbiologists and food manufacturers. The first consideration is the type of spoilage that may produce a public health hazard. Spoilage by mesophilic or thermophilic spore-forming organisms that represents economic loss is considered to be of secondary importance (Pflug, 1987a-e). The main public health hazard in sterilization of LACF are the spores of proteolytic strains of *Clostridium botulinum* characterized by a $D_{121.1°C}$-value of 0.2 min (as determined in phosphate buffer of neutral pH) and a z-value of 10°C. Years of canning practice allowed to conclude that for wet-heat sterilization of LACF, a $^{10°C}F_{121.1°C}$-value of 3 min for the slowest heating zone of a container will result in a product that is safe from a public health standpoint. To reduce economic losses due to spoilage by thermophilic microorganisms, in practice, most LACF are processed beyond the minimum *botulinum* cook because of the greater heat resistance of this kind of bacteria (Maesmans et al., 1990). $^{10°C}F_{121.1°C}$-values of the order used in the canning industry are in the range from 3 to 21 min (Pflug, 1987e).

In conclusion, the research objective of the present study can be formulated in kinetic terms: (1) to monitor the sterilization efficacy of LACF from a safety point of view, an enzymic system with a z-value of 10°C and thermostable so as to allow the measurement of an F_0-value between 3 min and 20 min should be looked for. For the evaluation of pasteurization processes, a separate TTI needs to be developed for each target safety aspect one wants to monitor. Hence, from a safety point of view, several enzymic systems with z-values ranging from 5 to 12°C are to be developed that allow the measurement of a lethal impact $^zF_{80°C}$ in the range from 0.002 min to 400 min.

5 Development of enzymic isolated extrinsic time–temperature integrators

Since the link between a target attribute and an appropriate TTI can only be made based on knowledge of its kinetic inactivation behavior, extensive kinetic studies of thermal enzyme inactivation were performed under isothermal conditions to determine the kinetic inactivation model and to estimate its kinetic parameters. Next, the potential indicators were evaluated on their integrating properties under time-variable temperature conditions.

5.1 Isothermal calibration of enzymic time–temperature integrators

5.1.1 Materials and methods

Enzymic systems. *Bacillus subtilis* α-amylase (BSA) was purchased from Fluka (Biochemika) as a powder with a specific activity of 390 units/mg of enzyme, where 1 unit corresponds to the amount of enzyme that liberates 1 mol of maltose from starch per minute at pH 6.0 and 25°C. *Bacillus licheniformis* α-amylase (BLA) (type XII-A, Sigma) was purchased as an aqueous solution containing 15% NaCl, 25% sucrose, and 31 mg protein per ml (Biuret) with a specific activity of 500 Sigma units/mg protein, where 1 Sigma unit will liberate 1.0 mg of maltose from starch in 3 min at pH 6.9 and 20°C. The BLA solution was extensively dialyzed (membrane 6-27/32) to minimize possible interferences from the additives and subsequently lyophilized. *Bacillus amyloliquefaciens* α-amylase (BAA) (type II-A, Sigma) was purchased as a lyophilized powder with a specific activity of 2500 Sigma units/mg protein (Biuret).

Enzymes were dissolved in Tris(hydroxymethyl)aminomethane buffer of pH 8.6 (at 25°C) at the desired concentration. Trehalose was purchased from Fluka (Biochemika). In order to obtain a certain water content in the enzymic sample, α-amylase was first dried above P_2O_5 at 4°C for 10 days and then equilibrated for at least 6 days above a saturated salt solution of an equilibrium relative humidity at 4°C in the range from 74 to 93 (Greenspan, 1977). For convenience, *Bacillus subtilis* α-amylase equilibrated above a saturated salt solution of a certain equilibrium relative humidity will be referred to as BSAxx, where xx indicates the equilibrium relative humidity of the saturated salt solution at 4°C.

Determination of response property. The degree of thermal enzyme inactivation was monitored either by measuring the (residual) enzymic activity or by determination of the (residual) heat of enzyme deterioration.

The α-amylase activity was measured spectrophotometrically (Philips-PU8680 VIS/NIR kinetics spectrophotometer) according to procedure n°577 of Sigma diagnostics. This procedure is based on the progressive hydrolysis of the α-1,4-glycosidic bonds in 4,6-ethylidene-p-nitrophenyl-α-D-maltohep-taside, thus releasing p-nitrophenol gradually, which has an absorption maximum at 405 nm. Temperature was kept constant at 30°C. Activities were expressed in terms of the change in optical density per minute calculated by linear regression from a plot of the absorption as a function of reaction time. Absorption was measured 15 times each 18 sec. Linearity between enzyme concentration and enzymic activity was investigated for each α-amylase and all activity measurements were verified to be in this working range.

Denaturation enthalpy was measured by use of a calibrated Power Compensation Differential Scanning Calorimeter (DSC) (Perkin-Elmer DSC7). DSC measures the enthalpy related to transition processes by recording the difference in heat flow between an inert reference material and the sample necessary to reach thermal equilibrium between both systems while the sample and the reference material are subjected to a controlled temperature program. A DSC-assay at a scan rate of 10°C/min from 50 to 150°C revealed an endothermic protein deterioration process, observed as a peak-shaped deviation from the baseline. The baseline of each peak was taken as a straight line between the two extremes of the peak. The denaturation enthalpy per unit sample weight is determined as the area between the peak and the baseline.

Heat inactivation experiments. The heat-induced enzyme inactivation kinetics was determined by monitoring the evolution of a response property of the enzymic system (X) as a function of inactivation time at constant temperature (steady-state procedure). To ensure isothermal inactivation conditions, small sample recipients were used to avoid heating and/or cooling lags: in case of activity measurement, enzyme solutions were heated in glass capillary tubes

(2-mm diameter and 150-mm length); in case of denaturation enthalpy measurement, 20 µL of enzyme solution or, in case of dry enzymic systems, approximately 10 mg of enzyme was hermetically sealed in a 75 µL capacity stainless steel volatile-sample capsules (Perkin-Elmer n° 0319-0218 Large Volume Capsules). Samples were either heated in a temperature-controlled water bath at constant temperature for predetermined heating times, or in the DSC (heating and cooling rate of 200°C/min). It was ensured that the heating times were sufficiently long to allow at least a 50% reduction in response value and hence facilitate a discrimination between zero-, first-, or second-order reactions. After withdrawal from the water bath, samples were cooled immediately in ice water to stop thermal inactivation. The possible reactivation after thermal treatment occurring during the storage period was studied, but no reactivation was observed in any of the enzymic systems considered.

5.1.2 Results and discussion

Data analysis of kinetic studies of isothermal heat inactivation of enzymes involves: (1) the choice of an appropriate kinetic inactivation model (e.g., first order, biphasic, n^{th} order), (2) the choice of an appropriate temperature dependence model (e.g., thermal death time model — TDT, Arrhenius model) and (3) the choice of a regression method so that kinetic parameter values are estimated with the highest probability of being correct. In this work, the integration approach was used to check from a loglinear plot of the response value (X- i.e., residual enzymic activity or residual heat of enzyme deterioration) as a function of inactivation time at constant temperature whether the heat-induced deterioration of the enzymic system studied could be adequately described by a first-order reaction. As a measure of the linearity, the coefficient of determination R^2, the F-value and the t-statistic of the slope were calculated (SAS, 1982). In addition, residuals of a linear regression analysis of $\log(X/X_0)$ vs. time were critically considered in a qualitative (i.e., generating plots of model residuals to represent visually trends and/or correlations) as well as in a quantitative way (i.e., performing a runs test — Mannervik, 1982). Based on the coefficients of determination R^2, the runs test for randomness of residu-plots, the F-value and t-statistic, the heat inactivation of all enzymic systems studied in this work could be adequately described by a first-order reaction. As an example, heat-induced inactivation of BSA 5 mg/mL is depicted in Figure 2.

In relation to the selection of a temperature dependence model, in the present study the Thermal Death Time (TDT) model was preferably applied when appropriate because, in general, kinetic data in the open literature of thermal processing and thermobacteriology are reported as D_T and z-values. Hence, the equality of the temperature coefficient of TTI and target attribute could be checked directly. The appropriateness of the TDT model to describe the temperature dependence of the decimal reduction time was checked by evaluation of the linear relationship of the logarithm of the decimal reduction times vs. inactivation temperatures using the same statistical criteria as for the evaluation of a first-order heat inactivation. By inspection of the linear

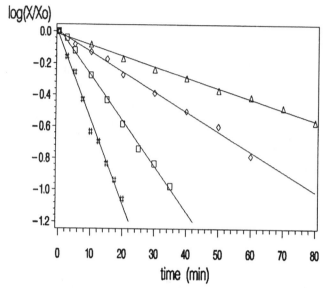

Figure 2   Thermal inactivation curves of BSA (5 mg/ml in TrisHCl buffer pH 8.6) thermally treated at 70°C (△), 72°C (◊), 75°C (□) and 78°C (#). Xo and X are initial α-amylase activity and activity after thermal treatment, respectively.

relationship of logD vs. temperature, the suitability of the TDT model was confirmed for all systems studied.

With regard to the regression method, kinetic parameters can be estimated from isothermal inactivation data using an individual approach (i.e., sequentially regressing X vs. time to obtain a D_T-value at each temperature and D_T vs. temperature to obtain an estimate of the z-value) or alternatively using a global regression approach that considers the dataset as a whole. Since the superiority of a global fit is proclaimed throughout literature (Cohen and Saguy, 1985; Haralampu et al., 1985; Doucet and Sloep, 1992; Johnson, 1992; van Boekel, 1996), in the present study kinetic parameters (D_{ref},z) were estimated in a global fit using nonlinear regression analysis on relative residual response values. As the assumption of a normally distributed error structure with mean zero and constant variance is inherent to least-squares fitting — for linear as well as for nonlinear least squares — in order to yield parameter estimates having the highest probability of being correct (Beechem, 1992, Johnson, 1992, van Boekel, 1996), it was decided not to apply a logarithmic transformation on the data prior to parameter estimation because the error structure of the nontransformed data were normally distributed (tested using the Shapiro-Wilk statistic, SAS, 1982). Hence, kinetic parameters (D_{ref},z) of heat-induced enzyme inactivation were estimated using Equation (1) (SAS, 1982) and are reported in Table 1.

$$\frac{X}{X_0} = 10^{-\frac{t}{D_{ref}}10^{\left(\frac{T-T_{ref}}{z}\right)}}$$

(1)

Table 1 Kinetic Parameter Estimates (D_T,z) of BSA, BAA and BLA at Different Environmental Conditions Obtained from Nonlinear Regression Analysis on Isothermal and Nonisothermal Inactivation Data, Assuming a First Order Heat Inactivation. Asymptotic Standard Errors Are Indicated.

		Isothermal data		Nonisothermal data	
		$D_{80°C}$ (min)	z (°C)	$D_{80°C}$ (min)	z (°C)
BSA	2 mg/mL	4.7 ± 0.3	9.2 ± 0.2	nd	nd
	5 mg/mL	10.3 ± 0.3	9.0 ± 0.2	10.7 ± 0.5	12.8 ± 0.7
	10 mg/mL	12.1 ± 0.4	9.8 ± 0.3	13.9 ± 0.5	10.2 ± 0.4
	20 mg/mL	18.3 ± 0.5	9.1 ± 0.3	18.5 ± 0.7	8.2 ± 0.3
	30 mg/mL	26.7 ± 0.8	10.4 ± 0.3	27.1 ± 0.7	10.2 ± 0.4
	200 mg/mL[a]	30.1 ± 0.9	8.8 ± 0.3	28.3 ± 0.7	8.4 ± 0.3
BAA	2 mg/mL	11.5 ± 0.6	12.3 ± 0.6	nd	nd
	5 mg/mL	14.4 ± 0.2	10.9 ± 0.2	nd	nd
	10 mg/mL	18.8 ± 0.6	9.7 ± 0.3	16.8 ± 1.0	8.6 ± 0.5
	30 mg/mL	26.2 ± 0.7	9.3 ± 0.2	34.7 ± 1.8	9.4 ± 0.9
	200 mg/mL[ab]	41.5 ± 0.8	7.5 ± 0.1	115.8 ± 34.4	7.7 ± 1.6
		$D_{100°C}$ (min)	z (°C)	$D_{100°C}$ (min)	z (°C)
BLA	2 mg/mL	7.2 ± 0.4	6.5 ± 0.2	nd	nd
	5 mg/mL	7.8 ± 0.3	6.7 ± 0.1	nd	nd
	30 mg/mL	3.7 ± 0.2	7.0 ± 0.2	5.3 ± 0.8	6.9 ± 0.6
	300 mg/mL[ab]	3.4 ± 0.4	6.2 ± 0.3	8.8 ± 2.9	6.8 ± 1.3
		$D_{100°C}$ (min)	z (°C)	$D_{100°C}$ (min)	z (°C)
BSA (200 mg/mL)	+trehalose[a] (500 mg/mL)	3.1 ± 0.4	7.5 ± 0.4	1.4 ± 0.2	6.0 ± 0.2
BAA (200 mg/mL)	+trehalose[ab] (500 mg/mL)	8.1 ± 0.3	6.6 ± 0.2	6.4 ± 0.6	6.8 ± 0.6
BLA (300 mg/mL)	+trehalose[a] (500 mg/mL)	172.3 ± 9.5	5.2 ± 0.2	195.4 ± 32.8	6.2 ± 0.5
		$D_{120°C}$ (min)	z (°C)	$D_{120°C}$ (min)	z (°C)
BSA	93[a]	0.05 ± 0.02	6.5 ± 0.4	nd	nd
	88[a]	0.25 ± 0.09	4.3 ± 0.4	nd	nd
	81[a]	13.4 ± 0.6	13.9 ± 0.7	nd	nd
	76[ab]	25.3 ± 0.9	9.4 ± 0.5	14.8 ± 0.4	9.6 ± 0.3
	74[a]	58.6 ± 3.2	6.7 ± 0.3	nd	nd

nd: not determined

[a] Denaturation enthalpy was measured as response property.

[b] Nonisothermal experiment performed in a DSC.

It can be observed from Table 1 that in the case of heat inactivation of BSA and BAA, the decimal reduction time at 80°C increases as enzyme concentration increases, indicating that dissolved α-amylase is more thermostable at higher enzyme concentrations. No systematic influence of the initial BSA concentration on the z-value can be noticed, whereas the z-value of thermal BAA inactivation slightly decreases as enzyme concentration increases. As dilution of an enzymic solution should drastically reduce the rates of the polymolecular reactions such as thermo-aggregation, while the monomolecular processes, generally speaking, will be hardly affected (Klibanov, 1983), the dependence of the decimal reduction time on initial enzyme concentration might be an indication that aggregation is the cause of thermo-inactivation of BAA and BSA in the concentration range studied (2 mg/mL up to 30 mg/mL). The fact that polymolecular aggregation likely occurs at high enzyme concentrations, whereas monomolecular conformational changes prevail at low enzyme concentrations, might explain the apparently contradictory results of Tomazic and Klibanov (1988a), who demonstrated that BAA in the concentration range from 0.1 to 1 mg/mL irreversibly inactivates due to a monomolecular conformational process. Indeed, at higher concentrations of BAA (10–30 mg/mL) enzyme aggregation upon heating was clearly observed, which diminished distinctly by lowering the enzyme concentration to 2 mg/mL.

BLA is much more thermostable than BSA or BAA at the same enzyme concentration (Table 1). The higher thermostability of BLA as compared to BAA and BSA has been extensively reported in literature (e.g., Morgan and Priest, 1981; Kindle, 1983; Tomazic and Klibanov, 1988b; Maassen, 1991; Weemaes et al., 1996). According to Tomazic and Klibanov (1988b) the higher thermostability of BLA was found to be mainly due to additional salt bridges involving a few specific lysine residues. They observed that BLA does not inactivate irreversibly due to the formation of incorrect structures, but as a consequence of deamidation of asparagine and/or glutamine residues. Although enzyme concentrations studied were not equal in cases where denaturation enthalpy was measured as response variable, the same conclusion with regard to the thermostability ranking can be made: thermostability of BSA and of BAA is of the same order of magnitude, BLA being by far more thermostable.

Addition of trehalose causes an increase in thermostability of α-amylase (see Table 1), which affords the possibility of using these systems to monitor pasteurization processes up to 105°C. Due to the addition of trehalose the z-value of the systems is lowered, which is noticeable from the steeper TDT curves as compared to the respective systems without addition of trehalose. It is well known that polyols and sugars protect against heat-induced protein deterioration (Gerlsma, 1968, 1970; Gerlsma and Stuur, 1972; Back et al., 1979; Arakawa and Timasheff, 1982; Gekko and Koga, 1983; Gray, 1988).

It can be observed in Table 1 that as moisture content decreases systems are more thermostable, as indicated by a higher D-value at 120°C. These experimental results are fully in agreement with the general assumption that the heat stability of an enzyme increases at low moisture content. Analog observations were made in literature by Mullaney (1966) (trypsin and ribonuclease), Multon and Guilbot (1975) (ribonuclease), Rüegg et al. (1975) (β-lactoglobuline), Hägerdal and Martens (1976) (myoglobine), Meerdink and van't Riet (1991) (*Bacillus licheniformis* α-amylase), Volkin et al. (1991) (ribonuclease),

Hendrickx et al. (1992) (horseradish peroxidase), and Saraiva (1994) (*Bacillus amyloliquefaciens* α-amylase and horseradish peroxidase). Dehydration of enzymes is a well-known method to increase its thermostability, but the mechanism of enhanced thermostabilization of enzymes upon dehydration is unknown. Since conformational mobility is necessary for partial unfolding, which in turn is the first step of the thermoinactivation process and water acts as a "plasticizer," the increased thermostability of dry enzymes may result from hindered unfolding of the protein macromolecule in a water-free solid phase (Klibanov, 1983; Zaks and Klibanov, 1984; Klibanov, 1986). Moreover, the enzyme undergoes irreversible reactions after unfolding, all of them requiring water. Therefore, they should not occur in a water-free environment (Klibanov, 1986). No systematic influence of the moisture content of BSA on the z-value of the system can be noticed in Table 1. Also, Multon and Guilbot (1975) observed that the activation energy of ribonuclease inactivation varies greatly with moisture content, and Saraiva (1994) observed a random like variation of the z-value of heat inactivation of dried BAA with water content. Due to the increased thermostability of α-amylase at reduced moisture content these systems can be used to monitor thermal processes at temperatures in the range from 100 to 130°C.

5.2 Evaluation of integrating properties under variable temperature conditions

5.2.1 Materials and methods

Heat inactivation experiments. The correct functioning of several isothermally calibrated potential TTIs was evaluated under nonsteady state conditions. To create variable temperature conditions two methods were applied. In most cases, a DSC capsule containing the enzymic sample was incorporated in the center of a sylgard (Sylgard 184, Dow Corning, Belgium) sphere (25 mm diameter), which was closed using a silicone stopper. The sensing junction of a calibrated copper-constantan thermocouple (Pyrindus) was inserted through the silicone stopper to make contact with the DSC pan. Time–temperature data were registrated each 15 sec using a calibrated MDP 82 series datalogger (Mess+Technik system GmbH). Spheres containing the DSC capsule were immersed in a water (oil) bath and processing times at various temperatures were chosen to cover a wide range of processing values. After a preset processing time the sphere was removed from the water (oil) bath and cooled and the TTI recovered and read-out as explained previously. In a second method, nonisothermal conditions were created using the nonisothermal mode of the DSC. After an equilibration of the enzymic sample at 40°C in the DSC, the sample was heated slowly at a constant rate until a desired holding temperature T_H was reached. The sample was kept at that constant holding temperature for a predetermined time t_H and then cooled at a cooling rate of 200°C/min to 0°C. The heating rate as well as the holding temperature and holding time were varied from sample to sample.

Data analysis. After heat treatment, processing values determined from the reading (activity or enthalpy measurement) of the TTIs (F_{TTI} – Equation (2) valid for first-order heat inactivation) were compared to processing values calculated from the nonisothermal time–temperature profiles to which TTIs were subjected (F_{t-T}). When time–temperature profiles were recorded, process values were calculated according to the general method (Bigelow, 1921) (Equation 3) using the numerical integration routine of Simpson's rule (Carnahan et al., 1969). When the nonisothermal heat treatment was performed in the DSC, process values were calculated from the programmed time–temperature profiles in the DSC using the analytical solution of the integrated impact value (Equation 4). The kinetic parameter estimates (D_{ref},z) obtained from nonlinear regression analysis on the isothermal inactivation data were used as initial estimates.

$$\left({}^{z}F_{T_{ref}}\right)_{TTI} = D_{ref}\log\left(\frac{X_0}{X}\right) \tag{2}$$

$$\left({}^{z}F_{T_{ref}}\right)_{t-T} = \int_0^t 10^{\left(\frac{T-T_{ref}}{z}\right)}dt \tag{3}$$

$$\left({}^{z}F_{T_{ref}}\right)_{t-T} = \frac{z}{2.303\beta}\left[10^{\left(\frac{T_H-T_{ref}}{z}\right)} - 10^{\left(\frac{T_0-T_{ref}}{z}\right)}\right] + 10^{\left(\frac{T_H-T_{ref}}{z}\right)}t_H \tag{4}$$

5.2.2   Results and discussion

In several cases, processing values calculated from the response of the TTI systems (F_{TTI}) differed systematically from the integrated process values (F_{t-T}), which were considered as true process values. The agreement between F_{TTI} and F_{t-T} for BSA (10 mg/mL) is depicted in Figure 3. The systematic over- or underestimation when using kinetic parameter estimates determined by nonlinear regression analysis of isothermal heat inactivation data lead to a reestimation of kinetic parameters using nonisothermal instead of isothermal inactivation data. Hereto, Equation (5) was derived from which the z- and D_{ref}-value were estimated by nonlinear regression analysis (SAS, 1982). The integral of Equation (5) was again approximated using a Simpson integration routine or solved analytically in the case of a linearly increasing temperature.

$$\frac{X}{X_0} = 10^{-\int_0^t \frac{dt}{D_{ref}} 10^{\left(\frac{T-T_{ref}}{z}\right)}} \tag{5}$$

Re-estimated kinetic parameters based on nonisothermal inactivation data are reported in Table 1 with an indication of asymptotic standard errors. It can be noticed from Table 1 that for several enzymic systems the kinetic

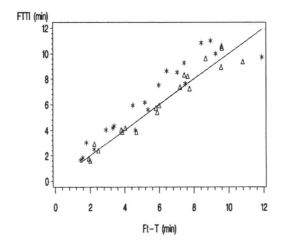

Figure 3 Comparison of processing values as determined from the reading (i.e., enzymic activity measurement) of the TTIs (BSA 10 mg/ml) (F_{TTI}) with processing values obtained by numerical integration of the recorded time–temperature profiles (F_{t-T}) using kinetic parameter estimates from isothermal data (*) and from nonisothermal data (Δ) ($T_{ref} = 80°C$).

parameters estimated based on isothermal inactivation data are different from the estimates based on nonisothermal inactivation data. Although it is not expected *a priori* that heat inactivation kinetics determined under steady-state and under non-steady-state conditions would differ, several potential explanations can be postulated: (1) from a mechanistic point of view, it could be that in different temperature domains, different thermal inactivation mechanisms, characterized by a different z-value, dominate. Hence, the fact that the isothermal inactivation experiments were conducted over a limited temperature range of about 8°C, whereas the nonisothermal inactivation experiments were conducted over a much broader temperature range, might explain the difference in heat inactivation kinetics observed. In addition, the history of the enzymic system prior to thermal denaturation, which is definitely different in the steady-state experiments as compared to the unsteady-state ones, might influence the irreversible inactivation mechanism and hence the observed heat inactivation kinetics; (2) from an experimental point of view, potential explanations can be traced back both to the design of kinetic experiments and as to their execution. A kinetic experiment should be designed so that the inactivation data contain enough information to find a unique and reliable set of kinetic parameter values. If this is not the case, the optimization routine used in the kinetic parameter estimation procedure will pose a problem. An often encountered optimization problem, which also occurs in case the jointly estimated kinetic parameters are highly correlated, is the existence of a "gutter-shaped" accumulation of minima instead of a global minimum so that no unique set of kinetic parameter values can be

Table 2   Average Percentual Error in the Prediction of the Relative
Retention after Heat Treatment and in Process Value Determination
by Using Kinetic Parameter Estimates from Regression of
Isothermal Inactivation Data (ISO) and of Nonisothermal
Inactivation Data (NonISO)

		$(X/X_0)_{exp}$ vs. $(X/X_0)_{pred}$ % error (Equation 6)		F_{TTI} vs. F_{t-T} % error (Equation 7)	
		ISO	NonISO	ISO	NonISO
BSA	5 mg/mL	22.9	9.1	28.7	7.8
	10 mg/mL	16.7	8.4	23.0	8.8
	20 mg/mL	13.3	9.8	12.2	10.0
	30 mg/mL	43.4	7.1	26.8	9.5
	200 mg/mL[a]	8.5	7.5	9.7	7.4
	+trehalose[a]	13.7	6.6	27.0	13.9
	76[ab]	19.2	6.1	69.5	11.4
BAA	10 mg/mL	7.5	7.1	15.6	15.1
	30 mg/mL	32.6	10.0	26.8	13.0
	200 mg/mL[ab]	373.3	3.2	65.5	6.7
	+trehalose[ab]	13.0	7.1	30.6	9.2
BLA	30 mg/mL	90.7	9.8	32.1	9.6
	300 mg/mL[ab]	240.7	1.5	75.4	4.7
	+trehalose[a]	138.6	6.8	41.0	7.4

[a] Denaturation enthalpy was measured as response property.
[b] Nonisothermal experiment performed by use of the DSC.

found. In this context, it would be worthwhile studying the optimal experimental design to estimate kinetic parameters, considering both steady-state
and non-steady-state conditions. With regard to the execution of the kinetic
experiments, it should be noted that small errors in temperature and/or time
registration might also contribute to a mismatching between F_{TTI} and F_{t-T}.

A measure of the reliability of the kinetic parameter estimates to predict
the residual relative retention after thermal treatment was obtained by comparing the experimentally observed relative retentions $(X/X_0)_{exp}$ to the ones
predicted by the model $(X/X_0)_{pred}$ and by calculating a percentage error using
Equation (6) (Table 2).

$$\text{|\% error|} = \left| \frac{(X/X_0)_{pred} - (X/X_0)_{exp}}{(X/X_0)_{pred}} \right| \times 100 \qquad (6)$$

In the context of TTI development, as a measure for the adequacy of an
enzymic system to measure the integrated impact of time and temperature
on a product factor with an equal z-value, an absolute percentage error was
calculated using Equation (7) and is reported in Table 2.

$$|\% \, error| = \left| \frac{F_{t-T} - F_{TTI}}{F_{t-T}} \right| \times 100 \tag{7}$$

The enhanced agreement between F_{TTI} and F_{t-T} when using parameter estimates obtained from regression on nonisothermal inactivation data of BSA (10 mg/mL) is clearly shown in Figure 3.

An average error of 15% in process-value determination corresponds to an accuracy of temperature measurement of 0.6°C, assuming a z-value of 10°C, and was considered allowable. Based on the satisfactory correlation between F_{TTI} and F_{t-T} using kinetic parameter estimates of nonisothermal inactivation data (Table 2) these enzymic systems can be used to monitor the thermal death of several pathogenic or spoilage microorganisms characterized by an equal z-value. A comparison of the kinetic parameters for α-amylase of *Bacillus* species at different environmental conditions with those of microorganisms relevant to pasteurized products reveals potential target attributes with z-values in the range from 5 to 12°C: e.g., inactivation of *Clostridium botulinum* nonproteolytic type B in cod (z = 8.6°C); inactivation of *Clostridium botulinum* nonproteolytic type E in water (z = 9.4°C) and in cod (z = 8.3°C); inactivation of *Streptococcus faecalis* in fish (z = 6.7°C); inactivation of *Bacillus cereus* in buffer (pH 7) (z = 10.5°C); inactivation of *Listeria monocytogenes* HAL 957E1 in egg yolk plus 10% NaCl (z = 7.7°C). As the European Chilled Food Federation (ECFF, 1996) stated that in the case of chilled pasteurized food products, heating must achieve the correct time–temperature conditions to destroy *Listeria monocytogenes* (z = 7.5°C) and/or *Clostridium botulinum* type B (z = 7°C at T<90°C and z = 10°C at T>90°C), the developed enzyme-based monitoring systems exhibiting a z-value of 7°C, 7.5°C, or 10°C are of particular importance to the food industry. Also BSA76 is especially important since its kinetic characteristics, i.e., its z-value of 9.6°C, allow the system to monitor heat destruction of spores of proteolytic strains of *Clostridium botulinum* (z = 10°C), the reference microorganism for sterilization of low acid canned foods. Taking into account that a one log reduction in response value can be measured accurately, BSA76 can be used to measure thermal impact values ($^{9.6°C}F_{120°C}$) up to 15 min.

Since TTIs are designed to determine an integrated time–temperature impact, careful calibration of the system under isothermal conditions and validation under nonisothermal conditions is highly recommended.

6 Application of enzymic time–temperature integrators for thermal process evaluation

Safe thermal process design and process evaluation should be based on the lethal impact achieved in the coldest point of a container located in the coldest zone in a retort. As the currently available process evaluation tech-

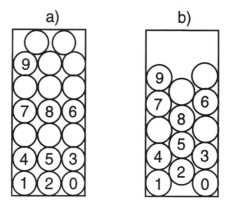

Figure 4 Spatial distribution of spheres in a glass jar (a) and in a can (b). Numbered spheres contain a TTI.

niques are often inadequate to monitor these critical control points, the use of TTIs as monitoring devices has been proposed. Hence, after careful calibration of enzymic systems in terms of their heat inactivation kinetics and evaluation on their integrating properties under variable temperature conditions, several relevant case studies to demonstrate the use of enzymic TTIs as wireless devices were elaborated.

6.1   Determination of coldest point in a particulate model food system

6.1.1   Materials and methods

For the evaluation of pasteurization processes, BSA 10 mg/mL (based on activity measurement) was selected as TTI, whereas for sterilization processes BSA76 (based on denaturation enthalpy measurement) was chosen. Heat inactivation kinetics of these systems was previously determined under isothermal and nonisothermal conditions.

As model systems of a particulate food, both Teflon (Eriks NV, Belgium) and silicon elastomer (Sylgard 184, Dow Corning) spheres of 25 mm diameter were used. An opening was drilled in the spheres allowing to embed the TTI at the particle center. As packaging material, cans (450 ml — 109.5 mm height, 36.5 mm radius, 0.23 mm wall thickness — CMB, Belgium) and glass jars (750 ml — 172 mm height, 40.5 mm radius, 2.6 mm wall thickness, Carnaud-Giralt Laporta S.A., Spain) were chosen. In each process, a TTI was incorporated in the center of ten spheres that were placed in between other spheres, as depicted in Figure 4. Loaded spheres were numbered to observe eventual changes in position. Distilled water or silicon oil of different viscosities (0.340 Ns/m2 — Pleuger, 0.545 Ns/m2 — Fluka, 1.070 Ns/m2 — Fluka, all measured at 20°C) were added as brine up to a predefined headspace (0–15 mm), measured before closing the container.

Still and end-over-end rotary processes (0–25 rpm) were simulated in a one-basket Steriflow simulator (microflow type 911R n°877, Barriquand,

France). Ambient temperatures were recorded every 15 sec. using calibrated copper-constantan thermocouples (Ellab, Denmark) and a calibrated CMC-92 data acquisition system (TR9216, Ellab, Denmark). For rotational processes a slipring contact (DCS85-12, Ellab, Denmark) was used. The container with the TTIs was always positioned at the central rotational axis of the processing installation. Due to the tight packaging of the particles in the container, rotational processes did not dislocate spheres with respect to their initial position. Before heat processing a container, an equilibration at 40°C was carried out to assure an homogeneous initial temperature. Processes consisted of a linear increase in retort temperature from 40°C to constant processing temperature (BSA 10 mg/mL: up to 78°C in 4 min and BSA76: up to 120°C in 10.5 min), followed by a holding period of variable length. Cooling was done as quickly as possible. After heat treatment, TTIs were recovered from the spheres and for each TTI the residual enzymic activity or denaturation enthalpy was determined and the process impact calculated using Equation (2).

6.1.2   Results and discussion

In still processing of a heterogeneous model food system with water as brine, the coldest point, defined as the position where the lethal impact is the smallest, was always located in the lower central part of the container (sphere n°2). A horizontal stratification of process values could be observed, with a cold central axis (spheres n° 2-5-8) (Figure 5). As the viscosity of the brine increased, the coldest spot was moved toward the center of the container (sphere n°5) and process values decreased significantly. By increasing the brine viscosity, the process value distribution increased, which is reflected by the ratio of minimal over maximal process value. In the case of still processing a can using water as brine F_{min}/F_{max} was 0.58, whereas for viscous silicon-oil (1.07 Ns/m^2) F_{min}/F_{max} was 0.13 as measured by BSA76. For still processes, the influence of the headspace on the process lethalities and on the lethality distribution was very small (Figure 6).

Increasing rotational speed resulted in increased processing values as compared to still processing for all brine viscosities studied (Figures 5 and 6). Analogous observations were made by the evaluation of the influence of rotational speed on the process lethality distribution in a glass jar, thermally processed at 105°C in a water cascading retort, using a BLA-based TTI, characterized by a z-value of 7°C (Maesmans et al., 1994). The means of increasing the rate of heat penetration in a filled, closed container by agitation of the contents during processing has been recognized for a long time (Wilbur, 1949). This effect is enhanced by the presence and movement of a headspace bubble and in the case of a particulate food by the movement of the particles. Besides the increase in process lethality due to agitation, end-over-end rotation leveled the spatial variation in processing values, especially for water and low-viscosity silicon oil as brine (see Figures 5 and 6). Few studies in literature, however, describe the influence of rotational speed on the in-pack lethality distribution. For higher brine viscosities the leveling

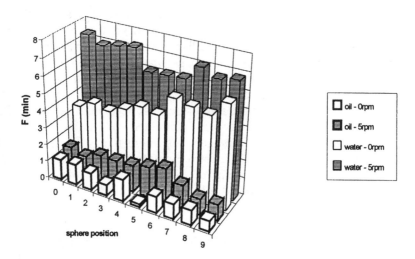

Figure 5 Lethality distribution (T_{ref} = 121°C) in a particulate model food system, measured by read-out of BSA76 (z = 9.6°C). CUT = 10.5 min, t_H = 10 min, T_H = 120°C, HS = 10 mm, brine = water or silicon-oil (viscosity = 1.07 Ns/m²). For sphere positions, refer to Figure 4b.

was less pronounced and often the presence of a coldest particle was still observed. As previously reported in literature (Conley et al., 1951) mixing of the fluid is easier for a lower-viscosity fluid. Hence, temperature differences are maintained for a longer time in viscous fluids, resulting in a broader lethality distribution. In contrast to the negligible effect of a headspace on the process lethality and the lethality distribution in processes without external agitation, in the case of a rotational process, the headspace influenced the process lethalities significantly and reduced the lethality distribution (Figure 6). This can be explained by the increased brine motion in the container and the extra mixing effect due to movement of the headspace bubble.

In this application case study, it was demonstrated that TTIs are sufficiently sensitive to distinguish between in-pack spatial variations of the lethal impact. They allow the determination of the coldest spot in a container in an easy and fast way, which can be very useful in case of, e.g., continuous thermal process design or process evaluation.

6.2   Determination of the coldest zone in a retort

6.2.1   Materials and methods

A lethality distribution study was performed in a pilot plant of a water cascading retort (Steriflow, Barriquand) using BSA76 as TTI. In a Steriflow retort, a low quantity of water is pumped through an external, steam-supplied heat exchanger and the superheated water is sprayed in a closed circuit at a high flow rate on the top tray of the basket. This water is continuously recycled through the heat exchanger. The overpressure is regulated independently from

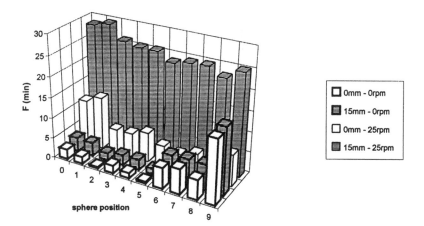

Figure 6 Lethality distribution (T_{ref} = 78°C) in a particulate model food system, measured by read-out of BSA 10 mg/mL (z = 10.2°C). CUT = 4 min, t_H = 40 min, T_H = 78°C, HS = 0 or 15 mm, brine = silicon-oil (viscosity = 0.545 Ns/m²). For sphere positions, refer to Figure 4a.

the heating temperature by injection of compressed air. Cooling is achieved by providing the heat exchanger with cold water, thus using the same water in the retort to cool the product as was used for heating (Figure 7a). The pilot retort is a single basket unit (height 370 mm, width 400 mm and depth 700 mm), which was filled with three layers of 45 cans filled with pet food (109.5 mm height, 36.5 mm radius) per layer. A divider plate with an open area of 52% and 3.1 mm thick was placed between each layer of cans. In each layer several cans filled with 23 sylgard spheres of which sphere 5 contained a TTI (Figure 4b) and water as brine up to a headspace of 10 mm, were placed between cans containing pet food as ballast load (Figure 7b). Initially, an equilibration phase at 40°C was performed to ensure an homogeneous initial temperature. Processes consisted of a linear increase in retort temperature from 40°C to constant processing temperature of 120°C, followed by a holding period of 10 minutes. Cooling was done as quickly as possible. After heat treatment, TTIs were recovered from the spheres, read-out, and the process impact calculated using Equation (2).

6.2.2 Results and discussion

The lethality distribution in the pilot plant of a Steriflow retort as measured by use of BSA76 due to a still process at 120°C is presented in Figure 8. Inspection of the vertical process-value distribution (position 2-6-8, 5-10, 4-9 and 1-7) allows the conclusion that the lethal impact in containers positioned in the retort bottom layer was the lowest and that the process impact increased upward (Figure 8). Taking the working principle of the retort into account, this observation is logical since the superheated water is sprayed on the top, over the containers, and hence the hot water gives up part of its

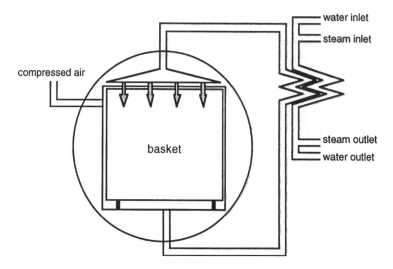

Figure 7a Front view of Steriflow retort (Barriquand). ⇓ represents (overheated) water flow.

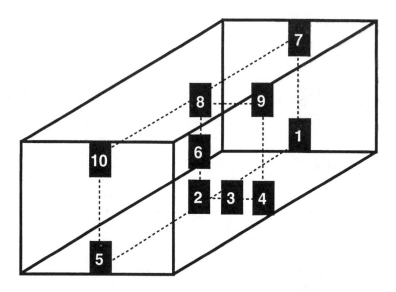

Figure 7b Graphical representation of basket lay-out. Numbered boxes represent cans loaded with TTIs.

heat as it trickles down onto the product on other layers. The occurrence of such a temperature gradient from top to bottom of the retort, especially at the beginning of the sterilization hold, was confirmed by Adams and Hardt-English (1990), who reported that the slowest heating area of the retort basket was on the bottom tray. Ramaswamy et al. (1991), who studied the heat distribution in a single-basket water-cascading retort, also found the smallest

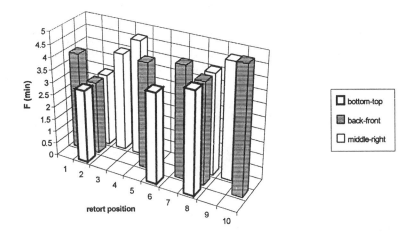

Figure 8 Lethality distribution (T_{ref} = 121°C) in a pilot plant of a water cascading retort, measured by use of BSA76 (z = 9.6°C) incorporated in sphere number 5 (Figure 4b) due to a still process at 120°C. For retort positions, refer to Figure 7b.

accumulated lethality in the Lexan bricks located in the bottom layer, though differences between top, middle, and bottom trays were small (5.6%). With regard to the horizontal lethality distribution, moving from a central position outward, the process impact increases (Figure 8). From the back of the retort to the front (position 1-2-5 and 7-8-10), it can be noticed that containers in the middle of the retort received the lowest lethal effect whereas the highest process impact was observed at the front positions (5 and 10). Going from a central position to the side positions (2-3-4 and 8-9) the process value increased.

From the lethality distribution study it can be concluded that the coldest zone in the retort under study was situated in the lower central part of the basket. Based on process lethality, therefore, the bottom layer appears to be the most conservative choice for process design and evaluation.

7   Conclusions

The development of TTIs to monitor thermal processes in terms of food safety was endeavored in the present study. Taking into account the merits and limitations of different types of TTIs, the development of single-component, isolated, extrinsic enzymatic TTIs seemed a suitable approach. As the main kinetic requirement for proper TTI functioning is the equality in z-value between TTI and target attribute, potential target attributes to monitor safety of thermal processes were identified that allowed the formulation of the objectives of the experimental work in kinetic terms.

For the development of suitable enzymic TTIs to monitor safety of thermal processes, α-amylases from different *Bacillus* species were selected and their degree of thermal inactivation was monitored by measuring the decrease in enzymic activity or the decrease in denaturation enthalpy. Thermal inactivation of BSA, BAA, and BLA at different environmental conditions could all be adequately described by a first-order decay. The dependence of the decimal reduction time on the initial enzyme concentration and the high thermostability of BLA as compared to BSA and BAA allowed developing several enzymic monitoring systems workable in a broad thermal impact range. Testing of several potential TTIs on their integrating properties under time-variable temperature conditions revealed that these systems can be reliably used to monitor with an accuracy of at least 85% the thermal impact on target attributes characterized by an equal z-value. Addition of trehalose to the enzymic environment as well as reduction of the enzymes' moisture content seemed successful approaches to increase thermal stability of enzymic systems, the latter being much more effective. The extensive kinetic studies revealed an extremely useful monitoring system to assess lethal efficacy of sterilization processes of low-acid canned foods, namely BSA equilibrated above a saturated salt solution of an equilibrium relative humidity of 76% at 4°C.

In the application experiments, TTIs were found to be sufficiently sensitive to distinguish between in-pack spatial variations of the lethal impact. It was demonstrated that TTIs can be used in monitoring critical control points (i.e., determination of the in-pack coldest point or evaluation of the coldest zone in a retort) as part of a hazard analysis and critical control points (HACCP) program.

Acknowledgments

This research has been supported by the Flemish Institute for the promotion of scientific–technological research in industry and by the European Commission as part of the Agriculture and Agro-industry, including Fisheries, project AIR1-CT92-0746.

References

Adams, H.W. and Hardt-English, P.K. (1990). Determining temperature distribution in cascading water retorts. *Food Technology,* 42 (12), 110.

Arakawa, T. and Timasheff, S.N. (1982). Stabilization of protein structure by sugars. *Biochemistry,* 21, 6536.

Back, J.F., Oakenfull, D., and Smith, M.B. (1979). Increased thermal stability of proteins in the presence of sugars and polyols. *Biochemistry,* 18 (23), 5191.

Beechem, J.M. (1992). Global analysis of biochemical and biophysical data. *Methods in Enzymology,* 210, 37.

Betts, G.D. (1992). The microbiologic safety of sous-vide processing. In *Technical manual n°39,* Campden and Chorleywood Food Research Association, Chipping Campden, U.K., p. 1–58.

Bigelow, W.D. (1921). The logarithmic nature of thermal death time curves. *Journal of Infectious Diseases,* 29 (5), 528.

Brown, B. (1992). Cooking vs. pasteurisation. In: *Pasteurised food products,* conference proceedings, Campden and Chorleywood Food Research Association, Chipping Campden, U.K.

Carnahan, B., Luther, H.A., and Wilkes, J.O. (1969). Numerical Integration. In: *Applied Numerical Methods,* John Wiley and Sons, New York, p. 69.

Cohen, E. and Saguy, I. (1985). Statistical evaluation of Arrhenius model and its applicability in prediction of food quality losses. *Journal of Food Processing and Preservation,* 9, 273.

Conley, W., Kaap, L., and Shuhmann, L. (1951). The application of "end-over-end" agitation to the heating and cooling of canned food products. *Food Technology,* 5 (11), 457.

De Cordt, S., Avila, I., Hendrickx, M., and Tobback, P. (1994). DSC and protein-based time–temperature integrators: case study of α-amylase stabilized by polyols and/or sugar. *Biotechnology and Bioengineering,* 44 (7), 859.

De Cordt, S., Hendrickx, M., Maesmans, G., and Tobback, P. (1992). Immobilized α-amylase from *Bacillus licheniformis*: a potential enzymic time–temperature integrator for thermal processing. *International Journal of Food Science and Technology,* 27, 661.

De Cordt, S., Hendrickx, M., Maesmans, G., and Tobback, P. (1993). The influence of polyalcohols and carbohydrates on the thermostability of α-amylase. *Biotechnology and Bioengineering,* 43 (2), 107.

Doucet, P. and Sloep, P.B. (1992). Curve fitting. In: *Mathematical modelling in the life sciences;* Bell, G.M. (Ed.), Ellis Horwood, New York, p. 169.

ECFF (1996). *Guidelines for the hygienic manufacture of chilled foods,* European Chilled Food Federation.

Gaze, J.E. (1992). Guidelines to the types of food products stabilised by pasteurisation treatments. In: *Food Pasteurisation treatments,* Technical manual n°27, Campden and Chorleywood Food Research Association, Chipping Campden, U.K.

Gekko, K. and Koga, S. (1983). Increased thermostability of collagen in the presence of sugars and polyols. *Journal of Biochemistry,* 94, 199.

Gerlsma, S.Y. (1968). Reversible denaturation of ribonuclease in aqueous solutions as influenced by polyhydric alcohols and some other additives. *The Journal of Biological Chemistry,* 243 (5), 957.

Gerlsma, S.Y. (1970). The effects of polyhydric and monohydric alcohols on the heat induced reversible denaturation of chymotrypsinogen A. *European Journal of Biochemistry,* 14, 150.

Gerlsma, S.Y. and Stuur, (1972). The effect of polyhydric and monohydric alcohols on the heat-induced reversible denaturation of lysozyme and ribonuclease. *International Journal of Peptide and Protein Research,* 4, 377.

Gray, C.J. (1988). Additives and enzyme stability. *Biocatalysis,* 1, 187.

Greenspan, L. (1977). Humidity fixed points of binary saturated aqueous solutions. *Journal of Research of the National Bureau of Standards-A.* Physics and Chemistry, 81 A(1), 89.

Hägerdal, B. and Martens, H. (1976). Influence of water content on the stability of myoglobin to heat treatment. *Journal of Food Science,* 41, 933.

Haralampu, S.G., Saguy, I., and Karel, M. (1985). Estimation of Arrhenius model parameters using three least squares methods. *Journal of Food Processing and Preservation,* 9, 129.

Hendrickx, M., Maesmans, G., De Cordt, S., Noronha, J., Van Loey, A., Willockx, F., and Tobback, P. (1993). Advances in process modelling and assessment: the physical mathematical approach and product history integrators. In: *Minimal Processing of Foods and Process Optimisation*, Singh, R.P. and Oliveira, F.A.R. (Eds.), CRC Press, Boca Raton, FL, p. 315.

Hendrickx, M., Maesmans, G., De Cordt, S., Noronha, J., Van Loey, A., and Tobback, P. (1995a). Evaluation of the integrated time–temperature effect in thermal processing of foods. *Critical Reviews in Food Science and Nutrition*, 35 (3), 231.

Hendrickx, M., Saraiva, J., Lyssens, J., Oliveira, J., and Tobback, P. (1992). The influence of water activity on thermal stability of horseradish peroxidase. *International Journal of Food Science and Technology*, 27, 33.

Hendrickx, M., Van Loey, A., and Tobback, P. (1995b). Assessment of safety of thermal processes using process indicators. In: *Proceedings of the International Symposium on New Shelf Life Technologies and Safety Assessment*, Ahvenainen, R., Mattila-Sandholm, T., and Ohlsson, T. (Eds.), VTT, Espoo, Finland, p. 35.

Hendrickx, M., Van Loey, A., Van den Broeck, I., Haentjens, T., and Tobback, P. (1995d). The development of new time–temperature integrators (product history indicators) for the quantification of thermal processes in terms of food safety and quality. In: *Proceedings of the First Main Meeting*, Project Process optimization and minimal processing of foods, Vol. 1 — Thermal Processing, Oliveira, J., Oliveira, F., and Hendrickx, M. (Eds.), Published by Escola Superior de Biotecnologia, Porto, Portugal, p. 1.

Hendrickx, M., Van Loey, A., Van den Broeck, I., and Tobback P. (1995c). Techniques to evaluate preservation processes: mathematical modelling and use of product history integrators. In: *Proceedings of the Conference on Technologias Avanzadas en Esterilizacion y Seguridad de Alimentos y Otros Productos*, Rodrigo, M., Martinez, A., Fiszman, S.M., Rodrigo, C., and Mateu, A. (Eds.), IATA, Valencia, Spain, p. 9.

Johnson, M.L. (1992). Review: why, when and how biochemists should use least squares. *Analytical Biochemistry*, 206 (2), 215.

Kindle, K.L. (1983). Review: Characteristics and production of thermostable α-amylase. *Applied Biochemistry and Biotechnology*, 8, 153.

Klibanov, A.M. (1983). Stabilisation of enzymes against thermal inactivation. *Advances in Applied Microbiology*, 29, 1.

Klibanov, A.M. (1986). Enzymes that work in organic solvents. *Chemtech*, June, 354.

Lund, B.M. and Notermans (1993). Potential hazards associated with REPFEDS. In: *Clostridium botulinum ecology and control in foods*, Hauschild, A.H.W. and Dodds, K.L. (Eds.) Marcel Dekker, New York, p. 279.

Maassen, A. (1991). Vergleich zweier α-amylasen aus *B. amyloliquefaciens* und *B. licheniformis. Biomed. Acta Biochim.*, 50 (2), 213. In German.

Maesmans, G. (1993). *Possibilities and limitations of thermal process evaluation techniques based on Time Temperature Integrators*. Ph. D. Thesis, Katholieke Universiteit te Leuven, Faculty of Agricultural Sciences, Centre for Food Science and Technology, Leuven, Belgium.

Maesmans, G., Hendrickx, M., De Cordt, S., Van Loey, A., Noronha, J., and Tobback, P. (1994). Evaluation of process value distribution with Time Temperature integrators. *Food Research International*, 27, 413.

Maesmans, G., Hendrickx, M., Weng, Z., Keteleer, A., and Tobback, P. (1990). Endpoint definition, determination and evaluation of thermal processes in food preservation. *Belgian Journal of Food Chemistry and Biotechnology*, 45 (5), 179.

Mannervik, B. (1982). Regression analysis, experimental error, and statistical criteria in the design and analysis of experiments for discrimination between rival kinetic models. *Methods in Enzymology,* 87, 370.

Meerdink, G. and van 't Riet, K. (1991). Inactivation of a thermostable α-amylase during drying. *Journal of Food Engineering,* 14, 83.

Morgan, F.J. and Priest, F.G. (1981). Characterization of a thermostable α-amylase from *Bacillus licheniformis* NCIB 6346. *Applied Bacteriology,* 50, 107.

Mullaney, P.F. (1966). Dry thermal inactivation of trypsin and ribonuclease. *Nature,* 210, 953.

Multon, J.L. and Guilbot, A. (1975). Water activity in relation to the thermal inactivation of enzymic proteins. In: *Water relations to foods,* Duckworth, R.B. (Ed.), Academic Press, London, p. 379.

Pflug, I.J. (1987a): *Textbook for an introduction course in the microbiology and engineering of sterilization processes.* Sixth edition, Environmental Sterilization Laboratory, Minneapolis, USA

Pflug, I.J. (1987b). Using the straight-line semilogarithmic microbial destruction model as an engineering design model for determining the F-value for heat processes. *Journal of Food Protection,* 50 (4), 342.

Pflug, I.J. (1987c). Endpoint of a preservation process. *Journal of Food Protection,* 50 (4), 347.

Pflug, I.J. (1987d). Factors important in determining the heat process value F for low-acid canned foods. *Journal of Food Protection,* 50 (6), 528.

Pflug, I.J. (1987e). Calculating F-values for heat preservation of shelf-stable, low-acid canned foods using the straight-line semilogarithmic model. *Journal of Food Protection,* 50 (7), 608.

Ramaswamy, H., Campbell, S., and Passey, C. (1991). Temperature distribution in a standard 1-basket water-cascade retort. *Canadian Institute of Science and Technology Journal,* 24 (1/2), 19.

Rüegg, M., Moor, U., and Blanc, B. (1975). Hydration and thermal denaturation of β-lactoglobulin: a calorimetric study. *Biochimica et Biophysica Acta,* 334.

Saraiva, J.M.A. (1994). *Effect of environmental aspects on enzyme heat stability and its application in the development of time–temperature integrators,* Ph.D. Thesis, Escola Superior de Biotecnologia, Porto, Portugal.

SAS (1982). *SAS User's guide: Statistics,* 1982 Edition, SAS Institute Inc., Cary, U.S.A.

Taoukis, P.S. and Labuza, T.P. (1989a). Applicability of time–temperature indicators as shelf life monitors of food products. *Journal of Food Science,* 54 (4), 783.

Taoukis, P.S. and Labuza, T.P. (1989b). Reliability of time–temperature indicators as food quality monitors under nonisothermal conditions. *Journal of Food Science,* 54 (4), 789.

Tomazic, S.J. and Klibanov, A.M. (1988a). Mechanisms of irreversible thermal inactivation of *Bacillus* α-amylases. *The Journal of Biological Chemistry,* 263 (7), 3086.

Tomazic, S.J. and Klibanov, A.M. (1988b). Why is one *Bacillus* α-amylase more resistant against irreversible thermoinactivation than another? *The Journal of Biological Chemistry,* 263 (7), 3092.

van Boekel, M.A.J.S. (1996). Statistical aspects of kinetic modeling for food science problems. *Journal of Food Science,* 61 (3), 477.

Van Loey, A., Ludikhuyze, L., Hendrickx, M., De Cordt, S., and Tobback, P. (1995) Theoretical consideration on the influence of the z-value of a single component time–temperature integrator on thermal process impact evaluation, *Journal of Food Protection,* 58 (1), 39.

Van Loey, A., Hendrickx, M., De Cordt, S., and Tobback, P. (1996a). Quantitative evaluation of thermal processes using time–temperature integrators. *Trends in Food Science and Technology,* 7 (1), 16.

Van Loey, A., Hendrickx, M., Ludikhuyze, L., Weemaes, C., Haentjens, T., De Cordt, S., and Tobback, P. (1996b). Potential *Bacillus subtilis* α-amylase based time–temperature integrators to evaluate pasteurization processes. *Journal of Food Protection,* 59 (3), 261.

Van Loey, A., Arthawan, A., Hendrickx, M., Haentjens, T., and Tobback, P. (1997a) The development and use of an α-amylase based Time Temperature Integrator to evaluate in pack pasteurisation processes. *Lebensmitteln Wissenschaft und Technologie,* 30, 94.

Van Loey, A., Haentjens, T., Hendrickx, M., and Tobback, P. (1997b). The development of an enzymic time–temperature integrator to assess thermal efficacy of sterilization of low-acid canned foods. *Food Biotechnology,* 11(2), 147–168.

Van Loey, A., Hendrickx, M., Haentjens, T., Smout, C., and Tobback, P. (1997c). The use of an enzymic of Time Temperature Integrator to monitor lethal efficacy of sterilization of low-acid canned foods. *Food Biotechnology,* 11(2), 169–188.

Volkin, D.B., Staubli, A., Langer, R., and Klibanov, A.M. (1991). Enzyme thermoinactivation in anhydrous organic solvents. *Biotechnology and Bioengineering,* 37, 843.

Weemaes, C., De Cordt, S., Goossens, K., Ludikhuyze, L., Hendrickx, M., Heremans, K., and Tobback, P. (1996). High pressure, thermal, and combined pressure-temperature stabilities of α-amylase from *Bacillus* species. *Biotechnology and Bioengineering,* 50, 49.

Weng, Z., Hendrickx, M., Maesmans, G., and Tobback, P. (1991a). Immobilized peroxidase: a potential bioindicator for evaluation of thermal processes. *Journal of Food Science,* 56 (2), 567.

Weng, Z., Hendrickx, M., Maesmans, G., Gebruers, K., and Tobback, P. (1991b). Thermostability of soluble and immobilized horseradish peroxidase. *Journal of Food Science,* 56 (2), 574.

Wilbur, P.C. (1949). Factors influencing process determination in agitating pressure cookers. *The Canner,* May 7, 13, 26.

Zaks, A. and Klibanov, A.M. (1984). Enzymatic catalysis in organic media at 100°C. *Science,* 224, 1249.

chapter three

Possibilities and limitations of computational fluid dynamics for thermal food process optimization

Pieter Verboven, Nico Scheerlinck, Josse De Baerdemaeker, and Bart M. Nicolaï

Contents

Summary 42
1 Introduction 42
2 Physical background 44
   2.1 Conservation equations 44
   2.2 Additional information 44
   2.3 Additional features of CFD 46
      2.3.1 Multicomponent multiphase flow 46
      2.3.2 Porous flow 48
      2.3.3 Turbulence 49
   2.4 Initial and boundary conditions 52
3 Numerical solution procedures 53
   3.1 Discretization 53
   3.2 Solution of the discretized equations 56
4 Examples in thermal food processing 59
   4.1 Forced convection oven 59
   4.2 Calculation of local surface heat transfer coefficients 60
5 Final remarks 61
Acknowledgments 64
References...64

Summary

The general principles of computational fluid dynamics (CFD) are covered in this text. The theoretical aspects and practical implications of the several assumptions that are made to generate a manageable set of equations are discussed. The text focuses on the application of CFD to thermal processing of foods, highlighting its potential for process optimization and design, with a particular concern with the assurance of product safety. A case study is presented on the modeling of a forced convection oven. It is shown that the method provides an accurate insight into the colder (stagnant) zones of the oven. A summary of the current advantages and limitations of CFD is given.

1 Introduction

In many thermal food processes, several product units are processed simultaneously, in batch or continuously. To assure a uniform quality and microbial inactivation in all units, the heat transfer must be as uniform as possible throughout the heating equipment. Especially in the case of minimal processes, small deviations of the processing variables may result in large deviations in food product temperatures. Nicolaï and De Baerdemaeker (1996) showed that a slab of 3 cm of immobilized water, initially at a temperature of 20°C, and heated in a medium of 100°C with a surface heat transfer coefficient of 80 W/m²°C or 4 W/m²°C, will have a change in center temperature at 1800s of processing of 0.2°C and 5°C per unit deviation of the surface heat transfer coefficient, respectively. These severe deviations in food temperature may result in unacceptable deviations of the quality attributes and the microbial load. In this context, the optimization of the design of the heat processing equipment is of great importance.

In traditional heat processing practice the processing medium is air, some form of water, or a mixture of both. This text will not deal with alternative technologies such as microwave heating or high pressure. In the former cases, the product is heated by convection from the heating medium, conduction from the shelves, radiation of any surrounding physical surfaces, and latent heat transfers due to evaporation and condensation at the product surface. Obviously, all the above processes are dependent on the design of the heat processing equipment and the properties of the product. One can imagine that the optimization of the design of the processing unit is a difficult task, because of the interaction of many physical phenomena on a relatively small scale in a complex geometry.

Of main interest is the food temperature and how it is influenced by the above processes. Therefore, the distribution of the processing variables (heating medium temperature, velocity, and water content) must be established. When the physical parameters of the materials involved (air, water, food, and processing equipment) and the geometry of the problem are known, the processing variables can be obtained from the conservation laws of mass, momentum, and energy. Unfortunately, these equations have inherited the complexity of the transfer processes they describe, and they are coupled and highly nonlinear. Analytical solutions exist only in a small number of simplified cases. The most common approach is to use simplified equations, which result from a combination of approximations and dimensional analysis and require considerable experimental input. This empirical approach is very efficient when global parameters, such as heat-transfer coefficients, are to be estimated for a specific geometry under specific conditions. Clearly, with the objective of design optimization in mind, details of the distribution of the processing variables are necessary. At this scale of detail, experiments become very time-consuming and costly.

Alternatively, with the increasing computational power and memory capacity of computers and the development of efficient numerical algorithms, the numerical solution of the governing fluid flow, mass and energy equations has become possible. The availability of user-friendly software-codes (*CFX*, AEA Technology, Harwell, UK; *Phoenix*, Flowsolve Ltd, London, UK; *Fluent/FIDAP*, Fluid Dynamics International, Evanston, USA; *STAR-CD*, Computational Dynamics Ltd, London, UK) has brought the method within reach of the engineer. This approach of using computers for solving fluid flow problems is known as computational fluid dynamics (CFD). Being used for many years in high-technology engineering (aeronautic, aerospace, and nuclear industries), the method has become increasingly popular in other fields as well, such as food engineering (Datta and Teixeira, 1987; Mirade et al., 1995; Verboven et al., 1997a).

This chapter will deal with the application of CFD to thermal food processing. The basic equations will be described and the numerical solution method briefly explained. The state of the art of the possibilities and the current limitations of CFD for use in food engineering applications will be discussed. One of the key questions throughout this chapter will be: "To what extent can CFD be an alternative and what else is needed in a complementary approach?" A first limitation for industrial use may certainly still be the investment cost. For a typical design study cycle, one needs a few man-months of a highly skilled CFD analysis, a fast computer (workstations are mostly used, although PCs are gaining on) with a high memory capacity (minimum 128 Mb RAM), and a commercial software code according to the personal needs and preferences. This may add up to a considerable amount of money, which probably is still a major drawback for many food-related companies. Nevertheless, the commercial benefits can be tremendous. Rhodes (1995) states a two-thirds reduction in the energy-cost of steel-making and a fuel cost saving of £500,000 per annum for a power station steam condenser, both through improvements resulting from the application of CFD. From a consumer point of view, if design optimization with CFD can considerably reduce the microbial risk and improve the quality of food products, the balance is quickly made.

2 Physical background

2.1 Conservation equations

Restricting the discussion to a single Newtonian fluid system, applying the conservation principle to a fixed infinitesimal control volume $dx_1.dx_2.dx_3$, leads to the continuity, momentum, and energy equations, written in index notation for Cartesian coordinates x_i ($i = 1,2,3$), and whenever an index appears twice in any term, summation over the range of that index is implied:

$$\frac{\partial \rho}{\partial t} + \frac{\partial(\rho u_j)}{\partial x_j} = 0 \tag{1}$$

$$\frac{\partial(\rho u_i)}{\partial t} + \frac{\partial(\rho u_j u_i)}{\partial x_j} = \frac{\partial}{\partial x_j}\mu\left(\frac{\partial u_i}{\partial x_j} + \frac{\partial u_j}{\partial x_i}\right) - \frac{\partial}{\partial x_i}\left(p + \frac{2}{3}\mu\frac{\partial u_j}{\partial x_j}\right) + f_i \tag{2}$$

$$\frac{\partial(\rho h)}{\partial t} + \frac{\partial(\rho u_j h)}{\partial x_j} = \frac{\partial}{\partial x_j}\left(k\frac{\partial T}{\partial x_j}\right) + \frac{\partial p}{\partial t} + q_h \tag{3}$$

where u_i ($i = 1,2,3$) are the Cartesian components of the velocity vector U, p the pressure, h the static enthalpy, and T the temperature. The fluid properties involved are the density ρ, the viscosity μ and the thermal conductivity k. The terms f_i and q_h represent sources of momentum (external forces) and energy (heat sources) per unit volume, respectively.

The gravitational force in Equation (2) is incorporated into the pressure p. Since only the pressure gradient appears in the equation, the absolute value of the pressure is not important, except in compressible flows at high velocities.

The energy equation is written in an enthalpy (h) formulation. It is obtained after some algebra from the total energy (E) equation based on the first law of thermodynamics. The relations $E = e + \rho U^2/2$, where e is the internal energy, and $h = e + p/\rho$ are used, and potential energy effects are neglected. The pressure term in (3) is often omitted when dealing with incompressible, low-velocity flows. For the same reason, the viscous work term in the original equation is neglected. A full derivation of these equations can be found in any text book on computational fluid dynamics (Versteeg and Malalasekera, 1995; Ferziger and Peric, 1996).

2.2 Additional information

The system of 5 equations contains 7 variables (u_i, p, h, T, ρ). Additional equations are therefore needed to find a closed solution. The thermodynamic equation of state gives the relation between density ρ, pressure p, and temperature T. The constitutive equation relates the enthalpy h to pressure and temperature. The following equations are mostly used, assuming perfect gas behavior:

$$\rho = \frac{pM}{RT} \tag{4}$$

$$C_p = \left(\frac{\partial h}{\partial T}\right)_p \tag{5}$$

with M the molecular weight of the fluid, R the universal gas constant, and C_p the specific heat capacity at constant pressure. When the heat capacity is assumed constant, the constitutive equation reduces to a linear relation between h and the difference between the actual temperature, T, and a reference temperature. Several formulas can however be used to expand C_p for temperature, and an integral formulation of Equation (5) applies.

Since only relatively low velocities are encountered in the food processes under consideration, the flow is often assumed incompressible. In the case of fully incompressible flow, the density ρ is simply constant and Equation (4) is not used. In this case, density gradients due to small temperature differences can be incorporated using the linear Boussinesq approximation, using a reference density ρ_0 at a reference temperature T_0 and the thermal expansion coefficient β of the fluid:

$$\rho g_i = \rho_0 g_i + (\rho - \rho_0)g_i = \rho_0 g_i - \rho_0 g_i \beta(T - T_0) \tag{6}$$

where g_i is the gravitational acceleration in the i-direction (i = 1,2,3). The first term of Equation (6) is usually absorbed into the pressure term. With this approximation, errors of the order of 1% are obtained if the temperature differences in air, for example, are kept below 15°C (Ferziger and Peric, 1996). In a "weakly" compressible flow assumption, formula 4 is used at the reference atmospheric pressure, ignoring small pressure fluctuations, making the model more suitable for larger temperature differences.

In incompressible or weakly compressible flows it is noted that there is no equation for the pressure, which cannot be obtained directly. Because the dominant variable in the continuity equation has disappeared (ρ = constant), the common approach is to construct the pressure field so as to assure global continuity. The continuity equation is used as a constraint on the momentum equations in order to produce an equation for pressure. This is called pressure-velocity coupling or pressure-correction and several algorithms for its numerical solution have been developed (Caretto et al., 1972; Patankar, 1980; Van Doormaal and Raithby, 1984; Issa, 1986).

2.3 Additional features of CFD

2.3.1 Multi-component multiphase flow

The concept of multiphase flow in CFD is wider than just the transport of multiple thermodynamic phases. Any mixture of species may be considered. Multiphase transport models describe the mixed flow on a macroscopic scale, where species may have different temperature and velocity. Multiphase flow may also include transport of small solid particles, when the number of particles is large. In this case, the solids are regarded as a fluid. In multiphase flow the mixing is assumed to occur on a smaller scale than the spatial resolution of the model, so that each phase occupies a certain volume fraction in each region considered. The general model for a multiphase system is:

$$\frac{\partial\left(r_\alpha \rho_\alpha\right)}{\partial t} + \frac{\partial\left(r_\alpha \rho_\alpha u_j^\alpha\right)}{\partial x_j} = \sum_{\beta=1}^{n_p}\left(\dot{m}_{\alpha\beta} - \dot{m}_{\beta\alpha}\right) \tag{7}$$

$$\frac{\partial\left(r_\alpha \rho_\alpha u_i^\alpha\right)}{\partial t} + \frac{\partial\left(r_\alpha \rho_\alpha u_j^\alpha u_i^\alpha\right)}{\partial x_j} =$$

$$\frac{\partial}{\partial x_j} r_\alpha \mu_\alpha\left(\frac{\partial u_i^\alpha}{\partial x_j} + \frac{\partial u_j^\alpha}{\partial x_i}\right) - \frac{\partial}{\partial x_i} r_\alpha\left(P_\alpha + \frac{2}{3}\mu_\alpha \frac{\partial u_j^\alpha}{\partial x_j}\right) + r_\alpha f_i \tag{8}$$

$$+ \sum_{\beta=1}^{n_p} c_{\alpha\beta}^d\left(u_i^\beta - u_i^\alpha\right) + \sum_{\beta=1}^{n_p}\left(\dot{m}_{\alpha\beta} u_i^\beta - \dot{m}_{\beta\alpha} u_i^\alpha\right)$$

$$\frac{\partial\left(r_\alpha \rho_\alpha h_\alpha\right)}{\partial t} + \frac{\partial\left(r_\alpha \rho_\alpha u_j^\alpha h_\alpha\right)}{\partial x_j} = \frac{\partial}{\partial x_j}\left(r_\alpha k_\alpha \frac{\partial T_\alpha}{\partial x_j}\right) + r_\alpha q_h$$

$$\tag{9}$$

$$+ \sum_{\beta=1}^{n_p} c_{\alpha\beta}^h\left(T_\beta - T_\alpha\right) + \sum_{\beta=1}^{n_p}\left(\dot{m}_{\alpha\beta} h_{\alpha\beta} - \dot{m}_{\beta\alpha} h_{\beta\alpha}\right)$$

These equations are similar to the original conservation equations for incompressible or weakly compressible flows, written for each of the phases α. The symbol r_α represents the volume fraction of phase α, ρ_α is the phase density, u_i^α ($i = 1,2,3$) the three Cartesian components of the velocity vector of phase α, p_α the partial pressure of phase α, h_α is the enthalpy, and T_α the temperature of phase α. Additional transport terms relate to interphase mass transport $m_{\alpha\beta}$ (phase transition from phase β into phase α), interphase drag, using the drag coefficients $c_{\alpha\beta}^d$ and interphase heat transfer, using the heat transfer coefficients $c_{\alpha\beta}^h$. The symbol $h_{\alpha\beta}$ in Equation (9) represents the heat needed for phase transition from phase β to phase α.

Sometimes some simplifications can be made when considering that the flow consists of multiple components instead of different phases. This assumption is valid when transport occurs on the microscopic level. Mass transfer takes place by diffusion and convection of the different species that move at the same speed and have the same temperature. This approach is particularly suited for moist air calculations. For a single component fluid, the transport equation for component a is:

$$\frac{\partial(\rho M_a)}{\partial t} + \frac{\partial(\rho u_j M_a)}{\partial x_j} - \frac{\partial}{\partial x_j}\left(\Gamma_a \frac{\partial}{\partial x_j} M_a\right) = 0 \tag{10}$$

$$\sum_{a=0}^{N_a} M_a = 1 \tag{11}$$

with M_a the mass fraction of component a, and Γ_a the diffusivity of a in the mixture. It should be noted that Γ_a is not constant in compressible flows and always depends on the gradients of all other species present. There are N_a+1 species in total. Species 0 is regarded as the carrier fluid and it determines the thermal conductivity, the flow viscosity, and the diffusivity of the fluid, which are used in the transport Equations 1 to 3. The evaporation/condensation at surfaces must be incorporated by suitable source terms. This approach is valid only for laminar flows with low concentrations of the additional species and for high turbulent flows, where molecular transport is negligible compared to turbulent transport (Anon., 1995). The density and the specific heat capacity are calculated from the mass fractions of the species.

The above equations can be applied to model mixtures of air/water/steam that are encountered in food heat-transfer processes. However, multiphase and multicomponent modeling is still a very active domain of research. The difficult task is the determination of the complex relation between heat and mass transfer, temperature, and pressure in the system. Multiphase flows require a considerable input of psychrometric data (Anon., 1993). The amount of parameters to be established in these types of models may well allow the user to fit every experimental dataset. Last but not least, when dealing with these complex relations, one must be able to handle the computational difficulties that may arise from them. Applied to heat exchanger design, progress is being made in multiphase modeling of boiling flows (Wadekar, 1997). Yan et al. (1997) worked out a multiphase model of bubbly water/air and water/steam flows, including mass transfer, drag, interphase heat transfer, and turbulence. A good example of a multicomponent model has been developed by Khankari et al. (1995) for the natural convection moisture transport in the storage of grain, which also involves porous flow.

2.3.2   *Porous flow*

Flow through porous media can be described using a CFD model. In food and post-harvest technology, this type of flow can occur in storage of packed foods or bulks of agricultural produce, where it is too difficult to model individual product units, and during the internal moisture transport in the food, where the food structure is hard to model explicitly. In these cases the problem is regarded to be porous on a large scale because the geometry is too complex to resolve on a small scale. The general conservation equations become the following, using an isotropic area porosity equal to the volume porosity γ:

$$\frac{\partial(\gamma\rho)}{\partial t} + \frac{\partial(\gamma\rho u_i)}{\partial x_j} = 0 \tag{12}$$

$$\frac{\partial(\gamma\rho u_i)}{\partial t} + \frac{\partial(\gamma\rho u_j u_i)}{\partial x_j} =$$

$$\frac{\partial}{\partial x_j}\gamma\mu\left(\frac{\partial u_i}{\partial x_j} + \frac{\partial u_j}{\partial x_i}\right) - \frac{\partial}{\partial x_i}\gamma\left(p + \frac{2}{3}\mu\frac{\partial u_j}{\partial x_j}\right) - R_i\bar{u}_i \tag{13}$$

$$\frac{\partial(\gamma\rho h)}{\partial t} + \frac{\partial(\gamma\rho u_j h)}{\partial x_j} = \frac{\partial}{\partial x_j}\left(\gamma k\frac{\partial T}{\partial x_j}\right) + \gamma\frac{\partial p}{\partial t} + \gamma q_h \tag{14}$$

The last term in the momentum equations reflects the resistance to the flow and is expressed in terms of the superficial velocity $\bar{u}_i = \gamma u_i$. The parameter R_i may consist of a constant and a velocity-dependent component:

$$R_i = R_i^c + R_i^u|\bar{U}| \tag{15}$$

A typical formula for the resistance is the Ergun-equation (Ergun, 1952), which depends on porosity and particle size. However, experimentation is usually required for the accurate determination of relation 13.

When the Reynolds number of the porous flow is low (low velocity, high viscosity, or small dimensions), the flow is dominated by viscous, pressure, and body forces. The convective term in the momentum equations becomes negligible and the flow is termed creeping flow. The resulting linear equations are called Stokes equations. In the limit of very low velocity, the unsteady terms can also be neglected. When the resistance is large, a further simplification becomes valid because the viscous term is then dominated by the pressure and resistance forces. In this case an anisotropic version of

Darcy's law applies, in terms of the actual flow velocity components rather than the continuous average velocity. It should be noted that the discontinuity in velocity at the porous region interface and the strong pressure gradients in porous flow could give rise to severe convergence problems (see below).

Porous flow modeling has received considerable attention over the last few years. Some examples include the modeling of air flows in the storage of agricultural products (Khankari et al., 1995; Xu and Burfoot, 1997; Bibby, 1997).

2.3.3 *Turbulence*

Forced convection heating and cooling of foods often occurs in highly turbulent conditions. The presence of flow obstructions such as baffles, shelves, and the foods themselves are the cause of the flow becoming seemingly unstructured. This feature of the flow should be considered advantageous because of the mixing capacity of swirling eddies, which cause a strong increase of the heat transfer across the main flow direction (Bejan, 1984). Because of this effect, turbulence is an important feature, which cannot be neglected in the model formulation.

In simplified cases (e.g., pipes, flat plates, cylinders or simple prismatic geometry) general properties of the flow and geometry can be used to decide whether the empirically established barrier for the onset of turbulence has been reached. This information can then be used to switch to another empirical formula for the calculation of flow and heat transfer properties or, in CFD, to switch to a turbulent version of the model. In most practical cases, however, defining a criterion for turbulence onset is not straightforward.

Turbulence is a complicated state of the flow: the flow variables fluctuate in a large range of scales, both in time and space. Typically, the size of the eddies can vary in a 1000-fold range or more (Moin and Kim, 1997). Furthermore, turbulence actually is a local phenomenon, rather than a global feature of the flow. It is evident that the mathematical description of turbulence is a difficult task. If both the entire region of interest and the smallest turbulence scale must be resolved, one is soon limited by computer resources. The amount of grid points required to resolve all scales is estimated to be proportional to the Reynolds number raised to the 9/4 power. Only for simplified cases and low Reynolds numbers is it currently possible to perform such direct numerical simulations on supercomputers (Leonard, 1995). Simulation shortcuts are possible at different levels of complexity and approximations. The least approximations are needed in large eddy simulations, in which case the largest eddies are resolved but the effects of smaller eddies are estimated by additional models (Ferziger and Peric, 1996). This approach is now being more widely used, since it is becoming within reach of the available computer power.

The least computationally demanding and still most popular approach is using the Reynolds averaged Navier-Stokes (RANS) equations, which result from averaging the governing Equations 1 to 3 and including the effect of the turbulent fluctuations by additional models for the new terms appearing in the RANS equations. The Boussinesq assumption is the basis of the classical and widely used "eddy viscosity" models. Here the turbulence is accounted for by a turbulent "viscosity," which is incorporated in the viscous and thermal diffusion transport terms. In k-ε models, originally proposed by Jones and Launder (1972), the turbulent viscosity is obtained as a function of the turbulent variables k, which represent the turbulent kinetic energy associated with the fluctuating components of the flow velocities, and ε, the turbulent energy dissipation rate.

$$\mu_t = \rho C_\mu \frac{k^2}{\varepsilon} \tag{16}$$

The constant C_μ may be assumed constant for equilibrium conditions, where the turbulence production nearly equals the turbulence dissipation.

Additional transport equations have been derived for these turbulent flow variables. Several undefined constants appear in the model equations, which together with several assumptions and the specific near-wall treatment render this model empirical. Although the standard version of the k-ε model is the most validated turbulence model available, it has been reported to perform poorly in a number of flows, including curved boundary layers, low Reynolds number flows, and recirculating flows, especially near the separation and reattachment points. Unfortunately, the air flow over complexly shaped foods involves these conditions, and a performance assessment of the k-ε model is necessary for each application under study.

The transport equation for the turbulent energy k is well established. The reader is referred to basic textbooks on CFD (e.g., Versteeg and Malalasekera, 1995). It involves only minor assumptions. The principal diffusive transport of k is modeled using the diffusion gradient idea as introduced in the basic eddy-viscosity hypothesis: the diffusivity of k is expressed as the turbulent viscosity divided by an empirical turbulent Prandtl number, σ_k. The pressure-transport term of the exact equation is absorbed into the diffusion gradient term of k. The exact ε-equation, on the other hand, contains many unmeasurable unknowns and its derivation is not as straightforward. A "modeled" transport equation similar to the k equation is therefore mostly used, and it contains different terms and values for constants depending on the approach of derivation. Basically, there are three main approaches to obtain k-ε models, resulting in a standard k-ε model, a RNG (renormalization group) k-ε model and a LRN (low Reynolds number) k-ε model.

In the standard k-ε model (Jones and Launder, 1972), the ε-equation is similar to the k-equation, containing comparable convective and diffusive transport terms. The complete standard k-ε model contains five constants, for which values have been obtained by comprehensive data fitting over a wide range of turbulent flows. Near surfaces, there is a direct viscous effect on the flow due to the no-slip condition at the wall, which is not reflected in the transport equations of the standard k-ε model. Wall functions are therefore used to track the variables from a point outside this viscous sub-layer toward the wall. This wall boundary condition is only valid in high Reynolds number equilibrium flows. Despite its restriction to such conditions, this approach is popular as it saves the computational efforts needed to resolve the model into the narrow viscous sublayer close to the surface.

The renormalization group (RNG) method describes turbulence from a statistical mechanics approach. Small-scale motions are systematically removed from the governing equations to the point that turbulence can be described with a large-scale model resolvable with available computer capacities, such as a k-ε model (Yakhot et al., 1992). The advantage of this approach is that the constants in the equations are calculated explicitly, rather than determined from experiments. Additionally, a new term appears in the ε-equation, which accounts for anisotropy in strongly strained turbulent flows. This term is incorporated through a modeled constant in the production term, based on the equilibrium assumption, restricting the RNG k-ε model to a coarse grid approach near walls, as was the case for the standard k-ε model. The same wall functions as in the standard k-ε model can be used.

In order to improve predictions of flows at low Reynolds numbers (5000<Re<30000), efforts have been invested to improve the k-ε model. In this case, the log-law is no longer valid and the model must be resolved into the viscous sublayer (Launder and Sharma, 1974). In this model, in order to ensure that viscous stresses take over from the turbulent ones near the wall, damping of the turbulent viscosity in formula 1 and the dissipation term in the ε equation have been applied. The exact form of the damping function depends on the flow geometry and has been determined experimentally by Launder and Sharma. These default functions have proven their success in a variety of boundary layer flows. Additionally, viscosity-dependent terms are introduced into the k and ε equations. A very fine grid is needed in this case to resolve the steep gradients of the variables near the wall in the linear viscous region and wall functions are no longer needed.

A performance assessment study of these three turbulence models has been performed by Verboven et al. (1997b) for a typical forced-convection heating process of complexly shaped foods. The conclusions of this work were that the boundary layers are badly represented by the wall function approach and the departure from local equilibrium is not accounted for. A correction function can be added to correct for the latter behavior in conjunction with the low-Reynolds-number model (see also Yap, 1987). Nevertheless, it was found that experimental input for these corrections is needed in order to determine important constants.

More complex closures for the RANS models are based on dynamic equations for the Reynolds stresses and fluxes themselves in the RANS equations. In addition to the equations for the mean flow, this approach results in seven more partial differential equations. These models are believed to be more accurate but require a better insight into the process of turbulence, and care must be taken with their numerical solution. Finally, it must be noted that new turbulence models are constantly proposed and tested.

2.4   Initial and boundary conditions

Because only a subdomain of the total physical domain is studied, conditions on the boundaries of the subdomain and initial conditions are required. Due to the complex mathematical nature of the Navier-Stokes equations, there are no conclusive general rules for the implementation of boundary conditions in order to have a well-posed problem. The needed boundary conditions depend mainly on the flow type. A detailed discussion has been given by Hirsch (1991). Practically, for incompressible and weakly compressible flows, the boundary conditions described below are often applied.

Because of the special relationship between pressure and velocity, these two variables cannot be supplied simultaneously at boundaries. Initially, values for all variables must be provided.

At *inlet boundaries* (upstream), values are specified for all the quantities involved in the model (Dirichlet boundary condition), except for pressure, which is extrapolated from downstream. If the values are not known accurately, the inlet boundary can be moved far upstream from the region of interest.

At *mass flow boundaries* (downstream), based on the assumption of fully developed flow, the normal gradients of the variables are given a zero value (Neumann boundary condition), except for pressure, which is extrapolated from upstream. In the normal direction, constants are added to the extrapolated velocity values in order to ensure global mass conservation, which is needed for a correct solution of the pressure-correction equation. Outlets should be as far as possible from the region of interest for the gradient assumption to be valid.

Velocities cannot be specified when the pressure is defined at a flow boundary (*pressure boundary*). At an inflow pressure boundary, normal gradients of velocity are set equal to 0 and other variables are specified. At outflow pressure boundaries, the normal gradients of all variables except pressure are set equal to zero. The velocities are not modified.

Symmetry planes can be applied to reduce the computational domain. At a symmetry plane, the convective fluxes are zero, the normal velocity gradients of the parallel velocity component is zero and the normal velocity is zero.

At *walls*, the boundary conditions can include the specification of velocity, temperature and other scalars and/or the shear stress and fluxes of heat and scalars. Special treatment is needed for turbulent flows (see above). At the interface of conducting solids and fluids, no specifications are required. The exact implementation of the boundary conditions depends on the discretization scheme and on the pressure-correction algorithm.

Difficulties arise when accurate knowledge about the exact conditions is lacking. This is especially true in turbulent flow designs, when the turbulence energy and energy dissipation rate can only be guessed, using information about the velocity and the flow geometry. Furthermore, the position of boundaries and the direction of flow at boundaries may be difficult to specify, but may have considerable influence when the flow is complex and phenomena occur on a small scale. For example, modeling boundary conditions for fans may involve defining significant swirls. The effect of pressure resistance in the computational domain on the fan flow rate may be considerable and cannot always be taken into account fully. Further, when the fan is part of the flow system, rather than a boundary condition, its action must be incorporated as body forces, making the equations more difficult to solve. In all these cases a sensitivity analysis is desirable in order to estimate the error associated with approximate or guessed boundary conditions.

3 Numerical solution procedures

3.1 Discretization

The coupled system of partial differential Equations 1 to 3 cannot be solved analytically; a numerical approximation must be made. The discussion here is concentrated on the finite volume method of discretization, which is the most widely used in commercial CFD codes at the moment. The finite volume method owes its popularity to the fact that it obeys the clear physical principle of conservation on the discrete scale. The concepts of the method are easy to understand and have physical meaning. Another popular discretization method is the finite element method, which is becoming increasingly popular in CFD applications, because of its ability to deal with very complex geometries easily and more accurately using unstructured grids. This text will not deal with grid structures, such as block-structured, body-fitted grids, or moving grids. The general features of the finite volume method will be discussed.

When the system of Equations 1 to 3 is written in coordinate-free notation, the equations have the following general convection-diffusion form, with ϕ the dominant quantity of the equation:

$$\frac{\partial(\rho\phi)}{\partial t} + \nabla\cdot(\rho U\phi) - \nabla\cdot(\Gamma\nabla\phi) = S_\phi \qquad (17)$$

This equation can be integrated over a finite control volume V with surface A, using Gauss's theorem to obtain the surface integral terms:

$$\int_{CV}\frac{\partial\rho\phi}{\partial t}dV + \int_A(\rho U\phi)\cdot ndA - \int_A(\Gamma\nabla\phi)\cdot ndA = \int_{CV}S_\phi dV \qquad (18)$$

Clearly, this equation expresses exactly the conservation principle for all relevant quantities in the system when the surface integrals are the same for volumes sharing a boundary. Moreover, the finite volume form of the model is not dependent on the coordinate system. The grid only defines the boundaries of the volumes, which renders the method applicable to complex geometries. The volume integrals are approximated in terms of the volume-centered value of ϕ. The surface integrals in Equation 18 are approximated using values at the volume faces, which require further interpolation in terms of volume-centered values.

The governing equations are nonlinear, so that interactions of the variables generate a broad range of scales, which is further complicated by the turbulence in the flow due to its wide spectrum of fluctuations and the fact that variables at one point may depend on values at many surrounding points through different transport mechanisms. These properties of the model may cause difficulties and even errors in the numerical solution. Special attention must therefore be paid to the solution method of the system of equations. Generally, the method must have the following properties: consistency of the discretization, stability of the numerical solution algorithm, and, most importantly, the solution must converge to the exact solution when the grid is refined. Apart from consistency, these concepts are hard to evaluate. Alternatively, the following properties of discretization schemes can be investigated:

Conservativeness: The conservative nature of the basic equations must be respected by the discretization method, both locally and globally. The amount of conserved quantity leaving a closed volume must equal the amount entering the volume, in a steady state when no sources are present. Conservative schemes can only distribute the quantity improperly across the domain of interest. Clearly, this property is automatically assured by the finite volume method.

Boundedness: The solution must lie within proper bounds. This property requires that in a linear problem without sources, an increase of a quantity in one point may not result in a decrease in the surrounding points. This property resembles the concept of stability.

Transportiveness: Convection transfer influences a point only from the upstream locations. Conduction effects are transported in all directions. These properties must be incorporated in the numerical scheme.

In practicing CFD, a lot of time must be invested to find the right balance between stability and accuracy of the discretization scheme or, in fact, of the whole solution algorithm, which is described below. In *spatial discretization*, difficulties mainly arise with the approximation of the convection terms, which includes values of the variables at the cell volume faces, whereas the diffusive gradients at the cell faces can be directly approximated with a second-order accurate linear interpolation in terms of the control volume

Table 1 Properties of Spatial Discretization Schemes

	Accuracy[1]	Conservativeness	Boundedness[2]	Transportiveness
UPWIND	1st order	Assured	Always	Assured
		Falsely diffusive transport across streamlines		
CENTRAL	2nd order	Assured	$F/D<2$	Neglected
QUICK	3rd order	Assured	$F/D<8/3$	Assured

[1] Order of the truncation error

[2] F = coefficient of the discretized convective term; D = coefficient of the discretized diffusive term

values. For example, in a one-dimensional flow, for a control volume with a west face w and an east face e, the convective part becomes:

$$\int_A (\rho U \phi) n dA = (\rho u)_e A \phi_e - (\rho u)_w A \phi_w \qquad (19)$$

Accurate solutions can usually be obtained on very fine grids with all interpolation approaches for the right-hand side terms of Equation (19), but this is often not feasible in a practical situation. Some discretization schemes may be more accurate, but produce unbounded solutions when grids are too coarse. Others are unconditionally stable no matter how coarse the grid, but have a low accuracy and produce erroneous results, called numerical or false diffusion. When the above properties of a discretization scheme can be shown, they can be used with more confidence. Such investigations have been performed by different authors (Patankar, 1980; Versteeg and Malalasekera, 1995) for simple convection-diffusion problems and their conclusions are summarized in Table 1, for the most popular discretization schemes UPWIND, CENTRAL, and QUICK.

From Table 1 it is seen that the possibility of unboundedness and false diffusion must be considered, although validity of the results can only be proven by experimentation. Unbounded schemes like CENTRAL may result in unphysical overshoots and undershoots of the solution. False diffusion with the UPWIND scheme results in smeared distributions of the quantities. A popular solution to the problems of UPWIND and CENTRAL is to blend the good properties of two schemes together. For example, in the HYBRID scheme, for $F/D>2$, the scheme switches from central differencing to the stable upwind scheme. Similar formulations can be defined for the QUICK scheme. Also, higher-order schemes have been proposed (Anon., 1995), but these are not as widely validated.

In practical problems, the error of discretization can only be investigated by multiple runs with systematically refined grids. It must be mentioned that obtaining a grid-independent solution is of the utmost importance. However, in complex geometries such studies are often limited by the computer memory. This makes the experimental validation of the simulations a stringent necessity.

Table 2   Properties of Time
Discretization Schemes

	Accuracy	Boundedness
Explicit	1st	$\Delta t < f(\Delta x^2)^1$
Crank–Nicolson	2nd	$\Delta t < f(\Delta x^2)^1$
Implicit	1st	always

[1] Indication of boundedness limit, based on a 1-dimensional heat conduction model

Discretization in time normally uses the following approximation of the transient term in Equation (19), written for a control volume with a central node P:

$$\frac{\partial(\rho\phi_P)}{\partial t} = \frac{\rho(\phi_P - \phi_P^0)}{\Delta t} \tag{20}$$

where the index 0 indicates the value of the previous time step. The accuracy of the transient solution depends on the evaluation of the other terms in Equation (19) as a function of time. In general, the terms can be evaluated in time as follows:

$$\int_t^{t+\Delta t} f(\phi_P)dt = \left[f(\theta\phi_P) + f\big((1-\theta)\phi_P^0\big)\right]\Delta t \tag{21}$$

One can evaluate them at the previous time t (explicit, $\theta = 0$), the next time $t+\Delta t$ (fully implicit, $\theta = 1$) or at some time in between ($0<\theta<1$), of which the Crank-Nicolson interpolation ($\theta = 1/2$) is the most popular. The properties of these schemes are given in Table 2. Because of its robustness, the implicit scheme is mostly used in CFD applications. Note that the total discretization accuracy of a transient model depends both on the spatial and temporal approximations, so that when the implicit scheme is used, only first-order accuracy can be achieved.

3.2   *Solution of the discretized equations*

Discretization results in the following set of equations for all the separate control volume nodes P:

$$A_P\phi_P + \sum_{np} A_{np}\phi_{np} = Q_P \tag{22}$$

The index np runs over the neighboring nodes involved in the discretization. The coefficients A involve fluid properties, flow variables, and geometrical quantities. Q_P contains all terms that do not involve the unknown variable. In matrix-vector notation, this equation can simply be expressed as:

$$\mathbf{A}\phi = \mathbf{Q} \tag{23}$$

where $\mathbf{A}$ is a square sparse matrix containing the coefficients, ϕ is a vector containing all the variable values for all the control volumes, and $\mathbf{Q}$ is a vector containing all the variable-independent source terms.

The resulting system of equations after discretization (23) is still nonlinear: the flow variables appear in the coefficients of the discretized terms. The equations therefore need to be solved by an iterative method, starting from an initial guess that is systematically improved using the equations. The nonlinear terms have to be approximated, i.e., linearized. The least expensive and most common approach is to linearize the terms using the Picard iteration. This means that wherever the flow variables appear in the coefficients of the nonlinear terms, values of a previous guess in the iterative procedure are used. This method has been found to need more iterations than Newton-like methods, which use a Taylor series expansion, but do not involve the computation of complex matrices and are found to be much more stable.

The iteration cycle in which the coefficients of the equations are updated using the most recent available values of the variables is known as the outer iteration. The solution of the linear equations themselves for the dominant variables is the inner cycle. This can be performed by direct methods, which are computationally very costly and generally do not benefit from the mathematical properties of the linear system. It is therefore advantageous to use an iterative method for the inner cycle, called the inner iteration. This approach benefits further from the fact that high accuracy is not required in the inner iterations, because the coefficients need to be adjusted in the outer cycle and the accuracy is limited by the error of the discretization. In order to achieve a solution satisfying all the equations, careful design is needed of the number of inner and outer iterations, which is problem-dependent.

The iterative method should have certain properties. The main requirement for convergence of the solution is that the matrix $\mathbf{A}$ must be diagonally dominant, which has been shown by Scarborough (1958):

$$\frac{\sum \left|A_{np}\right|}{\left|A_P\right|} \begin{cases} \leq 1 \text{ at all P} \\ < 1 \text{ at one P at least} \end{cases} \tag{24}$$

This criterion is closely related to the boundedness property of the discretization schemes, discussed above. Many iterative solvers with different properties are available, with a different computational cost of the iteration and differences in convergence rate. For a detailed discussion the reader is referred to Ferziger and Peric (1996). The selection of an appropriate iterative solver is based on how accurate and how fast one wants to solve the equations. Besides the selection of an appropriate solver, one can improve the convergence by carefully selecting a so-called under-relaxation factor. This is a factor between 0 and 1, by which the value of the variable is multiplied after each iteration to smoothen rapid changes in its value. The effect of under-relaxation is that the diagonal dominance of the matrix A is improved.

It is advantageous to monitor the convergence during the iteration, certainly in complex cases. One can then stop the iteration based on a pre-defined convergence criterion and be assured of a convergent solution of the discretized equations. The convergence error ε^n can be defined as (Ferziger and Peric, 1996):

$$\varepsilon^n = \phi - \phi^n \qquad (25)$$

where ϕ is the converged solution of Equation (23) and ϕ^n is the approximate solution after n iterations. Unfortunately, it is impossible to obtain ε^n directly and it is even hard to calculate a suitable estimation of the value. In practice, the residual ρ^n is often evaluated to test for convergence. It is the amount by which the equations are not satisfied after n iterations:

$$A\phi^n = Q - \rho^n \qquad (26)$$

When the residual is driven to zero, the convergence error will be forced to become zero as well, because from Equations (23), (25), and (26) it is found that:

$$A\varepsilon^n = \rho^n \qquad (27)$$

The reduction of the norm of the residual is often used as a convergence criterion to stop the outer iterations. Experienced users have shown that for convergence of the outer iterations, the residual should be reduced by three to five orders of magnitude, depending on the required accuracy, which also depends on the discretization scheme and grid refinement (Ferziger and Peric, 1996). If the grid is refined, the discretization error is reduced and the convergence must be tightened. Caution must be taken, however, because it has been observed that at the beginning of the iteration procedure the residual falls much faster than the actual convergence error. Later on in the iteration, it has been observed in standardized problems with a monotonic convergence rate, that the residual reduction rate will approach the true convergence rate. In this case, stopping criteria can be defined with more confidence, relying on an estimate of the order of magnitude of the initial error.

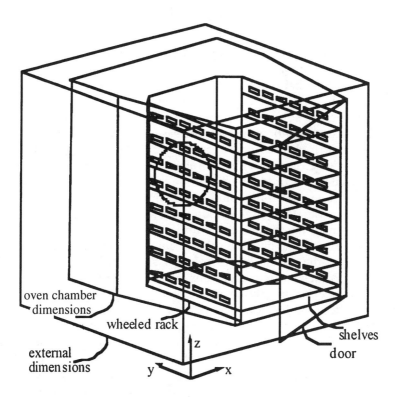

Figure 1 Geometry of the forced convection oven.

4 Examples in thermal food processing

4.1 Forced convection oven

In forced convection ovens, heated air is circulated over the stack of foods. A typical example of an oven for catering practice is given in Figure 1. The working dimensions of the cooking chamber of this device are 0.5 m wide, 0.4 m deep, and 0.7 m high, and it contains a wheeled rack with eight shelves. A fan of considerable dimensions (diameter = 0.26 m) is mounted at the back panel of the oven. Obviously, due to the complex geometry and the small scale, the flow pattern will be complicated. The air speed observed in the oven is in the range 1 to 6 m/s and is very turbulent. Cold spots may result from the fact that convection heating is restricted in regions of low velocity.

A CFD model has been developed to calculate the air flow in the device, using the code CFX4.1 (Harwell, UK). Because the relatively big fan has a large effect on the flow pattern, it is modeled as a body force on the flow. The body force represents the head-capacity curve of the fan. Additionally, based on further physical reasoning, a swirl is defined based on the flow rate and rotational speed of the fan. This makes it possible to observe the effect of the product stack on both the fan performance and the flow pattern, without a need for experimental determination of the exact boundary conditions for each product load. The turbulent RNG k-ε model was implemented. For the spatial discretization of the convection terms, the hybrid differencing scheme was used. Limited by computer memory, the grid could not be refined further than 55,188 control volumes, in which case cells with approximately 2 cm long edges are achieved. It was found that using a pseudo-transient solution toward steady state improved convergence. Indeed, transient solution methods increase the diagonal dominance of the coefficient matrix, as shown by Versteeg and Malalasekera (1995). The solution was advanced in time with small steps until the residuals of all equations were reduced by five orders of magnitude and constant values were achieved at a monitoring point in the solution domain.

The approximations in the model were simplifications in the geometry, the approximate fan model, the turbulence model, the first-order spatial discretization, and the iterative solution method. This should raise enough concern about the validity of the solution, so validation experiments were performed. Point measurements of the total velocity revealed that, using this model, the velocity magnitude can be predicted with an accuracy of about 30%. More importantly, however, the flow pattern agreed reasonably, as far as the identification of the zones of high and low velocities are concerned (Figure 2). In Figure 2 the magnitude of the velocity is plotted for the fourth shelf from the bottom. Values for velocity are only available at the intersection of the grid lines. The model can predict the regions of high velocity near the fan at the back (high y), and even the burst near the front left side of the wheeled rack (low x) is detected by the model. The low-velocity regions near the front, in the left corner at the back and on the right-hand side are also found. This property of the model assures that it can be used to search for overall deficiencies in the distribution of the heated air. A more detailed discussion can be found in Verboven et al. (1997c).

4.2 Calculation of local surface heat transfer coefficients

The coldest spot in the food is targeted when the microbial safety of the product during heating is concerned. For conduction heating symmetrical foods, the coldest spot is often chosen near the geometrical center, based on the assumption that the heat transfer coefficient is equal on all sides of the product. However, it is easily shown by boundary layer analysis that the surface heat transfer coefficient during convection is not constant across a surface.

CFD can be applied to solve the boundary layer near the surface of complexly shaped food products (Figure 3). Careful grid design and an accurate discretization scheme are required to resolve the complex separated, reattached, and recirculating flows close to the surface of the food. A further complication is introduced when the flow is turbulent. Again, with this type of problem, CFD cannot be used as a stand-alone solution, but requires experimental inputs to adjust the turbulence models and also careful validation of the results. Such validation has been performed by Verboven et al.

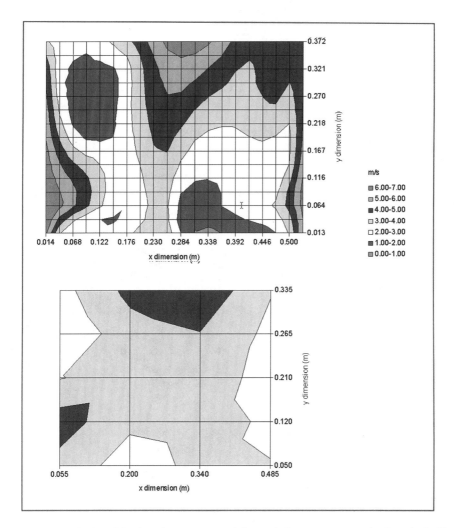

Figure 2 Calculated (*top*) and experimental (*bottom*) flow pattern on the fourth shelf of the oven.

(1996, 1997a, and 1997b) and for flows with a Reynolds number up to 10,000, good agreement was achieved between the model and experiments (Figure 4). The code CFX4.1 (Harwell, UK) was used.

5 Final remarks

This text has given an overview of computational fluid dynamics (CFD). The governing equations with their boundary conditions and the finite volume discretization method were discussed. Key features in the solution method were given. Some examples were given. The main advantage of CFD is the ease of use once a reliable model has been established. Then, CFD produces

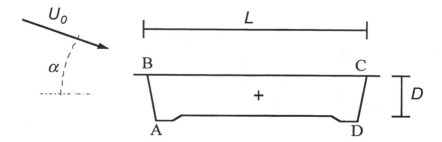

Figure 3 Geometry of a typical vacuum-packed food product.

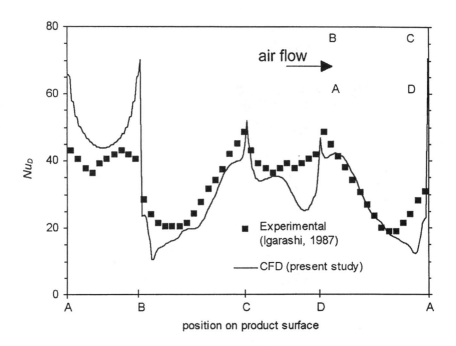

Figure 4 Local surface heat transfer to a rectangular shape for cross-flow heating with air, $L/D = 1.5$, $Re_D = 5{,}000$ (Reprinted from Verboven et al., 1997 with kind permission from Elsevier Science Ltd.).

quick answers to complex questions and gives details of all the variables involved. The availability of commercial codes with easy-to-use interfaces and flexible pre- and postprocessing facilities has cleared the path for the extensive use of CFD in many applications.

Preprocessing, where the computational grid is designed, is regarded as the most time-consuming component of a CFD study. Advances are being booked in grid generation with unstructured body-fitted multiblock meshes with nonhexahedral finite-element-like control volumes and the development of automatic adaptive grid generators interfaced with CFD packages (Gosman, 1997).

At the other end of a CFD study, interpretation of the results has benefited from the powerful 3D visualization tools for vector and color contour plots and particle tracking techniques.

However, CFD still has some drawbacks and points of concern, some of which were already raised in the text:

- In the model equations several unknown parameters need to be determined. A large amount of parameters can turn the model more into a fitting rather than a predictive tool based on physical principles. Many data often need to be measured, which limits the advantages of pure CFD over experiments. This is especially true when boundary conditions for complex problems need to be established. Currently, CFD is still regarded as too weak a foundation to base the whole design cycle on, but it is used successfully as a complementary tool to guide experiments that are even more expensive.
- Fitting has been widely used in turbulence modeling. Standard turbulence models are large-scale approximations of the complex physics of turbulence and have been identified to have severe limitations in many cases. It is not surprising that this limitation has led to a large number of variations of these models. Needless to say that for a large number of people, turbulence remains a phenomenon that has not yet been resolved properly, especially near surfaces and recirculation regions. It should also be noted that additional features such as natural convection and multiphase flows need additional turbulence modeling (Liu et al., 1997; Bessaih and Kadja, 1997; Yan et al., 1997).
- The drawback in turbulence modeling is related to the limitation in geometry-handling capabilities. If the mesh of control volumes could be sufficiently refined into complex geometries, no turbulence models would be needed. A considerable increase in computer speed and capacity as well as the development of more flexible grid generators is needed to achieve results with this approach. The development of the latter has received little attention in this text. Advanced meshes have been applied to large, complex geometries like full automobiles and trains. However, because computers cannot handle very large grids yet, only moderate results have been obtained up till now with these meshes in combination with turbulence models.
- The approximations that have to be made in the solution procedure cannot be overlooked. Erroneous results can still be produced when the properties and accuracy of the discretization schemes and of the iterative procedure are not clearly understood and investigated.
- It is clear that quality CFD requires highly skilled experts, which have a solid background in both fluid dynamics and numerical analysis. Not surprisingly, the learning period for CFD may be quite long.

Acknowledgments

Author Pieter Verboven acknowledges the scholarship from the Flemish Institute For Scientific Research in the Industry (I.W.T.). Author Bart Nicolaï is Postdoctoral Fellow with the Flemish Fund for Scientific Research (F.W.O.Vlaanderen). The Flemish Minister of Science and Technology and the E.U. (project AAIR-CT92-1519) are gratefully acknowledged for financial support.

References

Anon. (1993). *Psychrometrics. ASHRAE Handbook — Fundamentals.* ASHRAE, Atlanta, p. 6.1.

Anon. (1995). *CFX4.1-User Manual.* Computational Fluid Dynamics Services, AEAT, Harwell, UK.

Bejan, A. (1984). *Convection Heat Transfer.* John Wiley & Sons, Chichester, UK.

Bessaih, R. and Kadja, M. (1997). Three-dimensional turbulent natural convection in an enclosure. In: *Proceedings of the 5th U.K. National Heat Transfer Conference,* 17–18 September, Institute of Chemical Engineers, London, UK.

Bibby, I.P. (1997). A CFD approach to predicting airflow and heat transfer in bulk stored grains. In: *Engineering and Food at ICEF7,* Jowitt, R. (Ed.), Sheffield Academic Press, Sheffield, U.K., p.N-17.

Caretto, L.S., Gosman, A.D., Patankar, S.V., and Spalding, D.B. (1972). Two calculation procedures for steady, three-dimensional flows with recirculation. In: *Proceedings of the Third International Conference on Numerical Methods in Fluid Dynamics.* Paris, France.

Datta, A.K. and Teixeira, A.A. (1987). Numerical modelling of natural convection heating in canned liquid foods. *Transactions of the ASAE,* 30 (5), 1542.

Ergun, S. (1952). Fluid flow through packed columns. *Chemical Engineering Progress,* 48, 89.

Ferziger, J.H. and Peric, M. (1996). *Computational Methods for Fluid Dynamics.* Springer-Verlag, Heidelberg, Germany.

Gosman, A.D. (1997). Developments in Industrial Computational Fluid Dynamics. In: *Proceedings of the 5th U.K. National Heat Transfer Conference,* 17–18 September, Institute of Chemical Engineers, London, UK.

Hirsch, C. (1991). *Numerical Computation of Internal and External Flows.* Vol. 1 — Fundamentals of Numerical Discretisation. John Wiley & Sons, Chichester, UK.

Issa, R.I. (1986). Solution of implicitly discretised fluid flow equations by operator-splitting. *Journal of Computational Physics,* 62, 40.

Jones, W.P. and Launder, B.E. (1972). The prediction of laminarisation with two-equation model of turbulence. *International Journal of Heat and Mass Transfer,* 15, 301.

Khankari, K.K., Patankar, S.V., and Morey, R.V. (1995). A mathematical model for natural convection moisture migration in stored grain. *Transactions of the ASAE,* 38 (6), 1777.

Launder, B.E. and Sharma, B.I. (1974). Application of the energy dissipation model of turbulence to the calculation of flow near a spinning disc. *Letters of Heat and Mass Transfer,* 1, 131.

Leonard, A. (1995). Direct numerical simulation. In: *Turbulence and Its Simulation,* Gatski, T. (Ed.), Springer, New York.

Liu, F., Wen, J.X., Karayiannis, T.G., and Matthews, R.D. (1997). Computations of turbulent flows in a cavity with buoyancy-modified turbulence model. In: *Proceedings of the 5th U.K. National Heat Transfer Conference,* 17–18 September, Institute of Chemical Engineers, London, UK.

Mirade, P.S., Daudin, J.D., and Arnaud, G. (1995). Simulation en deux dimensions de l'aéraulique de deux tunnels de réfrigération de viandes. *International Journal of Refrigeration,* 18 (6), 403. In French.

Moin, P. and Kim, J. (1997). Tackling turbulence with supercomputers. *Scientific American,* 276 (1), 46.

Nicolaï, B.M. and De Baerdemaeker, J. (1996). Sensitivity analysis with respect to the surface heat transfer coefficient as applied to thermal process calculations. *Journal of Food Engineering,* 28, 21.

Patankar, S.V. (1980). *Numerical Heat Transfer and Fluid Flow.* Hemisphere Publishing Corporation, New York.

Rhodes, N. (1995). Making CFD pay for itself in the processing industries. In: *Computational Fluid Dynamics for Food Processing, Seminar Proceedings.* July 6, CCFRA, Chipping Campden, UK.

Scarborough, J.B. (1958). *Numerical Mathematical Analysis.* 4th Ed., Hopkins, J. (Ed.), University Press, Baltimore.

Van Doormaal, J.P. and Raithby, G.D. (1984). Enhancements of the SIMPLE method for predicting incompressible fluid flows. *Numerical Heat Transfer,* 7, 147.

Verboven, P., Scheerlinck, N., De Baerdemaeker, J., and Nicolaï, B.M. (1996). Importance of the local surface heat transfer coefficient in thermal food process calculations: a CFD approach. In: *Proceedings of the Second Main Meeting,* Project Process optimization and minimal processing of foods, Vol. 1 — Thermal Processing, Oliveira, J., Oliveira, F., and Hendrickx, M. (Eds.), Published by Escola Superior de Biotecnologia, Porto, Portugal.

Verboven, P., Nicolaï, B.M., De Baerdemaeker, J., and Scheerlinck, N. (1997a). The local and total surface heat transfer coefficient in thermal food process calculations: a CFD approach. *Journal of Food Engineering,* 33(1), 15–36.

Verboven, P., Scheerlinck, N., De Baerdemaeker, J., and Nicolaï, B.M. (1997b). Performance of different turbulence models for the calculation of the surface heat transfer coefficient. Presented at the Conference/Workshop on Thermal Properties and Behaviour of Foods during Production, Storage and Distribution, June 23–25, Prague, Czech Republic.

Verboven, P., Nicolaï, B., Scheerlinck, N., and De Baerdemaeker, J. (1997c). Variability of the air velocity in a forced convection oven: validation of the CFD model predictions. In: *Proceedings of the Third Main Meeting,* Project Process optimization and minimal processing of foods, Vol. 1 — Thermal Processing, Oliveira, J., Oliveira, F., and Hendrickx, M. (Eds.), Published by Escola Superior de Biotecnologia, Porto, Portugal, *In press.*

Versteeg, H.K. and Malalasekera, W. (1995). *An Introduction to Computational Fluid Dynamics: The Finite Volume Method,* Longman Scientific & Technical, Harlow, UK.

Wadekar, V.V. (1997). Boiling hot issues — Some resolved and some not-yet-resolved. In: *Proceedings of the 5th U.K. National Heat Transfer Conference,* 17–18 September, Institute of Chemical Engineers, London, UK.

Xu, Y. and Burfoot, D. (1997). Predicting air flows in potato stores. In: *Engineering and Food at ICEF7,* Jowitt, R. (Ed.), Sheffield Academic Press, Sheffield, U.K., p.N-13.

Yakhot, V., Orszag, S.A., Thangam, S., Gatski, T.B., and Speziale, C.G. (1992). Development of turbulence models for shear flows by a double expansion technique. *Physics of Fluids A*, 4 (7), 1510.

Yan, Y.Y., Smith, J.M., Müller-Steinhagen, H., and Malayeri, M.R. (1997). Numerical modelling of two-phase bubbly flow in a horizontal channel. In: *Proceedings of the 5th U.K. National Heat Transfer Conference*, 17–18 September, Institute of Chemical Engineers, London, UK.

Yap, C.R. (1987). *Turbulent Heat and Momentum in Recirculating and Impinging Flows*, Ph.D. Thesis, Faculty of Technology, University of Manchester, UK.

chapter four

Methodologies to optimize thermal processing conditions: an overview

Isabel M.L.B. Ávila and Cristina L.M. Silva

Contents

Summary 68
General introduction 68
1 Process modeling 69
   1.1 Transport phenomena 69
     1.1.1 Heat transfer 69
     1.1.2 Momentum transfer...71
     1.1.3 Mixed-mode heating 71
   1.2 Reaction kinetics...72
     1.2.1 Microorganisms' inactivation kinetics.............................73
     1.2.2 Quality attributes degradation kinetics...........................73
   1.3 Thermal processing applications..74
     1.3.1 Pasteurization 74
        1.3.1.1 In-pack pasteurization 74
        1.3.1.2 Continuous systems 75
     1.3.2 Sterilization 75
        1.3.2.1 In-pack sterilization..75
        1.3.2.2 UHT... 75
2 Process design 75
3 Process optimization 76
   3.1 Experimental validation..77
   3.2 Process assessment 77
4 Conclusions and recommendations for further work.........................78
References..78

0-8493-7905-9/99/$0.00+$.50
© 1999 by CRC Press LLC

Summary

A generalized methodology to optimize food thermal processes is presented. The approach involves the mathematical modeling of the phenomena and the design, optimization, and experimental assessment and validation of process conditions. The application of the proposed method is critically analyzed for continuous and in-pack pasteurization and sterilization of food products. Work in this field, particularly in pasteurization, is still quite limited. Relevant suggestions for further research are presented.

General introduction

It is well known that thermal processes are very important to stabilize foods and assure its microbiologic safety. Depending on the severity of the heat treatment and on the objectives to be accomplished, different thermal processes, such as pasteurization and sterilization, can be defined (Lund, 1975). Pasteurization is a relatively mild form of heat treatment designed to inactivate enzymes and destroy spoilage vegetative microorganisms (bacteria, yeast, and molds) present in low-acid foods (pH<4.6). On the other hand, if the pH of the food is high (pH>4.6) the main objective of this thermal treatment is to kill pathogenic microorganisms. This process is usually used in combination with other extrinsic factors (e.g., acidification, cooling, preservatives, and packaging) to minimize microbial growth and assure product safety during storage and distribution. Sterilization or "commercial sterilization" requires a severe heat-treatment to destroy heat-resistant bacterial spores. The severity of the heat treatment depends on the nature of the food (e.g., pH and water activity), the microorganism's heat resistance, the initial microbial load, the heat transfer characteristics of the food and heating medium, and on the storage conditions following the thermal processing (Lund, 1975; Fellows, 1988).

However, when a thermal process is applied, part of the original nutritional and sensorial quality is lost. Although the negative effect of thermal processes cannot be avoided, it can be minimized. After the identification of the process purpose, it is possible to optimize process conditions (Manvell, 1997). Currently, the main objective of an optimization of thermal processing conditions is the maximization of the final nutritional and/or sensorial product quality. The nutritional quality, such as vitamin content, is important in some specific products (e.g., baby foods) and in terms of public well being, but the consumer's perception goes to food sensory attributes, such as texture, color, and flavor (Villota and Hawkes, 1986). Thermal processing conditions can also be optimized in terms of economic aspects. The energy consumption can be minimized (Barreiro et al., 1984) or the productivity can be maximized (Banga et al., 1991; Noronha, 1996). Obviously, in this situation, a compromise must be attained in terms of the final product quality (Banga et al., 1991; Silva et al., 1993).

The quality optimization of thermally processed food products is possible due to the differences in temperature dependence of target microorganisms and quality attributes of thermal degradation kinetics. This is the basis for the high temperature short time (HTST) principle. In thermal processing, it is possible to find different time–temperature combinations with the same lethal effect, but leading to different quality losses. At higher temperatures the quality factors are relatively more thermal resistant. Therefore, a HTST process, when applicable, results in products with superior quality (Lund, 1977).

Five elements are common to all optimization problems: (1) the identification of the design variables that can be controlled, (2) the requirements that must be met (constraints or restrictions), (3) the definition of the objective function (the mathematical expression of what is to be optimized), (4) the mathematical model of the situation and (5) an optimization technique (Norback, 1980). The relevant parts of the optimization procedure are presented in detail in Figure 1. Mathematical modeling of heat, mass, and momentum transport phenomena and microorganisms, enzymes, and quality attributes of degradation kinetics have to be taken into consideration for process design. Here, the correct definition of the criteria to ensure the safety of the final product is required. The next step is to calculate the best conditions to process the product and design the implementation of the experiments to validate those optimal conditions. The optimization approach can be considered in terms of the two main alternatives: minimization of costs or maximization of quality. The experimental validation and process assessment are very important to evaluate the conditions achieved and especially to meet consumer requirements.

1   Process modelling

Mathematical models are essential tools to attempt the simulation of a thermal process impact on a food. The knowledge of the temperature evolution into the food combined with the kinetics of thermal destruction of quality attributes, microorganisms, and enzymes is of utmost importance for the final purpose that is the design and optimization of processes (Lund, 1975; Ohlsson, 1980a; Hallström et al., 1988).

1.1   Transport phenomena

The role of transport phenomena can involve heat, mass, and momentum transfer into a food. Mass transfer is a relatively slower process and therefore not relevant in sterilization or pasteurization processes. In these fields, heat and momentum transfer have a much wider importance.

1.1.1   Heat transfer

Heat can be transferred to a food product by conduction and/or convection. Conduction heating occurs in solids or semisolids by direct transfer

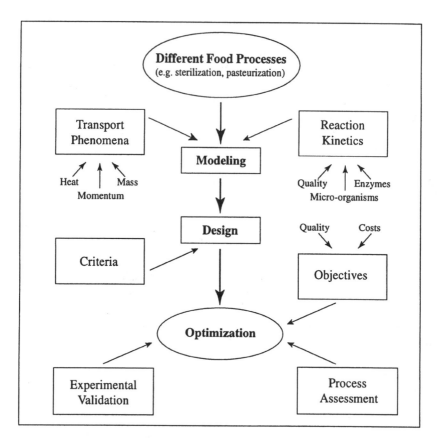

Figure 1 Schematic representation of optimization procedure.

of molecular energy and can be described by Fourier's laws. Transient heat conduction into a three-dimensional object is governed by the following partial differential equation using rectangular coordinates (Carslaw and Jaeger, 1959):

$$\rho C_p \left(\frac{\partial T}{\partial t} \right) = \frac{\partial}{\partial x} \left(\lambda x \frac{\partial T}{\partial x} \right) + \frac{\partial}{\partial y} \left(\lambda y \frac{\partial T}{\partial y} \right) + \frac{\partial}{\partial z} \left(\lambda z \frac{\partial T}{\partial z} \right) \tag{1}$$

where T is the product temperature (°C); t the time (s); λ_i the thermal conductivity in the i direction (W/(m K)), Cp the specific heat (J/(kg K)) and ρ the density (kg/m³).

Available solutions to this equation (analytical and finite difference or finite element numerical solutions) for conduction heating foods can be found in several literature references (e.g., Silva, 1993). A relatively large amount of research work on the simulation of heat transfer by conduction in in-packed thermal processing of foods is available (Silva et al., 1993).

Convection is the transfer of heat by groups of molecules that move as a result of differences in density or as a result of agitation (Fellows, 1988). To describe this phenomenon equations of momentum transfer have to be considered.

1.1.2 Momentum transfer

The modeling of heat together with momentum transfer involves the simultaneous solution of the three fundamental transport equations (Navier-Stokes equations): continuity, motion, and energy (Bird et al., 1960).

$$1.\ \text{Continuity:} \quad \nabla \bar{v} = 0 \tag{2}$$

$$2.\ \text{Motion:} \quad \rho \frac{D\bar{v}}{Dt} = \nabla p + \eta \nabla^2 \bar{v} + \rho g \tag{3}$$

$$3.\ \text{Energy:} \quad \rho C_p \frac{DT}{Dt} = \lambda \nabla^2 T \tag{4}$$

where g is the acceleration due to gravity (m/s^2), p the fluid pressure (Pa), $\bar{v}$ the mass average velocity vector (m/s), η the dynamic viscosity of a Newtonian liquid (Pa.s), and λ the thermal conductivity (W/m/K).

These equations are only valid for pure liquids and analytical solutions exist only for very simplified cases. A much more restrict number of simulation work is available in the case of prepackaged foods heated by convection (Rao and Anantheswaran, 1988). There is also a lack of experimental research concerning the sterilization of pre-packaged convection heating foods (Silva, 1993). Effects such as starch gelatinization can occur and cause transition from convection to conduction heat transfer, which makes the modeling very complex and with limited applicability.

In the case of continuous processing of fluid foods, where the flow behavior within the system is very complex, the residence time distribution (RTD) theory has to be applied. The knowledge of how long the different fluid elements stay in the system is necessary to ensure that the food is adequately processed (Singh and Lee, 1992). There are several research works on this field and a review can be found in Baptista (1995).

1.1.3 Mixed-mode heating

The previous two types of heat transfer can also occur simultaneously (mixed-mode heating) in particulate food products, such as soups or sauces. This makes the modeling even more complex, because it is necessary to consider both equations that describe heat transfer inside particles and heat transfer in the liquid (Silva, 1993; Baptista, 1995; Noronha, 1996).

For these processes also a very limited number of simulation work is available. Lenz and Lund (1979), Rumsey (1984), Deniston et al. (1987), and

Stoforos and Merson (1990a) attempted to model heat transfer into containers with liquid and particles. However, many assumptions (uniform fluid temperature, constant heat-transfer coefficients, uniform initial temperatures for both particles and fluid, particles with equal sizes and uniformly distributed, and constant physical and thermal properties for both the fluid and particles) had to be made in order to simplify the solution (Stoforos and Merson, 1990b). No research work was carried out considering more complex conditions such as: particle geometries other than spherical (e.g., cubical or cylindrical), mix of particles with different geometries, physical properties or size distribution (Silva, 1993). In terms of continuous processing of particulate foods the residence time distribution (RTD) of particles together with the estimation of heat transfer coefficients at the particle surface are the most important studies in this field (Baptista, 1995).

1.2   Reaction kinetics

It is not only important to know the extent of degradation or inactivation of nutritional and organoleptic factors or microbial and enzymic activity in foods submitted to thermal processing, but it is also crucial to know the rate at which these changes occur. The heat inactivation of microorganisms, enzymes, and quality attributes is frequently described by the zero order (Equation 5), first order (Equation 6) or second order reaction models (Equation 7):

$$C = C_0 - kt \tag{5}$$

$$C = C_0 \exp(-kt) \tag{6}$$

$$\frac{1}{C} = \frac{1}{C_0} + kt \tag{7}$$

where C is the measured concentration of microorganisms, enzymes, or quality attributes, C_0 the initial concentration, t the heating time (min), and k the reaction rate constant (min^{-1}).

The fractional conversion model (Equation 8) (Levenspiel, 1972) has also been recently applied to describe color (Ávila and Silva, 1996; Steet and Tong, 1996) and texture (Rizvi and Tong, 1997) changes.

$$\frac{C - C_f}{C_0 - C_f} = \exp(-kt) \tag{8}$$

where C_f is the final equilibrium concentration of the parameter.

The Arrhenius equation is usually applied to evaluate the effect of temperature on the reaction rate constant:

$$k = k_{ref} \exp\left[-\frac{E_a}{R}\left(\frac{1}{T} - \frac{1}{T_{ref}}\right)\right] \tag{9}$$

where Ea is the activation energy (J/mol), R the universal gas constant (8.314 J/(mol.K)), T the absolute temperature (K), and k_{ref} the rate constant at reference temperature (min^{-1}).

If kinetic data are available, it may be feasible to establish suitable processing conditions for optimizing food products. However, these studies are still very limited being an obstacle to the application of optimization techniques to the food industry (Villota and Hawkes, 1986).

1.2.1 Microorganisms' inactivation kinetics

In general, the thermal degradation of microorganisms is described by the first-order reaction model. However, in the microbiology field, the Bigelow model (thermal death time, TDT, model) (Ball and Olson, 1957) is more frequently used than the Arrhenius model:

$$D = \frac{\ln(10)}{k} \tag{10}$$

$$D = D_{T_{ref}} * 10^{(T_{ref}-T)/z} \tag{11}$$

where D is the decimal reduction time (time required for the number or concentration of microbial spores to be reduced by a factor of 10 at a given temperature, min), D_{Tref} the decimal reduction time at the reference temperature (min), T_{ref} the reference temperature (°C), and z the z-value (number of degrees Celsius required to reduce the D value by a factor of 10°C).

In food sterilization, the target microorganisms normally used to determine product safety are *Clostridium botulinum* spores (Richardson et al., 1988) that have a z-value of 10°C (Pflug and Odlaug, 1978).

For pasteurization processes the target microorganism depends on the specific product. Gaze (1992) reviewed relevant inactivation kinetic data in this field. However, for each new pasteurized product the target microorganism must be identified and its inactivation kinetics experimentally quantified.

1.2.2 Quality attributes degradation kinetics

A review on kinetic data of quality indicators is given by Lund (1975), Holdsworth (1985, 1990), Villota and Hawkes (1986), Hallström et al. (1988), and Silva and Igniatidis (1995). Quality attributes, such as nutrient content and sensory properties, are very important for the consumer acceptability of a certain food product. The sensorial quality is perhaps the most important characteristic of foods because it concerns attributes the consumer can readily assess (Lund, 1982). Appearance, flavor, and texture determine the initial acceptance of a food (Fennema, 1976). The influence of thermal processing on the sensory quality of foods has been studied by many researchers. However, kinetic data obtained using taste panels for sensory evaluation is limited. Hayakawa et al. (1977), Paulus and Saguy (1980), Ma et al. (1983), Harada et al. (1985) and Van Loey et al. (1994a) attempted to use trained taste panels for kinetic calculations.

Characterization and quantification of the appearance of a product should be determined by objective measurements and correlated with sensory analysis (Szczesniak, 1968; Kramer, 1969; Van Loey et al., 1994a). However, correlation between the two still requires additional research (Lund, 1982).

Mathematical modeling of real foods is always more difficult than of model systems because of the interactions between compounds. This is the reason why model systems under ideal conditions are commonly selected. However, due to the possible interactions in foods the mechanisms or the rate of degradation may change completely compared to the investigation of the isolated compound (Villota and Hawkes, 1986).

In addition, a degradation kinetics different from first order is not often considered in literature. Usually this mechanism of degradation is assumed as the result of a simplification of reality. However, if a first-order model instead of the real kinetics occurring in the product is applied, optimal temperatures can be affected (Silva, 1997). Therefore, extensive information is needed as a priority to obtain consistent, reliable, and comparable data on mathematical modeling of reaction kinetics (Villota and Hawkes, 1986).

1.3   Thermal processing applications

Different thermal processing systems are available. The suitability of each system depends on several factors, like type of food, production rates, availability of capital, and labor costs. Two basic methods of operation are being used: (1) in-pack processing and (2) continuous systems.

1.3.1   Pasteurization

Mathematical modeling of pasteurization processes with the aim of designing adequate and optimized processing conditions is relatively scarce. A pasteurization is a relatively mild thermal treatment with smaller impact on product quality. Furthermore, the identification of the process criteria is sometimes quite difficult. This explains the lack of relevant work in this field.

1.3.1.1   In-pack pasteurization.   In-pack pasteurization is normally applied to liquid products. Rao and Anantheswaran (1988) presented a detailed review on the modeling of convective heat transfer to fluid foods in cans.

The hot-fill technique is a simple pasteurization process, especially applied to acid fruit purees. This technique combines the two methods of processing. In order to achieve a required target pasteurization value, the food product is first heated, in a continuous system, until it reaches a certain hot filling temperature, and, after being introduced into a container of given dimensions, is cooled by conduction. Sandoval et al. (1994) and Silva and Silva (1997) presented studies on the modeling of this type of treatment. However, these authors took into consideration only the cooling phase for the lethality calculation. More sophisticated models would be welcome.

1.3.1.2 Continuous systems. Research work on the modeling of continuous pasteurization systems for fluid foods is mainly limited to the experimental determination of residence time distributions of fluids and particles (for particulate products) (Baptista, 1995). The development of modeling techniques to simulate heat and momentum phenomena together with inactivation and degradation kinetics would result in an excellent tool for further optimization of these thermal treatments.

1.3.2 Sterilization

As stated above a sterilization is a severe thermal treatment designed to inactivate all the microbial spores and thus extend the product shelf life for months and/or years. The need for rigorous process conditions design to meet consumer safety and minimize the impact on product quality is enormous.

1.3.2.1 In-pack sterilization. This method involves processing the product after packaging. The food product is placed in a container, sealed, and heated in pressurised retorts that may operate in batch or continuously. It is the simplest and most traditional thermal processing method but also the one that causes higher product quality losses. This explains the large amount of modeling work on this field (Silva et al., 1993). Although a lot of research already exists, there are still many needs. In particular, the modeling of sterilization of particulate products is still limited.

1.3.2.2 UHT. As for continuous pasteurization systems of fluid foods, modeling of UHT processes is mainly limited to the experimental determination of residence time distributions of fluids and particles (for particulate products).

2 *Process Design*

Process design requires the definition of the criteria to assure adequate lethality in all points of the food product (Silva, 1993), based on the definition of the equivalent processing time at a reference temperature.

For continuous pasteurization or sterilization processes, most of the work deals with residence time distribution studies. A minimum pasteurization or sterilization value must be assured in the fluid fraction with smaller residence time.

Considering in-pack processes, the most adequate criterion should be taken for the *worst case* conditions, which corresponds to a minimum lethality value identified at the coldest zone of the operation unit and at the coldest point of the containers (Pflug, 1975). Published research work refers mainly to the sterilization field. The experimental determination of those critical points is then obtained by carrying out heat-distribution studies and heat-penetration studies in the retort and container, respectively. On the other hand, during process calculations, parameters obtained from experimental procedures are subjected to errors in measurement and to inherent variability contributing to significant uncertainty on the calculated lethality (Hicks, 1961; Nicolai, 1994). The application of a statistical procedure is very important to identify the worst-case condition (Pflug, 1975). Therefore, the design of the process should be based on a complete statistical study in order to reduce excessive safety margins with consequent quality improvement and cost reduction. However, relatively little information has been reported on the magnitude of the uncertainties and errors involved in thermal process design (Robertson and Miller, 1984). Implementation of new processes should be based on a complete design involving all the above mentioned steps. Each design is only valid for the considered products, containers, fill weights, operation units, heating medium and processing conditions (Van Loey et al., 1994b). This means that the system should also be prepared for possible retesting if new variables are introduced in order to guarantee product safety and quality (May and Fletcher, 1995). From this point, optimization is the next stage to follow toward maximization of product quality.

3. Process optimization

In continuous pasteurization and sterilization systems for fluid foods the heat-penetration rate is high and the application of the HTST principle leads to products with high-quality retention. Although the mathematical modeling of heat and momentum transfer in these type of systems could be used to improve the quality of those products, research work in this field is very scarce.

For in-pack solid and semisolid foods the slow heat-transfer rate to the coldest region of the product limits the applicability of the HTST principle. If the principle would be applied, a severe quality thermal degradation would occur at the product surface. However, for these situations it is possible to calculate an optimum combination of time and temperature conditions leading to a microbiological safe food and minimized quality losses (Teixeira et al., 1969). In sterilization of prepackaged conduction heating foods several computer simulations have been performed to calculate suitable processing conditions leading to maximum product quality retention (Teixeira et al., 1969, 1975b; Thijssen et al., 1978; Saguy and Karel, 1979; Ohlsson, 1980a, 1980b, 1980c; Nadkarni and Hatton, 1985; Tucker and Holdsworth, 1990; Banga et al., 1991; Hendrickx et al., 1992; Silva et al., 1992). Optimization studies considering variable retort temperature are more limited (Teixeira et al., 1975a; Banga et al., 1991; Noronha et al., 1993). All authors concluded that when considering variable retort temperature on the optimization of thermal processing conditions for conduction heated foods, significant advantages were attained over the traditional constant temperature processes for surface quality retention. It is also feasible, although more complex, to simultaneously optimize more than one quality factor. Noronha (1996) simulated the simultaneous optimization of three surface quality factors of sterilized foods using constant- and variable-retort temperature profiles, proposing a new objective function that takes into account the relative thermal sensitivity of the different components.

For convective and mixed mode heating Van Loey et al. (1994b) reported a study with theoretical calculations on optimization of canned white beans.

For pasteurization in-pack processes very few optimization studies are found in research literature. Sarkin (1978) simulated the processing of cooked meat products and Silva and Silva (1997) modeled the quality optimization of hot filled pasteurized fruit purees.

3.1 Experimental validation

After the design and theoretical optimization of thermal processes, the experimental validation of the calculated conditions is of utmost importance (Silva, 1993). This procedure requires processing the product using different time–temperature profiles with equivalent lethal effect in order to confirm the best conditions.

Few authors attempted to validate experimentally the simulated optimal conditions for sterilized in-pack conduction heating products (Teixeira et al., 1975a; Banga et al., 1993; Nasri et al., 1993; Silva et al., 1994; Van Loey et al., 1994b). Van Loey et al. (1994b) and Ávila and Silva (1997) have validated optimum conditions experimentally using taste panels for mixed mode heating products.

The lack of research in this field hinders the relevance of theoretical optimization research works.

3.2 Process assessment

During the lifetime of a product, from its preparation, processing, and packaging, through its distribution chain, storage, and until the moment of consumption, the food is exposed to many adverse factors. Therefore, process evaluation is an indispensable tool to quantify product changes in terms of food safety and quality. This procedure, *Process Assessment,* is very important in process design and optimization and meets the increasing consumer demands for high-quality products.

The development of new heating technologies such as continuous processing in rotary retorts, aseptic processing, ohmic and microwave heating, and combined processes is limited by the applicability of currently used process evaluation methodologies (Van Loey et al., 1996). Therefore, considerable effort has been put into the development of new methodologies for process evaluation. The available procedures are the *in situ* method, the methods using time–temperature integrators (TTIs) and the physical–mathematical methods. These three methods were critically reviewed by Noronha (1996), and Van Loey et al., (1996) and can also be found in Chapter 2 of this book.

4 Conclusions and recommendations for further work

In order to design and experimentally validate optimal thermal processes for food products there is a need for further work in different fields such as modeling kinetics using sensory analysis evaluated by trained taste panels and heat-distribution and penetration studies. However, it is very difficult for the food industry to accomplish all these research steps, mainly due to economic and time-consuming aspects.

For the success of the application of a thermal process, some of the steps of the optimization approach such as the characterization of the heat-transfer model through heat-penetration studies and the calculation of the optimal time–temperature profile based on existing kinetics are indispensable. Fast analytical methods should also be chosen in order to rapidly verify the optimum.

References

Ávila, I.M.L.B. and Silva, C.L.M. (1996). Mathematical modelling of thermal degradation kinetics of peach puree total carotenoids and colour. Poster presented at the IFT '96 Annual Meeting, New Orleans, June 22–26.

Ávila, I.M.L.B. and Silva, C.L.M. (1997). Experimental validation of optimal sterilisation conditions for maximising quality of carotenoids and colour in canned peach puree. Oral presentation at the Institute of Food Technologists Annual Meeting, Orlando, June 14–18.

Ball, C. O. and Olson, F.C.W. (1957). *Sterilization in Food Technology,* McGraw-Hill, New-York.

Banga, J.R., Alonso, A.A., Gallardo, J.M., and Perez-Martin, R.I. (1993). Kinetics of thermal degradation of thiamin and surface colour in canned tuna. *Z. Lebensm Unters Forsch,* 197, 127.

Banga, J.R., Perez-Martin, R.I., Gallardo, J. M., and Casares, J.J. (1991). Optimisation of the thermal processing of conduction-heated canned foods: study of several objective functions, *Journal of Food Engineering,* 14, 25.

Baptista, P.N. (1995). *Flow and heat transfer analysis of two-phase systems with large solid particles moving in carrier fluids,* Ph.D. Thesis, Escola Superior de Biotecnologia, Porto, Portugal.

Barreiro, J.A., Perez, C.R., and Guariguata, C. (1984). Optimisation of energy consumption during the heat processing of canned foods, *Journal of Food Engineering,* 3, 27.

Bird, R. B., Stewart, W. E., and Lightfoot, E. N. (1960). *Transport Phenomena,* John Wiley and Sons, New York.

Carslaw, H. S. and Jaeger, J. C. (1959). *Conduction of Heat in Solids,* Oxford University Press, London.

Deniston, M. F., Hassan, B. H., and Merson, R. L. (1987). Heat transfer coefficients to liquids with food particles in axially rotating cans, *Journal of Food Science*, 52 (4), 962.

Fellows, P. (1988). *Food Processing Technology*, Elis Horwood, Ltd. Chichester, U.K.

Fennema, O. R. (1976). *Principles of Food Science*, Part I, Marcel Decker, New York.

Gaze, J. E. (1992). *Food Pasteurisation Treatments*, Campden & Chorleywood Food Research Association, Technical Manual, no. 27.

Hallström, B., Skjoldebrand, C., and Tragardh, C. (1988). *Heat Transfer and Food Products*, Elsevier Applied Science, New York.

Harada, T., Tirtohusodo, H., and Paulus, K. (1985). Influence of temperature and time on cooking kinetics of potatoes, *Journal of Food Science*, 42 (5), 1286.

Hayakawa, K. (1977). Mathematical methods for estimating proper thermal processes and their computer implementation. In: *Advances in Food Research*, Academic Press, New York, p. 23.

Hendrickx, M., Silva, C., Oliveira, F., and Tobback, P. (1992). Optimisation of heat transfer in thermal processing of conduction heated foods. In: *Advances in Food Engineering*, Singh, R.P. and Wirakartakusumah, A., CRC Press, Boca Raton, FL.

Hicks, E.W. (1961). Uncertainties in canning process calculations, *Journal of Food Science*, 26, 218.

Holdsworth, S.D. (1985). Optimisation of thermal processing — a review, *Journal of Food Engineering*, 4, 89.

Holdsworth, S.D. (1990). Kinetic data — What is available and what is necessary. In: *Processing and Quality of Foods*. Vol. 1 — HTST Processing, Field, R.W. and Howell, J.A. (Eds.), Elsevier Applied Science, London, p. 74.

Kramer, A. (1969). The relevance of correlating objective and subjective data, *Food Technology*, 23, 926.

Lenz, M. K. and Lund, D. B. (1979). The lethality-Fourier number method, heating rate variations and lethality confidence intervals for forced-convection heated foods in containers, *Journal of Food Process Engineering*, 2, 227.

Levenspiel, O. (1972). Interpretation of batch reactor data. In: *Chemical Reaction Engineering*, 2nd Ed., John Wiley & Sons, New York, p. 41.

Lund, D.B. (1975). Effects of blanching, pasteurisation, and sterilisation on nutrients. In: *Nutritional Evaluation of Food Processing*, Harris, R.S. and Karmas, E. (Eds.), AVI Publishing Co., New York, p. 205.

Lund, D. B. (1977). Design of thermal processes for maximising nutrient retention, *Food Technology*, 2, 71.

Lund, D. B. (1982). Applications of optimisation in heat processing, *Food Technology*, 2, 97.

Ma, L. Y., Deng, J. C., Ahmed, E. M., and Adams, J. P. (1983). Canned shrimp texture as a function of its heat history, *Journal of Food Science*, 48, 360.

Manvell, C. (1997). Minimal processing of food, *Food Science and Technology Today*, 11 (2), 107.

May, N. and Fletcher, P. (1995). Thermal process validation: the way forward. *Food Technology International Europe*, 150.

Nadkarni, M. M. and Hatton, T. A. (1985). Optimal nutrient retention during the thermal processing of conduction-heated canned foods: application of the distributed minimum principle, *Journal of Food Science*, 50, 1312.

Nasri, H., Simpson R., Bouzas, J., and Torres, J. A. (1993). An unsteady-state method to determine kinetic parameters for heat inactivation of quality factors: conduction-heated foods, *Journal of Food Engineering*, 19, 291.

Nicolaï, B. (1994). *Modelling and uncertainty propagation analysis of thermal food processes*, Ph.D. Thesis, Katholieke Universiteit Leuven, Belgium.

Norback, J. P. (1980). Techniques for optimisation of food processes, *Food Technology*, 2, 86.

Noronha, J. F. (1996). *Improved Procedures for Designing, Evaluating and Optimising In-Pack Thermal Processing of Foods*, Ph.D. Thesis, Katholieke Universiteit Leuven, Belgium.

Noronha, J. P., Hendrickx, M., Suys, J., and Tobback P. (1993). Optimisation of surface quality retention during the thermal processing of conduction heating foods using variable temperature retort profiles, *Journal of Food Processing and Preservation*, 17, 75.

Ohlsson, T. (1980a). Optimisation of heat sterilisation using C-values. In: *Food Process Engineering*, Applied Science Publishers, U.K., p. 137.

Ohlsson, T. (1980b). Optimal sterilisation temperatures for flat containers, *Journal of Food Science*, 45, 848.

Ohlsson, T. (1980c). Optimal sterilisation temperatures for sensory quality in cylindrical containers, *Journal of Food Science*, 45, 1517.

Paulus, K. and Saguy, I. (1980). Effect of heat treatment on the quality of cooked carrots, *Journal of Food Science*, 45, 239.

Pflug, I.J. (1975). *Procedures for carrying out a heat penetration test and analysis of the resulting data*, Published by the Department of Food Science and Nutrition, University of Minnesota, Minneapolis.

Pflug, I.J. and Odlaug, T.E. (1978). A review of z and F values used to ensure the safety of low-acid canned foods, *Food Technology*, 6, 63.

Rao, M.A. and Anantheswaran, R.C. (1988). Convective heat transfer to fluid foods in cans, *Advances in Food Research*, 32, 39.

Richardson, P.S., Kelly, P.T., and Holdsworth, S.D. (1988). Optimisation of in-container sterilisation processes. In: *Progress in Food Preservation Processes*, CERIA, Brussels, Belgium, p. 2.

Rizvi, A.F. and Tong, C.H. (1997). Fractional conversion for determining texture degradation kinetics of vegetables, *Journal of Food Science*, 62 (1), 1.

Robertson, G.L. and Miller, S.L. (1984). Uncertainties associated with the estimation of F_0 values in cans which heat by conduction, *Journal of Food Technology*, 19, 623.

Rumsey, T.R. (1984). Modelling heat transfer in cans containing liquid and particulates, *American Society of Agricultural Engineers*, Paper No. 84-6515.

Saguy, I. and Karel, M. (1979). Optimal retort temperature profile in optimising thiamine retention in conduction-type heating of canned foods, *Journal of Food Science*, 44, 1485.

Sandoval, A. L., Barreiro, J. A., and Mendonza, S. (1994). Prediction of hot-fill-air-cool sterilisation processes for tomato paste in glass jars, *Journal of Food Engineering*, 23, 33.

Sarkin, R. J. (1978). Computerised cooking simulation of meat products, *Journal of Food Science*, 43 (4), 1140.

Silva, C.L.M. (1993). *Optimisation of sterilised conduction heating foods: a generalised approach*, Ph.D. Thesis, Escola Superior de Biotecnologia, Porto, Portugal.

Silva, C.L.M. (1997). Optimisation of thermal processing conditions: Effect of reaction type kinetics. In: *Proceedings of the Conference Modelling of Thermal Properties and Behaviour of Foods During Production, Storage and Distribution*, Prague, Czech Republic, June 23–25.

Silva, C., Hendrickx M., Oliveira, F., and Tobback P. (1992). Optimal sterilisation temperatures for conduction heating foods considering finite surface transfer coefficients, *Journal of Food Science*, 57 (3), 743.

Silva, C.L.M. and Igniatidis, P. (1995). Modelling Food Colour Degradation Kinetics — A Review. In: *Proceedings of the First Main Meeting*, Project Process optimization and minimal processing of foods, Vol. 1 — Thermal Processing, Oliveira, J., Oliveira, F., and Hendrickx, M. (Eds.), Published by Escola Superior de Biotecnologia, Porto, Portugal.

Silva, C.L.M., Oliveira, F.A.R., and Hendrickx, M. (1993). Modelling optimum processing conditions for the sterilisation of pre-packaged foods, *Food Control*, 4 (2), 67.

Silva, C., Oliveira F., Lamb, J., Torres, A., and Hendrickx, M. (1994). Experimental validation of models for predicting optimal surface quality sterilisation temperatures, *International Journal of Food Science and Technology*, 28, 227.

Silva, F.V.M. and Silva, C.L.M. (1997). Quality optimisation of hot filled pasteurised fruit purees: container characteristics and filling temperatures, *Journal of Food Engineering*, 32, 351.

Singh, R.K. and Lee, J.H. (1992). Residence time distribution of foods with/without particles in aseptic processing systems. In: *Advances in Aseptic Processing Technologies*, Singh, R.K. and Nelson, P.E. (Eds.), Elsevier Applied Science, London, Chap. 1.

Steet, J.A. and Tong, C.H. (1996). Degradation kinetics of green color and chlorophylls in peas by colorimetry and HPLC, *Journal of Food Science*, 61 (5), 924.

Stoforos, N.G. and Merson, R.L. (1990a). Estimating heat transfer coefficients in liquid/particulate canned foods using only liquid temperature data, *Journal of Food Science*, 55 (2), 478.

Stoforos, N.G. and Merson, R.L. (1990b). An overview of heat transfer studies in rotated liquid/particulate canned foods. In: *Engineering and Food*, Spiess, W.E.L and Schubert, H. (Eds.), Elsevier Science Publishing, New York, p. 50.

Szczesniak, A. S. (1968). Correlations between objective and sensory texture measurements, *Food Technology*, 22, 981.

Teixeira, A. A., Dixon, J. R., Zahradnik, J. W., and Zinsmeister, G. E.(1969). Computer optimisation of nutrient retention in the thermal processing of conduction-heated foods, *Food Technology*, 23 (6), 137.

Teixeira, A. A., Stumbo, C. R., and Zahradnik, J. W. (1975a). Experimental evaluation of mathematical and computer models for thermal process evaluation, *Journal of Food Science*, 40, 653.

Teixeira, A. A., Zinsmeister, G. E., and Zahradnik, J. W. (1975b) Computer simulation of variable retort control and container geometry as a possible means of improving thiamin retention in thermally processed foods, *Journal of Food Science*, 40, 656.

Thijssen, H.A.C., Kerkhof, P.J.A.M., and Liefkens, A.A.A. (1978). Short-cut method for the calculation of sterilisation conditions yielding optimum quality retention for conduction-type heating of packaged foods, *Journal of Food Science*, 43, 1096.

Tucker, G. and Holdsworth, D. (1990). Optimisation of quality factors for foods thermally processed in rectangular containers. In: *Process Engineering in the Food Industry*. Vol. 2 — Convenience Foods and Quality Assurance, Field, R.N. and Howell, J.A. (Eds.), Elsevier Applied Science, London, p. 59.

Van Loey, A., Francis, A., Hendrickx, M., Maesmans, G., and Tobback, P. (1994a). Kinetics of thermal softening of white beans evaluated by a sensory panel and the FMC tenderometer, *Journal of Food Processing and Preservation*, 18, 407.

Van Loey, A., Francis A., Hendrickx, M., Maesmans, G., Noronha, J., and Tobback, P. (1994b). Optimising thermal process for canned white beans in water cascading retorts, *Journal of Food Science*, 59(4), 828.

Van Loey, A., Hendrickx, M., De Cordt, S., Haentjens, T., and Tobback, P. (1996). Quantitative evaluation of thermal processes using time–temperature integrators, *Trends in Food Science and Technology*, 7, 16.

Villota, R. and Hawkes, J. G. (1986). Kinetics of nutrients and organoleptic changes in foods during processing. In: *Physical and Chemical Properties of Foods*, Okos, M.R. (Ed.), American Society of Agricultural Engineering, St. Joseph, Michigan, Chap. 2.

chapter five

Microbial and quality assessment in thermal processing

Antonio Martínez, Carmen Rodrigo, Pablo S. Fernández, María J. Ocio, Francisco Rodrigo, and Miguel Rodrigo

Contents

Summary 83
1   Introduction 83
2   Microbiological assessment 84
3   Quality assessment 89
References 93

Summary

The optimization and assessment of thermal processes requires accurate and reliable kinetic models and parameters. The importance of predictive microbiology is evident, and the need for equally good models of enzyme inactivation is particularly important in novel HTST (high temperature short time treatments) processes. Aspects on the determination of microbial death parameters and enzymic inactivation are discussed in this text, highlighting the need for proper experimental methodologies.

1   Introduction

In recent years there have been important changes in habits and trends related to food consumption (the incorporation of women in the workplace, community meals, fast food, etc.), scientific and technological advances (interpretation of nonenzymatic browning, development of thermostable and highly impermeable plastics, HTST followed by aseptic packaging, automation and computer-controlled processing, etc.), and in increased safety requirements (application of hazard analysis and critical control point (HACCP) systems, adjustment of national legislations to harmonize at the European Union level, environmental protection, etc.). All these changes have acted as "driving forces" promoting advances in food preservation and an increased knowledge base for alerting the consumer to food quality and safety.

One of the most commonly used, economical, and versatile techniques for food preservation is heat sterilization. Thermal treatment is also a technique that is currently being proposed for combination with new technologies (e.g., high pressures, pulsed electric fields) for inactivation of enzymes and destruction of spores, aspects practically impossible to achieve by these new technologies alone or else involving unjustifiably high costs.

The concept of preservation by heat, the process itself, and the equipment used for its application have undergone major advances in the attempt to adapt to consumer demands (e.g., convenience, low cost, nutritive value) and health authority requirements, which call for greater control and microbiologic safety (e.g., application of the HACCP principle). In all processes in which heat is involved there are factors such as those related with microorganisms and quality factors that may affect microbiologic safety and the quality of the products manufactured. These factors include accuracy in the calculation of kinetic parameters and the deviations that may occur when values taken from the literature, obtained in reference media, or over a range of conventional temperatures, are extrapolated to high temperatures (HTST conditions).

In short, in order to achieve microbiologic safety and high quality in thermally processed foods it is necessary to have correct kinetic data for both microorganisms and quality factors, as well as inactivation or destruction predictive models, with which it is possible to use time–temperature integrators that are widely applicable tools to evaluate the impact of heat on microorganisms and quality factors in a variety of novel processing situations (see Chapter 2). This chapter explores the latest advances in these aspects of thermal processing.

2   Microbiological assessment

For appropriate design and evaluation of sterilization processes it is essential to have kinetic data for thermal inactivation of microorganisms that may potentially contaminate the food (see Chapters 2 and 4).

The differences between the D_T values obtained in reference media and those obtained in real food samples may be considerable (Rodrigo et al., 1993; Fernández et al., 1994). Figure 1 is an example that shows a comparison of the survival curves of *Bacillus stearothermophilus* in a reference medium and in unacidified mushroom extract. As can be seen, the extract perceptibly reduces the heat resistance of the spores. The same effect has been observed in *Clostridium botulinum* 213B (Brown and Martínez, 1992) and in *Clostridium sporogenes* PA 3679 (Ocio et al., 1994). Environmental factors around the spores also affect their heat resistance. Xezones and Hutchings (1965) studied

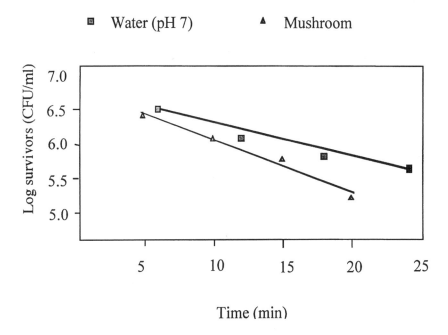

Figure 1 Survivors curves for *Bacillus stearothermophilus* spores at 115°C showing the effect of the food substrate.

the inactivation of *Clostridium botulinum* in various groups of foods with different pH values. The authors concluded that design, operation, and costs of sterilization were significantly affected by pH. Cameron et al. (1980) indicated that acidification to intermediate pH values could be used to achieve better utilization of sterilization for certain groups of food products. Rodrigo and Martínez (1988) studied the effect of an intermediate pH level of 5.2 on the heat resistance of *Clostridium sporogenes* spores in artichoke extract and observed D_T values lower than those obtained in phosphate buffer, the phenomenon being more evident as treatment temperature increased.

The pH value can affect the D_T value of microorganisms in various ways. On the one hand, acid pH may reduce the heat resistance of the spores, an effect directly reflected in the severity of the sterilization process since the lethality required for microbiologic safety depends, among other things, on the D_T value at a given temperature. On the other hand, the pH of the food product may inhibit or diminish microbial growth, especially of spores sensitized to pH after heat treatment (Fernández et al., 1995), and consequently reduce the apparent heat resistance of the spore. This phenomenon may be considered as a further safety factor in addition to the effect of heat treatment.

Although there is a general consensus on the reduction of heat resistance of spores as pH decreases, there are fairly recent studies in which it is shown that this effect could depend on the microorganism in question and on the treatment temperature. Brown and Martínez (1992) indicated that when

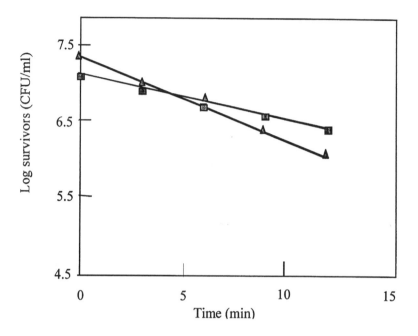

Figure 2 Survival curves for *Clostridium sporogenes* spores heated at 115°C in mushroom substrate with different pH values.

spores of *Clostridium botulinum* 213B were heated between 121 and 130°C in mushroom extract with various pH values and two different acidulants, they did not see a clear effect of reduction of pH on the heat resistance of the spores. In the case of *Clostridium sporogenes* PA 3679, it has been observed that the effect of acidification on pH depends on the specific pH value (Cameron et al., 1980; Ocio et al., 1994). Figure 2 is an example of survival curves for PA 3679 in acidified (pH 6.2) and unacidified (pH 6.7) mushroom extract. If we consider *Bacillus stearothermophilus*, much used in time–temperature integrators, there are studies that indicate reductions in D_T value as the pH of the mushroom extract decreases (Fernandez et al., 1994). In view of these and other studies that may be found in literature, it is fair to say that the effect of pH on the D_T value could depend on the microorganism group, aerobic or anaerobic.

An important aspect in heat inactivation kinetic studies is the form of heating and the method used for analyzing experimental data. Thermobacteriology studies have conventionally been carried out in isothermal conditions. Determination of kinetic parameters in these conditions is relatively simple and produces conservative sterilization or pasteurization processes. However, during heat treatment microorganisms are subjected to conditions that often differ substantially from isothermal experimental conditions. An alternative is to apply nonisothermal heating methods. These methods offer the advantage of subjecting the spores to dynamic temperature conditions, such as those that occur in real thermal sterilization processes. Tucker (1985) reviewed the various nonisothermal methods used in accelerated methods for the stability study of pharmaceutical products. These methods can also be used for microbial inactivation studies. Tucker (1985) discussed the following temperature increase programs: 1) uncontrolled temperature increase, 2) linear program, 3) linear program followed by an isothermal period, 4) polynomial program, and 5) hyperbolic program.

With nonisothermal heating processes it is possible to obtain much information (experimental data) with a single experiment, with consequent savings in material, time and labor costs. Moreover, they provide a wealth of data for subsequent development of predictive models.

Experimental data in microorganism inactivation studies are generally analyzed by means of two successive linear regressions. In the first, the logarithm of the concentration of the thermolabile factor remaining after heat treatment is plotted against time, giving the D_T value. In the second regression the logarithm of D_T is plotted against treatment temperature, giving the z value. This methodology generally provides high confidence ranges, due to the small number of degrees of freedom (Lund, 1983). Arabshahi and Lund (1985) presented a method for analyzing kinetic data for inactivation of thiamine that provided smaller confidence ranges than those obtained by using two linear regressions. To do this they calculated the activation energy, Ea, from the original experimental data taken as a whole, applying nonlinear regressions in a single step. A possible drawback of this methodology is the high correlation that may exist between the parameters estimated, D_T and z or k_T and Ea, causing convergence and accuracy problems. Nevertheless, the convergence can be improved if the equations are transformed by taking natural logarithms or reparameterizing.

Inactivation predictive models can be an excellent tool for ensuring the microbiologic safety of the thermal process. There is growing interest in predictive microbiology because of its many potential applications, such as estimating the effect of changes or of errors in estimating specific microorganism parameters (McNab, 1997), for example, D_T or z. The successful application of these models depends on developing and validating them in real conditions (McMeekin and Roos, 1996).

The parameters that define the inactivation of microorganisms by heat in nonisothermal conditions by analyzing experimental data obtained in real substrates by means of single-step nonlinear regressions are of great value for the development of inactivation models that relate the D_T value to environmental factors of importance in the canning industry, such as pH, sodium chloride concentration, anaerobiosis, or water activity. Figure 3 shows an example of a response surface relating the D_T value to pH and treatment temperature. These models can be incorporated into a hazard analysis and critical control point plan, to safeguard the sterilization process against any

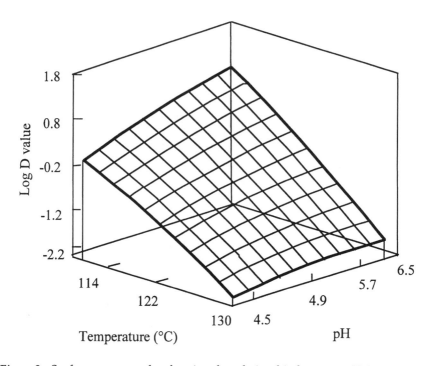

Figure 3 Surface response plot showing the relationship between pH, D value, and temperature of treatment for *Bacillus stearothermophilus* spores heated in acidified mushroom substrate.

eventuality during the manufacturing process which might affect the D_T value of the microorganisms, potentially causing spoilage of the food product.

Finally, process evaluation occupies an important place within the microbiologic safety of the sterilization process. Over the years, various techniques have been developed for evaluating sterilization processes, which take place in sealed cans or continuously by means of heat exchangers. Experimental inoculation of cans with *Clostridium sporogenes* was one of the first procedures used (NCA, 1968). Yawger (1978) developed the count reduction system, in which cans containing the product were inoculated with bacterial spores and subjected to heat treatment in the same way as in the experimental inoculation of cans. Subsequently, procedures were developed in which the spores were confined in small containers made of glass, metal, or plastic, which were inserted in the can together with the food product (Michiels, 1972; Pflug, 1982). Recently, time–temperature integrators have been developed by immobilizing spores in artificial food particles, polyacrylamide gel, or alginate containing food puree (Brown et al., 1984; Rönner, 1990, Ocio et al., 1997). This new generation of biological indicators is suitable for evaluating both conventional and heat exchanger sterilization processes, allowing interactions between the product and the spores immobilized in the particle (Rodrigo, 1997).

3   Quality assessment

Nowadays there is a demand for natural-processed foods, without additives and with characteristics that come closer to the fresh product — *fresh like product, invisible manufacturing* — without impairing microbiologic safety. Despite these consumer demands and the scientific and technological advances recorded in recent years, some foods are still sterilized by means of conventional systems and conditions or at temperatures that extract a high price in terms of quality and process yield. A typical example is canned asparagus, where heat has an appreciable impact on quality, revealed in loss of texture and weight, phenomena that are more evident in green asparagus.

Most of the studies related to the thermal destruction of enzymes and quality factors in high temperature short time conditions have been performed with dairy products, and there is little information on other foods (Holdsworth, 1992). Application of these sterilization processes by using novel heating technologies may partly solve the loss of quality in products such as asparagus, if we bear in mind the good results obtained in juices and milk. It has been verified that at high temperatures the rate of destruction of microorganisms is greater than the rate of loss of quality. It is this difference that allows optimization of thermal processes in terms of retention of nutrients and quality factors, including texture. However, this new situation may create a problem that does not appear in canned products sterilized by more conventional methods or at more conventional temperatures. Enzymes are also inactivated more slowly than microorganisms, so that there is a danger of residual enzymatic activity, which, by itself or by regeneration of the enzyme, may produce a degradation of food quality during storage. This kind of problem can be solved by application of optimized sterilization processes, for which it is necessary to ascertain the precise kinetic parameters for inactivation and regeneration of control enzymes (enzymes with greater heat resistance) and the parameters for the most appropriate quality factors, which in the case of asparagus would be destruction of texture and weight loss, all obtained at high temperatures.

Peroxidase has traditionally been chosen as a control for establishing heat processes for foods. It is one of the most heat-resistant enzymes, it is found in numerous vegetables, and the Food and Drug Administration (FDA) of the United States considers its inactivation necessary to reduce quality losses during storage of manufactured foods (Whitaker, 1992). One characteristic of peroxidase is its ability to regenerate after a sterilization process. The extent of the regeneration seems to depend on the type of vegetable, the isoenzyme considered, and basically on the heating time at a specific temperature (Adams, 1978). Consequently, HTST sterilization processes may be less effective for irreversible inactivation of peroxidase than conventional procedures. Therefore, in order to apply the new sterilization systems, it is necessary to study and develop models capable of interpreting the kinetics of inactivation and regeneration and of calculating the parameters that define them.

An aspect to be borne in mind when seeking to carry out studies on thermal inactivation kinetics is the equipment to be used and how the sample is to be heated. In order to study the heat resistance of enzymes such as peroxidase, direct or indirect heating methods may be used. In indirect heating methods the sample is inserted in a tube, can, etc., and receives heat through an intermediate barrier. A possible disadvantage of these methods is the relatively high period of heating inertia, which at high temperatures may affect the correct interpretation of experimental results. In direct heating methods the sample is in close contact with the heating medium, which in most cases is saturated steam. These methods have the advantage that the latent heat of steam is capable of producing extremely rapid heating. This group includes thermoresistometers, sophisticated devices in which it is possible to achieve very short periods of inertia, 0.3 to 0.4 seconds, thus obtaining greater accuracy in the results. Thermoresistometers have been mainly applied to the study of microorganism inactivation kinetics and little is known of their suitability for studying enzyme inactivation kinetics. Rodrigo et al. (1997) studied the inactivation kinetics of horseradish and asparagus peroxidase in a high-temperature thermoresistometer and concluded for the suitability of the equipment. The thermoresistometer enabled the study of the inactivation kinetics of the more labile fraction of horseradish peroxidase at temperatures above 100°C.

As mentioned earlier, in order to optimize HTST sterilization processes it is essential to have precise kinetic data for inactivation of microorganisms and enzymes and for quality losses over the high temperature range. The kinetic models employed for predicting and quantifying the impact of heat treatment on food constituents and the most reliable mathematical methods for determining the kinetic parameters for the degradation of those constituents are of the following types:

- First-order model
- n^{th} order model
- Consecutive reactions model
- Multifraction model

The first-order model is the most elementary chemical approach, consisting in considering the degradation of a quality factor or constituent as a first-order reaction in which the reaction rate is proportional to the concentration of the constituent in question:

$$E_N \xleftrightarrow{K_{eq}} E_R \xrightarrow{k_1} E_1$$

E_N is the native species, E_R is an inactive intermediate species, which, by heating, can be transformed reversibly into E_N and irreversibly into E_I, an inactive species; K_{eq} is the equilibrium constant between E_N and E_R, and k_I

is the reaction rate constant for the inactivation of E_R. This is the model most commonly employed for describing bacterial and enzymatic kinetics.

In the n-th order model it is assumed that the reaction rate is not proportional to the concentration of the constituent but to some power of it. The evolution of the concentration of a quality factor in relation to time corresponds to the following equation:

$$E^{(1-n)} = E_0^{(1-n)} - (1-n)k_T t$$

The model of consecutive reactions consists in assuming that the original component, E_1, is transformed into an intermediate component, E_2, which has a different thermostability and is in turn inactivated and transformed into E_3:

$$E_1 \xrightarrow{k_1} E_2 \xrightarrow{k_2} E_3$$

The inactivation of E_1 and E_2 follows a first-order kinetics with reaction rate constants k_1 and k_2, respectively. The intermediate component may have an activity that differs from that of the original. This model has been successfully applied to inactivation of various enzymes (Henley and Sadana, 1984a, 1984b, 1985) and also specifically to explain the inactivation kinetics of peroxidase (Clochard and Guern, 1973; Wang and Dimarco, 1972; Saraiva et al., 1996).

In the multifraction model it is assumed that the original component consists of several components (two in the simplest case), with different thermostabilities: a stable component (E_S) and a labile component (E_L), both following a first-order inactivation kinetic with reaction rate constants k_S and k_L respectively. This is the case with enzymes that consist of different isoenzymes (Ling and Lund, 1978), microorganisms that present various types of spores (Sapru and Labuza, 1993), or texture consisting of two substrates with differing thermal stability. An outline of this model could be the following:

$$E_s \xrightarrow{k_s} E_{I_S}$$

$$E_L \xrightarrow{k_L} E_{IL}$$

Texture has always been an important attribute of food, but in recent years it has begun to receive more attention. A decade ago texture was almost a forgotten attribute, but now it is recognized as a saleable characteristic that indicates the high quality sought by consumers; texture can lead to rejection or substantial differentiation of a food product that may increase its sales. It can be said that a good food product texture is an indicator of high quality and excellent preparation (Szczesniak, 1990). As has already been shown, the optimization of sterilization processes for green asparagus also calls for a knowledge of the kinetics of texture loss during heat treatment. However, there are few models relating texture loss to heat treatment similar to those that exist to represent the kinetics of microorganisms, enzymes, and food color (Lund, 1982).

As with chemical reactions, texture changes due to heating are studied by using kinetic models depending on time and temperature, the data being expressed in terms of their reaction rate constants k_T, or in terms of the D_T value (Bourne, 1989). The major property of fruit and vegetable texture is firmness. In general, texture is characterized by a physical quantity, such as, for example, the maximum force needed to cut an asparagus spear, and this property is measured in relation to time and temperature.

Historically, two models have been applied to analyze experimental data of texture degradation: the first-order model (Nagel and Vaughn, 1954, Paulus and Saguy, 1980), and the two-fraction model (Huang and Bourne, 1983). In the latter model it is assumed that the vegetable tissue consists of two substrates, A and B, with different thermal resistances, each with its kinetic parameters and following first-order kinetics. For short times, the faster process, corresponding to degradation of the labile substrate A, is dominant, and for longer times the dominant process is the degradation of the more resistant substrate B, which has the slower kinetic process. Rodrigo et al. (1997) studied the thermal degradation of green asparagus texture when asparagus spears were heated between 70 and 100°C, and in the softening curve they observed a rapid initial reduction in firmness, which was almost linear and coincided with first-order kinetics, but the softening rate decreased with time and a second straight line with a new slope appeared for longer heating times. The slopes of these two straight lines represent the reaction rate constants for mechanisms 1 and 2, respectively. However, when the samples were heated over a temperature range of 100 to 130°C, a first-order kinetics was observed that could correspond to the degradation of the more resistant component (Rodrigo et al., 1997).

In order to determine the kinetic model of thermal destruction of a food component during heating, traditionally two different approaches have been used (Lenz and Lund, 1980): the *steady-state approach* (heating at constant temperature, or isothermal methods), and the *unsteady-state approach* (heating at a nonconstant temperature, or nonisothermal methods). The selection of the steady-state or unsteady-state procedure depends on the features of the heating equipment and the type of experiment being performed, i.e., whether heating and cooling inertia time can be discounted or not.

Sadeghi and Swartzel (1990) found large differences in the values of kinetic parameters obtained by the steady-state and unsteady-state methods. De Cordt et al. (1992) also observed differences in the parameters obtained by making an isothermal and a nonisothermal analysis. In HTST kinetic studies performed with relatively large samples, the theoretical temperature of the assay is not the temperature of the entire sample during the heating period, so that in these cases it is particularly important to consider the thermal profile of the sample and estimate kinetic parameters by using unsteady-state methods. Most studies connected with the calculation of kinetic parameters are carried out using the steady-state approach due to the difficulty of obtaining prediction models for sample temperature profiles as a result of the natural complexity of foods. However, some researchers have used unsteady-state methods, as in the case of Svensson and Eriksson (1974), who studied thermal inactivation of lipoxygenase in the skin and cotyledons of whole peas by considering the pea divided into 17 concentric elements. Naveh et al. (1982) and Luna et al. (1986) studied thermal destruction of peroxidase during blanching of corn-on-the-cob by considering the cob as a finite homogeneous cylinder, and Nasri et al. (1993) studied thermal destruction of thiamine using a mathematical method for temperature prediction based on finite differences (Nasri et al., 1993). In the case of the heating of whole asparagus it is clear that, given its characteristics, the temperature of the spear will not be constant, at least during a good part of the heating and cooling. This distribution of temperatures varying with position and time affects the sensory quality of the asparagus (texture) and its weight. Therefore, more precise unsteady-state methods are needed for the calculation of kinetics and estimation of texture destruction and weight loss parameters. The use of mathematical heat-transfer models for estimating the evolution of temperatures inside a cylinder and the incorporation of the experimentally calculated thermal properties of asparagus have made it possible to improve the estimations of the kinetic parameters (Rodrigo et al., 1998).

The data found in literature on weight loss during heating are mainly for food of animal origin, types of meat, which generally have first-order reaction kinetics and activation energies of the order of 13 kcal/mol for veal muscle (Bertola et al., 1994). Rodrigo et al. (1997) studied the kinetics of weight loss in asparagus heated between 100 and 130°C, obtaining an E_a of 8.9 kcal/mol.

References

Adams, J.B. (1978). The inactivation and regeneration of peroxidase in relation to the high temperature-short time processing of vegetables, *Journal of Food Technology*, 13, 281.

Arabshahi, A. and Lund, D.B. (1985). Considerations in calculating kinetics parameters from experimental data, *Journal of Food Process Engineering*, 7, 239.

Bertola, N.C., Bevilaqua, A.E., and Zaritzky, N.E. (1994). Heat treatment effect on texture changes and thermal denaturation of proteins in beef muscle, *Journal of Food Processing and Preservation*, 18, 31.

Bourne, M.C. (1989). Applications of chemical kinetic theory to the rate of thermal softening of vegetable tissue. In: *Quality Factors of Fruit and Vegetables*, Jen, J.J., (Ed.), American Chemical Society, Washington, D.C., p. 98.

Brown, K.L. and Martinez, A. (1992). The heat resistance of spores of *Clostridium botulinum* 213B heated at 121–130°C in acidified mushroom extract, *Journal of Food Protection*, 55, 913.

Brown, K.L., Ayres, C.A., Gaze, J.E., and Newman, M.E. (1984). Thermal destruction of bacterial spores immobilized in food-alginate particles, *Food Microbiology*, 1, 178.

Cameron, M.S., Leonard, S.J., and Barret, E.L. (1980). Effect of moderately acidic pH on heat resistance of *Clostridium sporogenes* spores in phosphate buffer and in buffered pea puree, *Applied and Environmental Microbiology*, 39, 943.

Clochard, A. and Guern, J. (1973). Destruction thermique de l'activité peroxidasique. Interpretation des curves experimentales, *Revue Generale du Froid*, 8, 860. In French.

De Cordt, S., Hendrickx, M., Maesmans G., and Tobback P. (1992). Immobilized α-amilase from *Bacillus licheniformis*: a potential enzymic time–temperature integrator for thermal processing, *International Journal of Food Science and Technology*, 27, 661.

Fernandez, P.S., Gomez, F.J., Ocio, M.J., Rodrigo, M., Sanchez, T., and Martinez, A. (1995). D-values of *Bacillus stearothermophilus* spores as a function of pH and recovery medium acidulant, *Journal of Food Protection*, 58, 628.

Fernandez, P.S., Ocio, M.J., Sanchez, T., and Martinez, A. (1994). Thermal resistance of *Bacillus stearothermophilus* spores heated in acidified mushroom extract, *Journal of Food Protection*, 57, 37.

Henley, J.P. and Sadana, A. (1984a). Mathematical analysis of enzyme stabilization by a series-type mechanism: influence of chemical modifiers, *Biotechnology and Bioengineering*, XXVI, 959.

Henley, J.P. and Sadana, A. (1984b). Series-type enzyme deactivations: influence of intermediate activity on deactivation kinetics, *Enzyme and Microbial Technology*, 6, 35.

Henley, J.P. and Sadana, A. (1985). Categorization of enzyme deactivation using a series-type mechanism, *Enzyme and Microbial Technology*, 7, 50.

Holdsworth, S.D. (1992). *Processing and aseptic packaging of food products*, Elsevier Applied Science, London.

Huang, Y.T. and Bourne, M.C. (1983). Kinetics of thermal softening of vegetables, *Journal of Texture Studies*, 14, 1.

Lenz, M.K. and Lund, D.B. (1980). Experimental procedures for determining destruction kinetics of food components, *Food Technology*, 34, 51.

Ling, A.C. and Lund, D.B. (1978). Determining kinetic parameters for thermal inactivation of heat-resistant and heat-labile isozymes from thermal destruction curves, *Journal of Food Science*, 43, 1307.

Luna, J.A., Garrote, R.L., and Bressan, J.A. (1986). Thermo-Kinetic modelling of peroxidase inactivation during blanching-cooling of corn on the cob, *Journal of Food Science*, 51, 141.

Lund, D.B. (1982). Quantifying reactions influencing quality of foods: texture, flavor and appearance, *Journal of Food Processing and Preservation*, 6, 133.

Lund, D.B. (1983). Considerations in modeling food processes, *Food Technology*, 37, 92.

McMeekin, T.A. and Roos, T. (1996). Modelling applications, *Journal of Food Protection*, Sup., 37.

McNab, B.W. (1997). A literature review linking microbial risk assesment predictive microbiology and dose response modeling, *Dairy Food and Environmental Sanitation*, 17, 405.

Michiels, L. (1972). Methode biologique de determination de la valeur sterilisatrice des conserves appertisées, *Industrie Alimentaire Agricole*, 1349.

Nagel, C.W. and Vaughn, R.H. (1954). Sterilization of cucumbers for studies on microbial spoilage, *Food Research*, 19, 612.

Nasri, H., Simpson, R., Bouzas, J., and Torres, J.A. (1993). An unsteady-state method to determine kinetic parameters for heat inactivation of quality factors: conduction-heated foods, *Journal of Food Science*, 19, 291.

Naveh, D., Mizrahi, S., and Kopelman, I.J. (1982). Kinetics of peroxidase deactivation in blanching of corn on the cob, *Journal of Agricultural and Food Chemistry*, 30, 967.

NCA (1968). *Laboratory Manual for Food Canners and Processors*, Vol. 1, Avi Publishing Company, Westport, CT, Chap. 10

Ocio, M.J., Fernandez P.S., Rodrigo M., Periago P., and Martinez A. (1997). A time–temperature integrator for particulated foods: thermal process evaluation, *Z Lebensm Unters Forsch A*, 205, 325.

Ocio, M. J., Sanchez, T., Fernandez, P. S., Rodrigo, M., and Martinez, A. (1994). Thermal resistance characteristics of PA 3679 in the temperature range of 110–121°C as affected by pH, type of acidulant and substrate, *International Journal of Food Microbiology*, 22, 239.

Paulus, K. and Saguy, I. (1980). Effect of heat treatment on the quality of cooked carrots, *Journal of Food Science*, 45, 239.

Pflug, I.J. (1982). Measuring the integrated time–temperature effect of a heat sterilization process using bacterial spores, *Food Process Engineering*, 78, 68.

Rodrigo F. (1977). *Establecimiento de las condiciones de utilización de un ITT microbiológico que utiliza como elemento sensor esporas de Bacillus stearothermophilus.*, PhD. Thesis Dissertation. Universidad de Valencia, Spain. In Spanish.

Rodrigo, M. and Martinez, A. (1988). Determination of a process time for a new product: canned low artichoke hearts, *International Journal of Food Science and Technology*, 23, 31.

Rodrigo, C., Alvarruiz, A., Martínez, A., Frígola, A., and Rodrigo, M. (1997). High-temperature short-time inactivation of peroxidase by direct heating with a five-channel computer-controlled thermoresistometer, *Journal of Food Protection*, 60, 967.

Rodrigo, M., Martinez, A., Sanchez, T., Peris, M.J., and Safon, J. (1993). Kinetics of *Clostridium sporogenes* PA 3679 spore destruction using computer controlled thermorresistometer, *Journal of Food Science*, 58, 649.

Rodrigo, C., Mateu, A., Alvarruiz, A., Chinesta, F., and Rodrigo, M. (1998). Estimation of kinetic parameters for thermal degradation of green asparagus texture by an unsteady-state method, *Journal of Food Science*, 63, 126.

Rodrigo, C., Rodrigo, M., and Fiszman, S.M. (1997). The impact of high-temperature, short-time thermal treatment on texture and weight loss of green asparagus, *Lebensmittel Wissenschaft und Technologie*, 205, 53.

Rodrigo, C., Rodrigo, M., Fiszman, S.M., and Sanchez T. (1997). Thermal degradation of green asparagus texture, *Journal of Food Protection*, 60, 315.

Rönner, U. (1990). Bioindicator for control of sterility, *Food Laboratory News*, 22, 51.

Sadeghi, F. and Swartzel, K.R. (1990). Generating kinetic data for use in design and evaluation of high temperature food processing systems, *Journal of Food Science*, 55, 851.

Sapru, V. and Labuza, T.P. (1993). Temperature dependence of thermal inactivation rate constant of bacterial spores as predicted by glass transition theory, *Journal of Industrial Microbiology*, 12, 247.

Saraiva, J., Oliveira, J.C., Oliveira, S., and Hendrickx, M. (1996). Inactivation kinetics of horseradish peroxidase in organic solvents of different hydrophobicity at different water contents, *International Journal of Food Science and Technology* 31, 233.

Svensson, S.G. and Eriksson, C.E. (1974). Thermal inactivation of lipoxygenase from peas. IV. Inactivation in whole peas, *Lebensmittel Wissenschaft und Technologie, 7,* 145.

Szczesniak, A.S. (1990). Texture: Is it still an overlooked food attribute?, *Food Technology, 44,* 86.

Tucker, I. (1985). Nonisothermal stability testing, *Pharmaceutical Technology, 6,* 78.

Wang, S.S. and Dimarco, G.R. (1972). Isolation and characterization of the native, thermally inactivated and regenerated horseradish peroxidase isozyme, *Journal of Food Science, 37,* 574.

Whitaker, J.R. (1992). Importance of enzymes to value-added quality of foods. (Review), *Food Structure, 11,* 201.

Xezones, H. and Hutchings, I.J. (1965). Thermal resistance of *Clostridium botulinum* (62A) spores as affected by fundamental food constituents, *Food Technology, 19,* 113.

Yawger, E.S. (1978). Bacteriological evaluation for thermal process design, *Food Technology, 32,* 59.

chapter six

Minimal processing of foods with electric heating methods

Thomas Ohlsson

Contents

Summary 97
1   The thermal heating approach to minimal processing............................98
2   Electric volume heating methods for foods 98
3   Electric resistance/OHMC heating...99
   3.1   Fundamentals 99
   3.2   Equipment...99
   3.3   Applications..100
4   High-frequency heating 100
   4.1   Fundamentals 100
   4.2   Equipment...101
   4.3   Applications..101
5   Microwave heating 101
   5.1   Fundamentals 101
   5.2   Equipment...102
   5.3   Applications..103
6   Comparing the frequencies 103
7   Safety aspects..104
References...104

Summary

A general overview of electric heating methods is given, analyzing the process fundamentals, equipment, and industrial applications. Three major types of processes are covered: ohmic heating, high frequency heating, and microwave heating. General advantages and limitations are summarized.

0-8493-7905-9/99/$0.00+$.50
© 1999 by CRC Press LLC

1 The thermal heating approach to minimal processing

Thermal methods are extensively used for the preservation and preparation of foods. Thermal treatment leads to desirable changes, such as protein coagulation, starch swelling, textural softening, and formation of aroma components. However, undesirable changes also occur, such as loss of vitamins and minerals, formation of thermal reaction components of biopolymers, and, in minimal processing terms, loss of fresh appearance, flavor, and texture.

The classical approach to overcome or at least minimize these undesirable quality changes in thermal processing is the HTST (high temperature short time) concept. It is based on the fact that the inactivation of microorganisms depends primarily on the time of the heat treatment, whereas the quality changes or deterioration depend mostly on the time duration of the heat treatment (Ohlsson, 1980). High temperatures will give rapid inactivation of microorganisms and enzymes, aimed for in pasteurization or sterilization, and short times will give less undesired quality changes. The problem in applying this principle to solid and high viscosity foods is that the parts of the food in contact with the hot surfaces will be overheated during the time needed for the heat to transfer to the interior or coldest spot of the food. The surface overheating will give quality losses that in severe cases will counterbalance the advantages of the HTST concept. Electric heating that directly heats the whole volume of the food is a method that may overcome these limitations caused by the low heat diffusivity of foods. Thus, direct volume heating methods are seen as minimal processing methods, where the thermal processing is applied to minimize the quality changes of the process (Ohlsson, 1994).

Another quality aspect of thermal processes is that the juice losses of meat and fish strongly depend on the temperatures reached. Only a few degrees will give large differences in juice losses, which is important both to the economical yield of the process and to the consumer's perceived juiciness of the product. Avoiding temperatures exceeding the desired culinary and bacteriologically determined final core temperature is then important. The thermal processing is then better done with the LTLT concept, using low temperatures over long times.

2 Electric volume heating methods for foods

In the food industry, thermal processing using electric heating is done by applying electromagnetic energy for producing temperature increases in the food. These temperature increases will in their turn cause the desired changes in the food, such as inactivation of microorganisms and enzymes or production of the desired flavor and texture that consumers associate with ready-made foods. Many of the electric heating methods that we will be talking about are not in themselves new, but the knowledge about them as well as their application in the food industry is limited at present (UIE, 1996).

The electric heating methods directly transfer the energy from the electromagnetic source to the food, without heating up heat transfer surfaces, etc., in the heat processing equipment. This direct energy transfer is of major advantage, as it gives excellent opportunities for high-energy utilization. Looking through the electromagnetic spectrum, we can identify three frequency areas that are employed today in the industry for direct heating of food. The 50/60 Hz, the electric power in the household, is used for electric resistant heating, sometimes called ohmic heating. In this application, the food itself acts as a conductor between a ground and a charged electrode, normally at 220 or 380 volt. In the high-frequency area of 10 to 60 MHz, foods are placed between electrodes — one of them again being grounded — and energy is transferred to the food placed between the electrodes. In the microwave region of 1 to 3 GHz, energy is transferred to the food through the air by guided waves, controlled by electromagnetic devices called applicators.

In all these electric heating methods, it is important to have an understanding of the interaction between the electromagnetic field at the frequency in question and the material being subjected to the energy. Electric and dielectric properties of foods and other materials used in equipment construction, etc., are important to know in order to better understand and control the application of electric energy for heating foods (Ohlsson, 1987).

3 Electric resistance/ohmic heating

3.1 Fundamentals

In electric resistance heating, the food itself acts as a conductor of electricity, taken from mains that are 50 Hz in Europe and 60 Hz in the United States. The food may also be immersed in a conducting liquid, normally a weak salt solution of similar conductivity to the food. Heating is accomplished according to Ohm's law, where the conductivity, or the inverse, the resistivity, of the food will determine the current that will flow between the ground and the electrode. Normally, voltages up to 5000 V are applied. The conductivity of foods increases considerably with increasing temperature. To reach high temperatures it is therefore necessary to increase the voltage current or to use longer distances between the electrodes and ground.

3.2 Equipment

The best known electric resistance heating system is the APV ohmic heating column, where electrodes immersed into the food are transported in a vertical concentric tube. Electrodes, normally four, are connected to earth and line voltage. The inside of the tube is lined with high-temperature and electrically inert plastic material. As the electrodes often are of highly conductive metals, with corresponding ions not being desirable in foods, the isolation of electrodes against the food components is of major importance.

3.3 Applications

The ohmic system of APV has been installed for pasteurization and sterilization of a number of food products with resulting excellent quality. The majority of these installations are found in Japan for the production of fruit products (Tempest, 1996). There are also other installations, e.g., for prepared food in the UK. The ohmic heating system is showing excellent retention of particle integrity due to the absence of mechanical agitation, typical for traditional heat-exchange-based heating systems. A special reciprocal piston pump is used to accomplish the high particle integrity. A long traditional tubular heat exchanger is used for the cooling.

Extensive microbiologic evaluation of the system has shown that the method can produce sterile products reliably.

Other industrial cooking operations for electric resistance heating involve rapid cooking of potatoes and vegetables for blanching in the industry and for preparing foods in institutional kitchens. One of the major problems with these applications is ensuring that the electrode materials are inert and do not release metal ions into the conducting solutions and eventually into the foods.

As mentioned, ensuring that the electrode material is isolated from the food, is one of the problems of electric resistive heating. Another problem is the need to properly control the electric conductivity of all constituents of the food product as this determines the rate of heating. This often requires well-controlled pretreatments to eliminate air in foods and control of salt levels in foods and sauces, etc. (Zoltai and Swearingen, 1996).

4   High-frequency heating

4.1 Fundamentals

High-frequency heating is done in the MHz region of the electromagnetic spectrum. The frequencies of 13,56 and 27,12 MHz are set aside for industrial heating applications. Foods are heated by transmitting electromagnetic energy through the food placed between an electrode and the ground. The high frequency energy used will allow for transfer of energy over air gaps and through nonconducting packaging materials. To achieve sufficiently rapid heating in foods, high electric field intensities are needed.

High-frequency heating is accomplished by a combination of dipole heating, when the water dipole tries to align itself with the alternating electric field, and electric resistance heating from the movement of the dissolved ions of the foods. In the lower temperature range, including temperatures below the freezing point of foods, dielectric heating is important, whereas for elevated temperatures electric conductivity heating dominates. The conductivity losses or the dielectric loss factor increases with increasing temperature, which may lead to problems of runaway heating when already hot parts of the food will absorb a majority of the supplied energy. Dielectric properties of foods are reasonably abundant in the low temperature range, but little data is available in temperatures above normal room temperature.

4.2   Equipment

An important part of high-frequency heating equipment is the design of the electrodes. A number of different configurations are being used, depending on the field strength needed, the configuration of the sample, etc. For high moisture applications, the traditional electrode configuration is mostly used. For low moisture applications, such as dried foods and biscuits, electrodes in the form of rods, giving stray fields for foods placed on a conveyor belt, are often used. Electrodes can be designed to create uniform electric field patterns and, thus, heating patterns for different types of food geometry, today supported by computer simulation techniques, such as FEM software packages (Metaxas, 1996).

The high-frequency power is generated in a circuit, where the food is part of the circuit. It consists of a coil, the condensator plates with the food in between, an amplifier in the form of a triode, and an energy source. Obviously, modern electronic control devices are also employed, among other things to maintain a given frequency, as this may vary as the food is heated; bear in mind that the food itself is part of the oscillating circuit. The control function is today improved by the introduction of the so-called 50 ohms technology, which allows a separate control for tuning the load circuit.

4.3   Applications

The largest application in the food industry for high-frequency heating is in the finish drying or postbaking of biscuits and other cereal products. Another application is the drying of products such as expanded cereals and potato strips. Previously, defrosting of frozen food using high frequency was a major application, but problems of uniformity with foods of mixed composition limited the actual use. The interest in high-frequency defrosting has increased again in the last number of years.

Recently, high-frequency cooking equipment for pumpable foods has been developed. These devices involve pumping a food through a plastic tube placed between two electrodes, shaped to give a uniform heating. Excellent temperature uniformity has been demonstrated in these applications, e.g., for continuous cooking of ham and sausage emulsions (Tempest, 1996).

5   Microwave heating

5.1   Fundamentals

Microwaves used in the food industry for heating are of the ISM (industrial, scientific and medical) frequencies 2 450 MHz or 900 MHz, corresponding to 12 or 34 cm in wavelength. In this frequency range, the dielectric heating mechanism dominates up to moderated temperatures. Polar molecules, the dominant water, try to align themselves to the rapidly changing direction of the electric field. This alignment requires energy that is taken from the electric field. When the field changes direction, the molecule "relaxes" and the energy

Table 1   Factors Influencing
Microwave Heating Uniformity

Food composition and geometry
Packaging geometry and composition
Microwave energy feed system

previously absorbed is dissipated to the surroundings, that is, directly inside the food. This means that the water content of the food is an important factor for the microwave heating performance of foods. The penetration ability of the microwaves in foods is limited. For normal "wet" foods the penetration depth from one side is approximately 1 to 2 cm at 2450 MHz. At higher temperatures, the electric resistance heating from the dissolved ions will also play a role in the heating mechanisms, normally further reducing the penetration depth of the microwave energy. The limited penetration depth of microwaves implies that the distribution of energy within the food can vary. The control of the heating uniformity of the microwave heating is difficult, as the objects to be heated are of the same size as the wavelength in the material. The difficulties to control heating uniformity must be seen as the major limitation for industrial application of microwave heating. Thus, an important requirement on microwave equipment and microwave energy application in the food industry is the ability to properly control the heating uniformity (Ohlsson, 1983).

5.2   Equipment

The transfer of microwave energy to food is done by contactless wave transmission. The microwave energy feed system is designed to control the uniformity during the heating operation. Many different designs are used in industrial applications, starting from the traditional multimode cavity oven, via direct radiation waveguide applicators to sophisticated periodic structures (Metaxas, 1996). Design of applicators needs to be done with proper care taken to the interaction between parameters important to the heating uniformity, as shown in Table 1.

As pointed out by Ryynänen and Ohlsson (1995), the importance of the food geometry and the actual layout of the components in a plate for reaching good heating uniformity is often poorly understood.

The microwave energy feed system controls the electric field polarization. This in its turn affects the tendencies for overheating of food edges, which is one of the most severe problems of uneven microwave heating of foods (Sundberg et al. 1996).

The very high frequency used in microwave heating allows for very rapid energy transfer and thus high rates of heating. This is a major advantage, but can also lead to problems of nonuniform heating when too high energy transfer rates are used.

5.3 Applications

Industrial applications of microwave heating are found for most of the heat treatment operations in the food processing industries. For many years the largest application has been defrosting or thawing of frozen foods, such as blocks of meat, prior to further processing. Often, meat is only partially defrosted (tempered) before it can be further processed. Another large application area is for pasteurization, and now also sterilization, of packaged foods. It is primarily ready-made foods that are processed. The objective of these operations is to pasteurize the food to temperatures in the range of 75 to 80°C, in order to prolong the shelf life to approximately 3 to 4 weeks. Sterilization using microwaves has been investigated for many years, but the commercial introduction has only come in the last few years in Europe and Japan. Microwave pasteurization and sterilization promise to give very quick heat processing, which should lead to small quality changes due to the thermal treatment, according to the HTST principle. However, it has turned out that very high requirements on heating uniformity must be met in order to fulfil these quality advantages (Ohlsson, 1991).

Pasteurization with microwave heating can also be done for pumpable foods, as pointed out already in 1964 by Püschner. Microwaves are directed to the tube where the food is transported, and heating is accomplished directly across the tube cross section. Again, uniformity of heating must be ensured, which requires selection of the correct dimensions of the tube diameter and the proper design of the applicators (Ohlsson, 1993). Systems where the food is transported through the heating zone by a screw are also available (Berteaud, 1995).

Further application of microwave heating is for drying in combination with conventional hot-air drying. Often, microwaves are primarily used for moving water from the wet interior of solid food pieces to the surfaces, relying on the preferential heating of water by microwaves. Applications can be found for pasta, vegetables, and various cereal products, where also puffing by rapid expansion of the interior of the food matrix can be accomplished using microwave energy (Tempest, 1996).

Microwave energy is also used for various cooking and coagulation processes from meat products, chicken, and fish, often in combination with other conventional cooking operations. A number of new applications in the microwave heating area have been reported recently in patent literature, often involving the use of the unique heating properties of microwaves with higher energy fluxes and direct heating of the interior.

6 Comparing the frequencies

The advantages and limitations of the various frequencies are listed in Table 2.

Table 2　Comparison between the Frequencies of Electric Heating

Ohmic and High Frequency	Microwave
Advantages	**Advantages**
+ Better for large, thick foods	+ Higher heating rate
+ Lower investment costs	+ Design freedom
+ Easier to understand and control	+ Less sensitive to food heterogeneity
	+ Much R&D available
Disadvantages	**Disadvantages**
– Risk of arching in HF	– Limited penetration
– Larger floor space	– Higher investment costs
– Narrow frequency bands	– More engineering needed
– Limited R&D support	

7　Safety aspects

In the application of electromagnetic energy for the heating of foods, questions on the safety of the food, nutritional value, and the existence of non-thermal heating effects are often raised. There have been extensive studies regarding the changes of chemical constituents in the foods as a result of electric heating. It has been demonstrated that the effect of the electromagnetic heating in all practical aspects is the same as for conventional heating to the same temperature.

Electric heating equipment for the food industry has to be designed and operated according to international and national safety standards. The levels of allowable leakage vary over the frequency range according to these standards. Equipment for measuring and monitoring electromagnetic energy leakage from electric equipment is readily available (IEC, 1982).

References

Berteaud, A-J. (1995). Thermo-Star. *Bulletin from MES*, 15, Rue des Solets, Rungis, France.

IEC (1982). *Safety in electroheat installations.* Publication 519. IEC, Geneva, Switzerland.

Metaxas, A.C. (1996). *Foundation of electroheat: A unified approach.* J. Wiley and Sons, Chichester, UK.

Ohlsson, T. (1980). Temperature dependence of sensory quality changes during thermal processing. *Journal of Food Science*, 45 (4), 836.

Ohlsson, T. (1983). Fundamentals of microwave cooking. *Microwave World*, 4 (2), 4.

Ohlsson, T. (1987). Dielectric properties — industrial use. In: *Physical Properties of Foods*, Vol. 2, Elsevier Applied Science Pub., London, p. 199.

Ohlsson, T. (1991). Microwave processing in the food industry. *European Food and Drink Review*, 3.

Ohlsson, T. (1993). In-flow microwave heating of pumpable foods. Presented at the International Congress on Food and Engineering, Chiba, Japan, May 23–27, 1993.

Ohlsson, T. (1994). Minimal Processing — preservation methods of the future: an overview. *Trends in Food Science and Technology*, 5 (11), 341.

Püsschner, H. (1964). Wärme durch Mikrowellen. Philips Techn. Bibliotek, Eindhoven, Netherlands.

Ryynänen, S. and Ohlsson, T. (1995). Microwave heating uniformity of ready meals as affected by placement, composition, and geometry. *Journal of Food Science 61* (3), 620.

Sundberg, M., Risman, P.O., Kildal, P-S., and Ohlsson, T. (1996). Analysis and design of industrial microwave ovens using the finite difference time domain method. *Journal of Microwave Power and Electromagnetic Energy* 31 (3), 142.

Tempest, P. (1996). *Electroheat Technologies for Food Processing*, Bulletin of APV Processed Food Sector, PO Box 4, Crawley, W. Sussex, England.

UIE (1996). *Electricity in the Food and Drinks Industry* UIE, B.P. 10., Place de la Defense, Paris, France.

Zoltai, P. and Swearingen, P. (1996). Product Development Considerations for Ohmic Heating. *Food Technology*, 50 (5), 263.

chapter seven

Application of the concepts of biomaterials science to the quality optimization of frozen foods

Jorge C. Oliveira, Pedro M. Pereira, Jesus M. Frias, Isabel B. Cruz, and William M. MacInnes

Contents

Summary 108
1 Introduction 108
2 Basic concepts 109
   2.1 Glass transition and Tg 109
   2.2 Factors that affect Tg 111
   2.3 Experimental measurement of Tg 112
   2.4 Glass transition below the freezing point (T_g', T_m', and T_f) 113
3 Application to frozen foods 114
   3.1 Glass transition in frozen food products 114
   3.2 Complexity of food matrixes 116
   3.3 Deteriorative phenomena between (T_g' (T_m') and T_f) 117
      3.3.1 Recrystallization 117
      3.3.2 Oxidation, coalescence, and microbial growth 118
      3.3.3 Modeling of the matrix mobility 119
4 Novel developments in the analysis of T_g in frozen foods 122
   4.1 Dynamic mechanical thermal analysis (DMTA) 122
   4.2 Measurement of T_g in DMT tests 123
   4.3 Application to the determination of T_g in frozen sugar
      solutions 125
5 Conclusions 126
List of symbols 128
References 128

107

Summary

The concepts of glass transition theory that were developed by materials science, with special relevance in polymer engineering, can be applied to the analysis of water and general molecular mobility in foods. In recent years, this approach has been finding increasing applications in biotechnology and food science. This text presents a general review of these concepts applied to frozen matrixes. How they can assist in the improvement of product formulation and logistics with a view to optimize the maintenance of the quality attributes achieved during improved freezing processes throughout the frozen food chain is highlighted. Some novel approaches concerning the measurement of water mobility in frozen foods are also discussed.

1. Introduction

Water molecules play several essential roles in all systems of interest to biotechnology and food and therefore its mobility within the system is a matter of utmost importance in the design, control, and improvement of industrial processes and products.

Conventionally, water is divided in free water and bound water. The bonds between water molecules and the solid matrix can have very diverse intensities (physical adsorption, hydrogen bonds, polar bonds, ionic bonds, etc.) and therefore some of the bound water can actually be made available for use by endogenous metabolisms or microbial activity. If one would assume that the capacity of microorganisms or enzymatic metabolisms to dispute water molecules with the specific structural bonds is similar to that of air below its saturation point, then the concept of water activity would be an excellent form of quantifying the mobility of water in the system. In its thermodynamic concept, water activity (the ratio between the partial pressure of water and its vapor pressure at the temperature in question) is roughly equal to the relative humidity of air in equilibrium with the solid matrix. For instance, a water activity of 0.5 means that the water present in the structure of the solid matrix is so strongly bound to it that air with a relative humidity of 50% is in equilibrium with this system and cannot capture water molecules from it (Karel, 1975).

For many decades, this simple concept has been sufficient to describe the stability of food products, especially dried foods. However, it proved to be inadequate in some cases: for instance, some solutions having the same water activity but other relevant factors (such as concentration of some specific macromolecules) exhibited quite different microbial growth rates (Slade and Levine, 1991). A different form of quantifying water mobility and food stability was necessary and the application of the concept of glass transition became popular in the late 1980s. Earlier suggestions had been made (White and Cakebread, 1966), but it was mostly the work of Slade and Levine and of other researchers from 1985 onward (Slade and Levine, 1986) that established the references for the developments in food science. Several applications of the concept of glass transition in food systems can be found in Blanshard and Lillford (1993).

In this text, a brief review of the major concepts pertaining to frozen foods is given, covering some specific applications.

2 Basic concepts

2.1 Glass transition and Tg

Most food constituents are molecules that can be organized in crystalline or amorphous structures.

In general, a crystalline structure is highly organized and comprises almost exclusively bonds between molecules of the specific component: that is, it has a high degree of purity. The high level of molecular organization corresponds to a stable matrix with minimum entropy, which is relatively little affected by temperature. Different types of crystalline structures may coexist, depending on the molecular bonds that are created, and therefore some crystalline structures have lower entropy than others. Obviously, the systems tend toward the form of lower entropy, but depending on the matrix mobility the process can be very time consuming under natural conditions. The essentially pure nature of crystalline structures implies that almost no water molecules are present, except at the surface of crystals due to hydration (in solution) or physical adsorption (in solids).

Amorphous structures have a much lower level of organization, that is, higher entropy. They are composed not only of molecules of the specific component, but involve many other molecules that are linked to the structure at several points and may even create intermolecular links. Of these "impurities" that may be trapped in an amorphous matrix, water is the most generally relevant and can be involved in different types of bonds (e.g., ionic, covalent, polar, hydrogen bonds, physical adsorption). Due to this variety of possible links, amorphous structures are much more sensitive to temperature, and entropy decreases significantly with cooling.

Complex macromolecules may show partly crystalline, partly amorphous structures. The most typical case is starch, broadly described as composed of a (partly) crystalline amylopectin part and an amorphous amylose part.

When a component cools down from liquid state below its fusion temperature, in many cases the resulting solid structure may be crystalline or amorphous depending on the cooling rate. If the cooling is slow, there is "sufficient time" for the molecular configuration to evolve to the more stable crystalline form, whereas if cooling is fast (called "quenching"), the molecules practically "freeze" in whatever conformation they were in, trapping other molecules, and particularly water, in the resulting matrix. It is obvious that the natural tendency of the system is to evolve from the amorphous to the crystalline structure, a process that can take seconds or centuries, depending on the molecular mobility.

In relation to water itself, this process is not found at normal conditions. Amorphous solid water can only be generated under special conditions, namely high pressure (Mishima et al. 1984), hyperquenching (Bruegeller and Mayer, 1980), or vapor deposition (Sceats and Rice, 1982). Furthermore, solid water would rapidly evolve into the usual crystalline ice I form above −135°C.

If one considers an amorphous structure created by fast cooling, that keeps being cooled down, entropy and the whole internal energy of the system keep decreasing. It is obvious that there has to be a physical limit for this decrease, which must evidently have an entropy well above that of crystalline structures. This discontinuity must correspond to a change of state. Below it, the relatively flexible and temperature-sensitive amorphous structure becomes more rigid and relatively insensitive to temperature changes, similar characteristics to those of crystalline structures. The structure between liquid state and this transition is called rubbery state and the rigid structure below it is known as glassy state. This process of conversion from one to the other is the glass transition.

There are several theories available to interpret this phenomenon theoretically (Mansfield, 1993), but the most important are the free-volume theory, according to which glass transition takes place when the molecular free volume approaches zero (Doolittle, 1951), and the entropic models, which explain the reason for it to be a second-order transition (Gibbs and DiMarzio, 1958).

Second-order transitions refer to the classification proposed by Ehrenfest (1933). In first-order transitions (such as phase transitions), there is a discontinuity of the free energy (for instance, enthalpy of fusion for a liquid-solid phase transition at constant temperature and pressure), while in a second-order phase transition there is no discontinuity of the free energy, but there is of its derivatives (for instance, the specific heat). This concept is represented in Figure 1. The major implication is that even for a pure component the glass transition does not occur at a specific (constant) temperature (which would depend on pressure), but during a given range of temperatures. In food matrixes, the difference between the onset and endset temperatures of glass transition is usually around 10 to 30°C (Slade and Levine, 1991; Roos, 1995).

From this description, it is evident that a DSC (differential scanning calorimetry) scan can identify the onset and offset of glass transition from deviations in an otherwise linear plot. Figure 2 sketches a typical DSC scan over glass transition. However, glass transition is usually indicated simply by the mid-temperature of the range detected in a DSC, which is called "glass transition temperature" (T_g). Recently, some authors have suggested that T_g should be taken as the onset temperature, especially for stability studies, as above it part of the structure is already in rubbery state, but this reasoning is not necessarily correct.

The concept of T_g can be related to water activity. In a range of values of a_w (around 0.1 to 0.8 for amorphous matrixes), there is an approximately linear relationship between T_g and a_w (Roos, 1995). This relationship obviously depends on temperature.

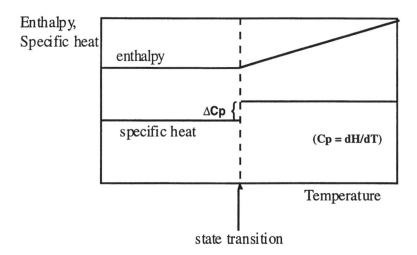

Figure 1 Variation of energy variables and their derivatives in a second-order transition.

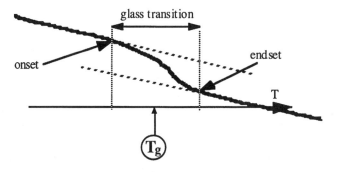

Figure 2 Sketch of a typical DSC thermogram around glass transition.

2.2 Factors that affect T_g

Glass transition and T_g are strongly affected by the composition of the matrix. The molecule that has the greatest impact is water, as shown in Figure 3 (Roos and Karel, 1991; Kokini et al., 1994). A significant decrease of T_g by a given molecule is called its plasticizing effect. Lowering T_g is important in polymer processing for improving the materials processability and hence the origin of the name. As water is present in virtually all biochemical structures it is sometimes referred to as an intrinsic plasticizer. Other components also have a similar effect, such as polyols. In the polymer industry, such additives are used as extrinsic plastifiers, of which glycerol is the most common.

Another important factor is the average length of the molecular chains. In general, T_g increases with the average molecular weight (Roos, 1995). The

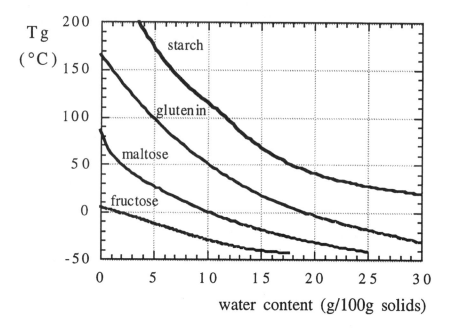

Figure 3   Influence of water content on the glass transition temperature (model fits of the data reported by Roos and Karel, 1991 and Kokini et al., 1994).

addition of macromolecules, of which polysaccharides are the most common, is the major way of increasing T_g.

In general, any effect that is known in conventional analysis to increase the water-holding capacity (WHC) will increase T_g, which is evident as it decreases the amount of water molecules that are available to plastify the structure. Similarly, it is obvious that a structure having amorphous components rather than crystalline (for instance, sugars) has a significantly higher T_g.

It is therefore possible to design the formulation of a matrix to change T_g. It is necessary to consider the effect of several components and additives and to have a good knowledge of polymorphism, as each conformation has a different effect on the T_g of the whole structure. This dependence on molecular conformation implies that intrinsic or extrinsic factors that affect molecular conformation (such as pH) may affect T_g.

2.3   Experimental measurement of T_g

The most usual and standard method for determining T_g, the onset and endset of glass transition is DSC. However, there are several alternatives and those that involve an understanding of molecular vibration or mobility would be particularly advantageous, such as NMR (nuclear magnetic resonance) and ESR (electron spin resonance), but their high cost makes them unusual. Other more common alternatives are based on the measurement of changes in a given physical property, such as dilatometry, TMA (thermo-mechanical analysis), DETA (dielectric thermal analysis), and DMTA (dynamic mechanical thermal analysis). In fact, it is theoretically possible to use DMTA to obtain a measurement of molecular vibration and mobility, as will be later discussed (Section 4). A brief overview of these techniques can be found in Roos (1995).

2.4 Glass transition below the freezing point (T_g', T_m', and T_f)

Most foodstuffs have high amounts of water (with the exception of dried and dehydrated food products), which implies that their T_g is well below freezing. Therefore, to visualize the glass transition in most foods as a result of lowering temperature, one has to consider first the effect of freezing. The formation of ice crystals that begins at the freezing point has an obvious dramatic effect on the T_g of the matrix, as it corresponds to a cryoconcentration. As water molecules change to the solid phase, the unfrozen solution becomes more concentrated and therefore its freezing temperature (T_f) decreases while its T_g increases. Inevitably, there will be a point when the unfrozen solution is so cryo-concentrated that its T_f equals its T_g and the whole unfrozen matrix (including the remaining water) will then undergo glass transition. This point therefore corresponds to the maximum concentration for ice formation. The temperature is designated as T_g' (glass transition temperature of the maximum cryo-concentrated solution) and the corresponding concentration as C_s' (maximum concentration for ice formation).

This evolution is sketched in Figure 4. It should be noted that in a complex matrix there may be differences between slow and fast cooling. If the cooling rate is low, some components that can crystallize, such as sugars and lipids, may precipitate in crystalline form when reaching its solubility point, releasing water molecules that previously hydrated the molecules. If the cooling rate is sufficiently high to minimize crystallization, amorphous solids will precipitate, which entrap some of the water molecules, and the T_g evolution will be faster.

Roos and Karel (1991) and other authors argue that the maximum viscosity for formation or melting of ice crystals is known to be around 10^8 Pa.s while the viscosity of a glassy matrix is around 10^9 to 10^{12} Pa.s. Therefore, there must be a range of temperatures above T_g' in which the formation of crystals is no longer possible. There would be a higher temperature, designated by T_m', which would be the limit for crystal formation/melting, identified to be around 5 to 8°C above T_g (Roos, 1995). This notion is depicted in Figure 5. However, it is noted that the argument is not universally accepted. Slade and Levine (1991) remind that glass transition is a phenomenon that occurs over a range of temperatures and that T_g is the mid-temperature of the transition. Therefore, what is called T_m' could simply be the endset of glass transition. This reasoning seems theoretically correct and later in this text (Section 4.3) some experimental evidence will be presented that would also be more in agreement with it than with the definition of a T_m'.

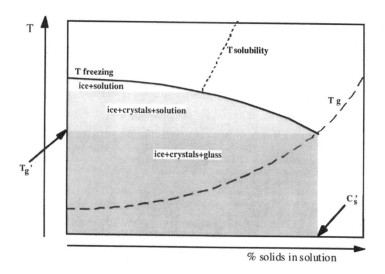

Figure 4   Sketch of a typical state diagram.

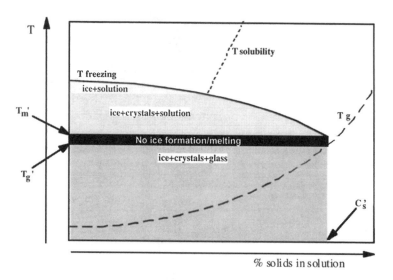

Figure 5   Concept of T'_m in a state diagram.

3   Application to frozen foods

3.1   Glass transition in frozen food products

There are two major aspects to consider in frozen foods: microbial safety and sensory quality. The latter is particularly relevant for products that will be consumed in the frozen state (such as ice creams) where texture is one of

A - cottage cheese
B - cream
C - ice cream
D - skim milk
E - skimmed milk + HL
F - whipped toppings
G - whole milk
H - apple
I - banana
J - blueberry
K - peach
L - strawberry
M- apple juice
N - lemon juice
O - orange juice
P - pear juice
Q - pineapple
R - prune juice
S - strawberry juice
T - white grape juice
U - broccoli
V - cauliflower
W- potato
X - spinach
Y - sweetcorn
Z - tomato

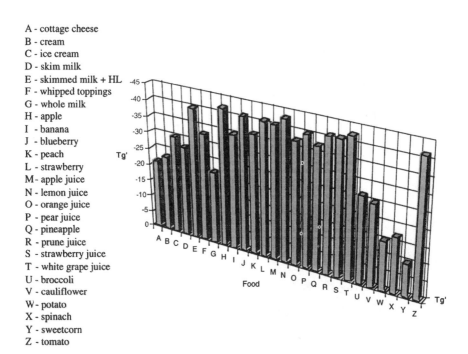

Figure 6 T'_g of several food products.

the major attributes. In any case, if the products were stored below T'_g, the rate of any type of change would be severely decreased and the product would be virtually stable: in the glassy state, molecular mobility is restricted to vibration and mutarotation. Unfortunately, most food products have such large amounts of water that their T'_g are below normal frozen storage temperatures. Figure 6 shows the values reported by Levine and Slade (1989). This implies that most frozen foods are stored with a more or less significant amount of unfrozen cryo-concentrated aqueous solution and there is a fair degree of molecular mobility. In these matrixes, there are several detrimental reactions that can take place and affect product quality, such as recrystallization of ice, crystallization of amorphous sugars, coalescence of the organic phase, transformation of lipid molecular conformations to the more stable forms, enzymatic and nonenzymatic oxidation, and possibly even microbial growth.

It follows that the higher quality achieved with improved freezing processes, such as cryogenic freezing, may be totally lost during storage and distribution, depending on the temperatures to which it is subjected, their fluctuation (phenomena such as recrystallization are enhanced by temperature fluctuations), and the resulting degradation rates. In general, molecular mobility and therefore the rates of quality loss are higher the further from T_g. Therefore, one would wish that the food matrix would have a T_g as high as possible.

3.2   Complexity of food matrixes

In its simplest form, food matrixes are a complex mixture of a large diversity of molecules (e.g., milk, dough). The complexity increases in cases where the matrix has a high degree of structurization (e.g., fruits and vegetables), with the cellular microstructure affecting significantly the properties of the matrix. In addition, food matrixes are rarely stable and continuously evolve as a result of natural endogenous metabolisms (e.g., senescence of vegetable tissues) or externally induced changes (e.g., oxidation, microbial activity).

It is however possible to consider in general terms the influence of the major specific components on T_g.

In relation to carbohydrates, there are three major groups to analyze: sugars, polysaccharides, and starch (which is individualized from the other polysaccharides due to its special role in food matrixes).

Sugars may be present in amorphous or crystalline form and have a plasticizing effect. Obviously, amorphous sugars lead to matrixes with higher T_g, but it is important to analyze whether the storage conditions might be sufficient to allow for reasonable crystallization rates, as this would have a major effect on lowering T_g below the design value and further increase crystallization rates. Replacing sugars by others with higher T_g will increase the T_g of the whole matrix. The T_g' values vary from around $-50°C$ (pentoses) to $-25°C$ (disaccharides and oligosaccharides), with *circa* $-40°C$ for hexoses (Slade and Levine, 1991).

Polysaccharides increase the T_g of the matrix mainly as a result of increasing the average molecular chain length, and therefore their effect is generally higher the longer the molecular chain. Notwithstanding, some particular phenomena can be observed, especially with molecules that act as cryoprotectants and form complex structures with proteins.

Starch is a special polysaccharide in this context. It is a major food ingredient, of a quite complex structure, that usually presents in its native form a semicrystalline conformation. The amylose/amylopectin ratio obviously affects the T_g of starch significantly, because amylose only forms amorphous structures. However, the T_g' of starches varies in a narrow band between -7 and $-5°C$ (Levine and Slade, 1989). In addition, starch also has a structuring effect on the whole matrix, particularly after gelatinization. Changing the type of starch (wheat, corn, potato, modified) allows for significant changes in T_g. The selection of the most adequate type of starch must therefore be based on the knowledge of the specific types of interactions with the other components and of the type of structure and of links that are formed, which is much more important than the actual individual T_g' value of the starch in question.

In relation to proteins, it is known that their structuring effect on the food matrix is as or more relevant than that of starch molecules. Therefore, their role in forming specific molecular structures in the food matrix is also more relevant than their individual T_g values. The proteins that have been more studied are those of cereals, whose T_g' varies around -12 to $-6°C$ (Slade et al., 1988; Kokini et al., 1994).

In lipids, the most important aspect concerning glass transition is polymorphism. Lipids are generally insoluble and usually crystallize with cooling. In the more important case of triglycerides there are three major conformations: α, β', and β, by decreased order of stability. The most stable form is usually undesirable for food purposes: the α crystals are normally too large and texture is considered generally poor. The sensorially preferred form for butters, margarines, and spreads is β'. Increasing T_g is important to lower the rates of transformation from β' to α. In frozen products, the major problem concerning lipids is the coalescence of the organic phase and therefore the diffusion of lipid molecules is the major factor of concern. It is evident that emulsion stabilizers play an important role, regardless of their effect on the T_g of the matrix.

3.3 Deteriorative phenomena between T_g' (T_m') and T_f

3.3.1 Recrystallization

It is known from freezing technology that the effect of freezing on food texture is greater the larger the size of the ice crystals. Therefore, slow freezing affects the product texture much more than fast freezing.

Cryogenic freezing is a high-quality producing technology, as the immersion in liquid nitrogen or liquid carbon dioxide gives very high freezing rates. An even more interesting technology currently under development is high-pressure-assisted freezing. However, the fact that such technologies can lead to high-quality products depends on the ability of the product to maintain such quality throughout storage and distribution, up to the final consumer. If temperatures are above T_g' (which is usual), ice recrystallization is inevitable. In fruits and vegetables this will lead to cell rupture and loss of texture. In products that are consumed frozen, such as ice creams, the size of ice crystals is also very important for texture: Blanshard and Franks (1987) indicate that the maximum size is 40 μm, above which consumers complain of grainy texture. In a study carried out by the Campden & Chorleywood Food Research Association (summarized in Richardson (1995) and in Chapter 9 of this book) with the objective of comparing the quality of thawed foods frozen by different technologies, including still air freezing, blast freezing, and cryogenic freezing, after several months of frozen storage no quality differences were found by consumer panels. This implies that the average consumer would not find any quality gain resulting from the improved freezing technologies, and the storage conditions in this study (constant, $-20°C$) were better than average real-life frozen food chains.

Two recrystallization mechanisms are particularly destructive: migration and aggregation, as these result in an increase of the average size of crystals at the expense of their number. In order to ensure that the improved product quality is not lost completely during storage and distribution, it is necessary to know well the recrystallization rates and which factors can be manipulated to influence it. Additives that inhibit recrystallization can be used (sugars, polysaccharides, cryoprotectants — Roos, 1995). The way by which many of these additives work is not entirely known and cannot be explained solely on the basis of the effect on increasing T_g. It is evident than any additive that increases T_g would decrease the recrystallization rate, but some additives have a much larger effect than what would be predicted by simple analysis of the effect on T_g. The physical–chemical interactions between the molecules of the additives and ice are equally important (Goff et al., 1993). Another possible beneficial effect of such additives would be an increase of the viscosity of the cryo-concentrated solution, as this obviously decreases the rate of recrystallization.

Another important aspect concerning crystallization phenomena is related to sugars. In ice creams the formation of lactose crystals is the major cause of sandiness. In this case, it is important to consider how crystals are formed. The process can be divided in several steps, of which nucleation and propagation (the first two) are highlighted. The crystallization rate increases with nucleation rate or with propagation rate. At temperatures somewhat above T_m' the rate of nucleation is maximum, but the rate of propagation is very low as viscosity is high. At temperatures close to the freezing point the rate of propagation is high as the matrix has its maximum mobility, but the rate of nucleation is low as the concentration is minimum. Consequently, the maximum overall rate of crystallization occurs somewhere in rubbery state. This means that lowering the storage temperature may not necessarily be an improvement, and it is important to know the crystallization rates in the specific product.

The above discussion highlights the need for describing the rates of crystallization/recrystallization and how temperature and water content affect them. The most widely used empirical equation is:

$$L = L_0 + \psi t^n \tag{1}$$

where L is the average crystal size at time t, L_0 is the initial average crystal size, and n and ψ are system parameters, with n usually being temperature- and water-independent in a reasonable range and ψ is considered to vary with temperature according to different models (WLF, Arrhenius, linear — Roos, 1995).

3.3.2   Oxidation, coalescence, and microbial growth

In many frozen products, other phenomena are also important, mainly the oxidation of sugars or lipids (especially nonenzymatic browning), the coalescence of an organic phase, and microbial growth. In general, the microbial growth rate is too low compared to the others. These phenomena have one important thing in common: they are strongly affected by molecular mobility, not only of water but also of other molecules. The rates of these reactions/metabolisms are usually well correlated to diffusion, which in turn is related to viscosity. Therefore, the type of quantitative methods applied to water mobility can usually be applied to the mobility of other molecules and hence to the overall rate of these deteriorative effects (Nelson, 1993).

3.3.3 Modeling of the matrix mobility

The kinetics of oxidation reactions, phase coalescence, microbial growth, crystallization, and recrystallization are normally well correlated to the molecular mobility in the cryo-concentrated solution above T_g', which in turn is related to the average molecular vibration time of the conformation arrangement at the temperature in question.

Williams, Landel, and Ferry (1955) have proposed an equation that relates the average molecular vibration time above T_g to temperature, known as WLF equation, which can actually be reduced to that proposed by Doolittle (1951) from the free volume theory:

$$\log \frac{\tau}{\tau_g} = \frac{-C_1\left(T-T_g\right)}{C_2+\left(T-T_g\right)} \tag{2}$$

where τ is the average molecular vibration time at temperature T and τ_g the respective value at T_g. C_1 and C_2 are model parameters. C_1 would be, in principle, the difference in order of magnitude between the average atomic vibration time (circa 10^{-13}) and the average molecular vibration time in glassy state (circa 10^3), that is, 16. Williams, Landel, and Ferry (1955) proposed the universal values of 17.44 for C_1 and 51.6°C for C_2 by fitting data from several polymers. However, some authors have cautioned against the use of universal constants, as specific systems may show quite a variation. Peleg (1992) collected C_1 values from literature ranging from 13.7 to 34 and C_2 values between 24 and 80°C, though even lower C_1 values were reported by Roos (1995). From the comprehensive work of Angell and co-workers (Angell et al., 1994), it is clear that while C_1 should indeed be a universal constant with the value of 16, C_2 is a system-dependent value and should be determined for the specific matrix in question. It should be noted that the WLF equation expresses both temperature and water-content dependency, because T_g varies with water-content, and this is a particularly interesting aspect for describing quality loss in drying processes (Frias and Oliveira, 1997).

Evidently, viscosity will show a similar type of relation, as it can be correlated very well to the average molecular vibration time. For viscosity, the VTF (Vogel, Tamman, and Fulcher) equation is more commonly used in physical chemistry. It is noted that the mathematical form of the VTF equation is interconvertible to the WLF equation (Angell et al., 1994). Therefore, diffusion-dependent phenomena can also be expected to follow a WLF model, given the relationship between viscosity and diffusivity.

However, the WLF equation is rarely applied in this form, because the system property being measured (viscosity, diffusivity, reaction rate) has extremely high/low values around and at T_g. Therefore, experimental data will usually concern a range of temperatures above T_g and the system property at T_g would not be a good model parameter, as it would have a very different order of magnitude compared to the experimentally measured values. In their original work, Williams, Landel, and Ferry (1955) suggested replacing T_g by a reference temperature that could be determined by regression of the experimental data. Writing Equation (2) for the reference temperature and manipulating both, one would obtain:

$$\log\frac{\tau}{\tau_r} = \frac{-C_1'(T-T_r)}{C_2'+(T-T_r)} \tag{3}$$

where τ_r is the system property (average molecular vibration time in this case) at the temperature T_r and the modified constants C_1' and C_2' are equal to:

$$C_1' = \frac{C_1 C_2}{C_2 + T_r - T_g}$$

$$C_2' = C_2 + T_r - T_g$$

For several inorganic and organic compounds and synthetic polymers, Williams, Landell, and Ferry (1955) verified that T_r was around 50°C above T_g (which, probably not coincidentally, is roughly the value of C_2). Later works have used this type of equation in a more or less semiempirical way, suggesting basically different forms of writing the same expression. Thus, Ferry (1980) and Peleg (1992) suggested:

$$\log\frac{\tau}{\tau_r} = \frac{-C_1'(T-T_r)}{C_2'+(T-T_\infty)} \tag{4}$$

where T_∞ would be another reference temperature located below T_g, to be determined by data regression (Peleg, 1992). Ferry (1980) considered that T_∞ should be around 50°C below T_g.

The WLF equation has been applied in several contexts with variable success (Kerr et al., 1993). The most common area of application is nonenzymatic browning in dried products (Roos and Karel, 1991). In our research work, we have found that in a large variety of situations the fit of experimental data to Equation (3) is particularly sensitive to the value of T_r, that should be considered a system parameter, rather than being fixed at a given value within the range of experimental data. Good results were obtained by analyzing the statistical significance, collinearity, and error region of the parameters and selecting for T_r the value that resulted in the highest statistical significance and lower error region for the other parameters — this also

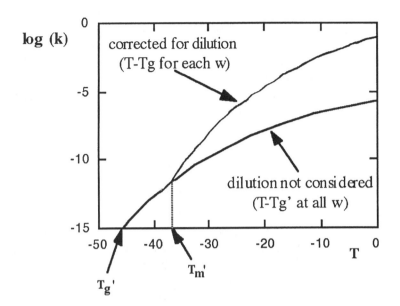

Figure 7 Example of the effect of crystal melting on the increase of the rate of deteriorative reactions, according to the WLF model (adapted from the concept presented by Roos, 1995, for data on the thermal degradation of ascorbic acid in maltodextrin with T_g as a function of water activity being given by the Gordon-Taylor equation).

resulted in the minimum collinearity between C_1 and C_2 (Frias and Oliveira, 1997). It is also possible to decrease problems of collinearity and large error regions by fixing the value of C_1 in 16.

It should be noted that describing the mobility in unfrozen matrixes between T_f and T_g' requires the knowledge of the whole phase diagram, not just only of T_g'. Above T_g' the cryo-concentrated solution contains more water the higher the temperature and therefore its T_g is lower than T_g' and depends on the actual temperature. This dilution effect is sketched in Figure 7. The mobility/stability of the matrix is related to the difference between T and T_g, not T and T_g'. The phase diagram allows to convert from T_g as a function of concentration (water content) to T_g as a function of the system temperature.

The plasticizing effect of water can be estimated from the Gordon-Taylor equation (Gordon and Taylor, 1952), which in this case becomes:

$$T_g = \frac{kwT_{g(water)} + (1-w)T_{g(drymatrix)}}{kw + (1-w)} \tag{5}$$

where w is the water content and k is a system parameter, to be obtained from model fitting of experimental data for several water contents. This is in fact a derivation of the theoretically based Couchman-Karasz equation

(Couchman, 1978), with the empirical parameter k replacing the ratio of the specific heat increase of water due to glass transition and that of the dry matrix. Although the equation becomes empirical, the added flexibility allows for a wider application. Both the Couchman-Karasz and the Gordon-Taylor equations were proposed for mixtures of polymers, but with the empirical constant k allowed to have any value, it has been applied with success to several food systems (Roos and Karel, 1991; Frias et al., 1998), with values ranging roughly between 2 and 10.

4 Novel developments in the analysis of T_g in frozen foods

4.1 Dynamic mechanical thermal analysis (DMTA)

The analysis of frozen matrixes has been largely performed by DSC, with some recent developments in NMR. However, DSC does not give much information, other than identifying when glass transition is beginning and when it is ending. Although the temperature scan rate can potentially affect the results, this has been little studied.

An interesting technique that can provide some more insight into the system and is relatively simple and less costly than NMR is DMTA. The added value of these measurements ultimately results from the fact that DMTA is a dynamic analysis, which can be theoretically interpreted in terms of system dynamics theory, well known in process engineering.

DMTA measures the rheologic properties of the system continuously, while the temperature is changed, usually by linear increase of decrease scans. Most usually, the rheologic measurements are made by compression or bending tests, that quantify the modulus of elasticity of the matrix by measuring the stress caused by a given input strain.

The concept of dynamic or frequency-response tests is to cause a sinusoidal input strain. It is fairly simple to demonstrate theoretically that in the linear viscoelastic range stable systems will tend to an output sinusoidal stress that has the same frequency as the input strain (Luyben, 1990), though with a phase angle, which can vary from 0° for an ideal solid body to 90° for an ideal liquid (Darby, 1975). Furthermore, it can also be demonstrated that in the linear viscoelastic range the phase angle (δ) and the ratio of the amplitudes of the output and input sinusoidal waves, which are independent of the amplitude of the input wave, allow for the quantification of the percentage of mechanical energy input that is stored (elastic behavior) and the percentage that is dissipated (plastic behavior). The relationships between the measuring and system variables are the following:

Input strain wave:	$\varepsilon = \varepsilon_0 \sin(\omega t)$
Output stress wave:	$\sigma = \sigma_0 \sin(\omega t + \delta)$
Complex elastic modulus:	$E^* = \sigma_0/\varepsilon_0 = E' + i\,E''$
Storage modulus:	$E' = E^* \cos \delta$
Loss modulus:	$E'' = E^* \sin \delta$
Phase angle:	$\tan \delta = E''/E'$

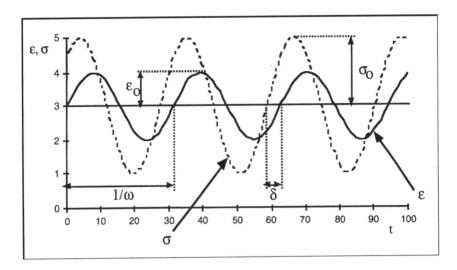

Figure 8 Input (strain) and output (stress) variables in a frequency-response test. The output signal only reaches the sinusoid represented after some time.

where ε is the strain at time t, ε_0 the strain amplitude, σ the stress at time t, σ_0 the stress amplitude, ω the frequency, δ the phase angle (tanδ is the viscoelastic ratio), E^* the complex elastic modulus, E' the storage elastic modulus (mechanical energy storage), E'' the loss elastic modulus (mechanical energy dissipation), and i is the imaginary unit (square root of –1). These variables are sketched in Figure 8.

Glass transition is easily detected from the graphs of E', E'', and tanδ as a function of the measuring temperature. In glassy state E' is constant and therefore the onset of glass transition corresponds to the initial drop in E'. The end of glass transition is identified by a plateau of the E' graph (called rubbery plateau), which corresponds to a peak of the tanδ plot. In between, E'' shows a peak that has been reasonably correlated to T_g, the mid-temperature of DSC measurements (MacInnes, 1993). For glass transition of a fully amorphous matrix, the drop in E' between the glassy state and the rubbery plateau is around two orders of magnitude (Roos, 1995). Figure 9 sketches a typical DMTA scan. Phase separation, crystal melting, recrystallization, annealing, and other effects can also be identified more or less precisely from peaks, shoulders, and/or break points in the E' or E'' plots. Peaks in E' and E'' are sometimes better visualized in semilog graphs.

4.2   Measurement of T_g in DMTA tests

These measurements are much more sensitive than DSC. In our research work, we were able to determine glass transition temperatures in native flours with DMTA, which cannot be done by simple DSC, and this allowed us to propose a new methodology to determine glass transition in native

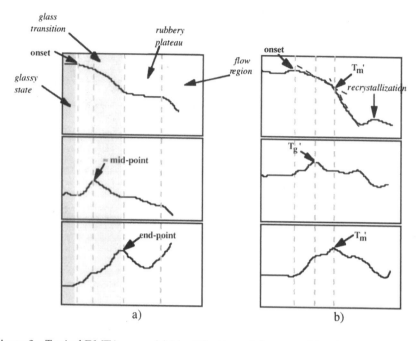

Figure 9 Typical DMTA scans. (a) Identification of glass transition from the variation of the viscoelastic parameters with temperature; (b) Typical scan with slow heating for a solution containing a crystallizable solute.

flour without disrupting the natural conformation of the starch molecules. DMTA analysis has however other remarkable advantages. As a dynamic test, the measuring parameters (E′, E″, and tanδ) will vary with the input frequency, but not with the strain amplitude (if one has indeed remained in the linear viscoelastic range, which can be ensured with preliminary strain sweep tests). This could seem a drawback because it would be necessary to establish a standard frequency of measurement and some authors have even thought about performing measurements at different frequencies and extrapolate for zero frequency — in this context, Arrhenius-type fits have been proposed (MacInnes, 1993). However, it is in fact a valuable information, because the variation with frequency in the frequency-response domain is a mirror image of the variation with time in the time domain (Luyben, 1990).

　　Considering therefore in greater depth the system dynamics theory, it is noted that the maximum of the energy dissipation (peak of E″) should be the point when the average molecular vibration time is in resonance with the input strain frequency. This result is easy to obtain for a first-order system (Allen, 1993 — note that E* is the transfer function of the system in the frequency-response domain, that is, the complex number obtained by replacing the Laplace variable s by ωi in the transfer function E(s) of the Laplace domain). Hence, the peaks of E″ obtained in tests at different frequencies should be good indications of the average molecular vibration times, τ, at

Table 1 T'_g Values of Sugar Solutions Determined
by DMTA and Reported in Literature[1], (Results obtained
from 30% w/w solutions for a strain amplitude of 10^{-4} m,
between 0.1 and 12.8 Hz, with $C_1 = 16$.)

Sugar	τ'_g (s)	C_2 (°C)	DMTA[2] T'_g (°C)	DSC1 T'_g (°C)
Glucose	100	44 ± 3	−54.3 ± 0.3	−53
Fructose	100	42 ± 2	−54.2 ± 0.2	−53
Sucrose	100	34 ± 2	−41.2 ± 0.2	−41
Lactose	100	28 ± 4	−36.2 ± 0.4	−36

[1] Roos (1995). Values are arithmetic midpoints between onset and end-set.

[2] Values obtained by fitting the WLF equation for E″ peak temperatures at several frequencies.

that specific temperature. Thus, Equation (1) could be applied to the data and yield the value of T_g in a much more precise and objective way than DSC measurements.

4.3 Application to the determination of T_g in frozen sugar solutions

In our research, we have applied this reasoning to several sugar solutions (sucrose, lactose, fructose, glucose, and also mixed solutions of these sugars), for frequencies varying between 0.1 and 12.8 Hz. Typical DMTA scans are shown in Figures 10 a and b, showing the influence of the frequency (Figure 10b also shows the existence of recrystallization in a slow heating test). The data fitted Equation (2) very well. The T_g and C_2 values were relatively insensitive to the values of τ_g in the physical range of interest (100 to 1000 s). C_1 was fixed at 16 and τ_g at 100 s, which avoided problems of lack of confidence (if τ_g were also a regression result, the error region of the three parameters would be physically unacceptable). It is noted that using $\tau_g = 1000$ s would result in a difference of *circa* 2°C in the estimated T_g. There is quite a lot of physical evidence for many liquids showing that the value of τ_g is close to 200 s (Angell et al., 1991), but this cannot be generalized to complex food matrixes without experimental verification. Fixing the value of τ_g at 100 s has been used by Moynihan et al. (1971) and Angell et al., 1994. The results for the four sugars are shown in Figure 11, and Table 1 gives the several parameters, comparing to published DSC T_g measurements. It can be seen that the above reasoning was indeed verified. This opens good possibilities for the study of frozen matrixes, as a precise and objective way of determining T_g was established, relating to molecular mobility.

This study also allowed for an analysis of the concept of T'_m vs. endset of glass transition (see Section 2.4). Crystal melting is a first-order transition and as such T'_m should be independent of the measuring frequency in a dynamic test, while the end of glass transition, which is a second-order transition, varies with the frequency. The peaks of tanδ correlated well for the four sugars studied with published T'_m values. However, they varied with

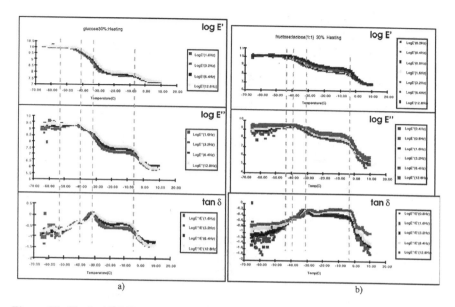

Figure 10 Typical DMTA scans of sugar solutions obtained with a rheometrics RSA II at several frequencies: (a) Sucrose, 30% w/w; (b) Fructose 15% + Lactose 15% w/w in slow heating, to favor crystallization.

the frequency as much as the peaks in E" did. Arrhenius-type plots of E" and tanδ peaks vs. the frequency yielded practically the same activation energy. This clearly established that the peak of tanδ is a measurement of glass transition, not of crystal melting phenomena, and its correlation to published T_m' values would therefore suggest that T_m' should be understood as the endset of glass transition, as suggested by Slade and Levine (1991).

5 Conclusions

The quality of frozen foods requires an integrated policy. Investments in improved freezing technologies can be wasted if the products are not designed to meet the problems that the storage and distribution chain can cause. Storing below T_g' is unfortunately unfeasible for most foods and it is inevitable that they will be subjected to temperature fluctuations (in and out of storage rooms, transport trucks, store displays, etc.), which can greatly enhance deteriorative reactions. The higher the product quality after freezing, the greater the need to protect such quality by minimizing the rate of the degradative phenomena of greater relevance to the specific food product.

It is important to seek a proper understanding of glass transition and of crystallization/recrystallization phenomena in the matrix. Ideally, one would like to increase T_g and inhibit and control crystal nucleation and growth (especially of ice). The product formulation can be designed with these concerns in mind, replacing some ingredients and/or using adequate

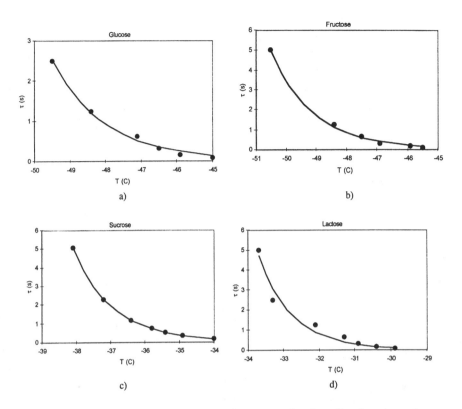

Figure 11 Fit of the WLF equation to the average molecular vibration time of sugar solutions, taken from the peaks of E″ DMTA scans (model parameters are given in Table 1): (a) glucose; (b) fructose; (c) sucrose; (d) lactose.

additives. In this context, the application of osmotic pretreatments using cryoprotectants in the osmotic solution is a particularly interesting option, that has barely been explored until recently. It is important to note that some specific synergies can greatly enhance the protective effects. The best results therefore tend to be system-specific. Thus, the development of effective and reliable ways of measuring glass transition is important.

A greater understanding of the importance of microstructure and molecular conformation to food quality control has been leading to significant advances in the pursuit of high-added-value products. The application of glass transition concepts and methodologies has improved the capacity to understand and thus improve food systems. The combination of such knowledge with novel technologies that have enormous potential in the creation of new structural conformations, such as high pressure, can lead to totally innovative food products that can be explored in high-added-value market niches.

It is evident that such concepts can find other important applications in high-added-value industrial sectors, such as biotechnology in the production of cell cultures and enzymatic preparations and even in advanced medicine applications, such as protection of organ viability for human transplants. An example can be found in Franks and Hatley (1992), which produced enzymatic solutions of increased shelf life applying glass transition concepts.

List of symbols

a_w Water activity
C_1 WLF constant
C_1' Modified WLF constant
C_2 WLF constant (°C)
C_2' Modified WLF constant (°C)
Cs'   Maximum concentration for ice formation (g water/g solid matrix)
E^* Complex modulus (Pa/m)
E' Storage modulus (Pa/m)
E'' Loss modulus (Pa/m)
i   Imaginary unit ($\sqrt{-1}$)
k   Gordon-Taylor constant
L   Average crystal dimension at storage time t (mm)
Lo   Average crystal dimension after freezing (mm)
n   Power-law parameter for the growth of crystals with storage time
t   Time (s; days for storage)
T_f Freezing temperature (°C)
T_g Glass transition temperature (°C)
T_g' Glass transition temperature of the maximum cryo-concentrated solution (°C)
T_m' Ice crystals melting temperature when heating from glassy state (°C)
T_r Reference temperature above Tg (°C)
T^∞ Reference temperature below Tg (°C)
w   water content (g water/g water+dry matrix)
δ Phase angle (°)
ε Strain (m)
ε_0 Amplitude of the sinusoidal strain variation (m)
σ Stress (Pa)
σ_0 Amplitude of the sinusoidal stress variation (m)
τ Average molecular vibration time at temperature T (s)
τ_g Average molecular vibration time at Tg (s)
τ_r Average molecular vibration time at Tr (s)
ω Input frequency (Hz)
ψ Growth constant for the growth of crystals with storage time (mm/days)

References

Allen, G. (1993). A history of the glassy state. In: *The Glassy State in Foods*, Blanshard, J. and Lillford, P. (Eds.), Nottingham University Press, Nottingham, U.K., p. 1.

Angell, C.A., Bressel, R.D., Green, J.L., Kanno, H., Oguni, M., and Sare, E.J. (1994). Liquid fragility and the glass transition in water and aqueous solutions, *Journal of Food Engineering*, 22, 115.

Angell, C.A., Monnerie, L., and Torell, L.M. (1991). Strong and fragile behavior in liquid polymers. *Materials Research Society Symposium Proceedings*, 215, 3.

Blanshard, J. and Franks, F. (1987). Ice crystallisation and its control in frozen food systems. In: *Food Structure and Behaviour*, Blanshard, J. and Lillford, P. (Eds.), Academic Press, Orlando, FL, p. 51.

Blanshard, J. and Lillford, P. (1993). *The Glassy State in Foods*, Nottingham University Press, Nottingham, U.K.

Brueggeller, P. and Mayer, E. (1980). Complete vitrification in pure liquid water and dilute aqueous solutions. *Nature*, 288, 569.

Couchman, P.R. (1978). Compositional variation of glass transition temperatures. 2. Application of the thermodynamic theory to compatible polymer blends. *Macromolecules*, 11, 1156.

Darby, R. (1975). *Viscoelastic Fluids*. Marcel Dekker, New York.

Doolittle, A. (1951). Studies in Newtonian flow. II — The dependence of the viscosity of liquids on free space. *Journal of Applied Physics*, 22, 1471.

Ehrenfest, P. (1933). Proceedings of the Academy of Sciences of Amsterdam, 36: 153. Cited by Wundelich, B. (1981). The basis of thermal analysis. In: *Thermal Characterisation of Polymeric Materials*, Turi, E. (Ed.), Academic Press, New York, p. 91.

Ferry, J. (1980). *Viscoelastic Properties of Polymers*. John Wiley & Sons., New York, USA

Franks, F. and Hatley, R. (1992). Stable enzymes by water removal. In: *Stability and Stabilisation of Enzymes*, van den Tweel, W., Harder, A., and Buitelaar, R. (Eds.), Elsevier Applied Science Pub., Chichester, U.K., p. 459.

Frias, J. and Oliveira, J.C. (1997). Application of dynamic experiments to the study of the kinetics of ascorbic acid degradation during drying. In: *Engineering and Food at ICEF 7*, Jowitt, R. (Ed.), Sheffield Academic Press, Sheffield, U.K., p.G-147.

Frias, J., Oliveira, J.C., Cunha, L.M., and Oliveira, F.A.R. (1998). Application of D-optimal design for determination of the influence of water content on the thermal degradation kinetics of ascorbic acid at low water contents. *Journal of Food Engineering*. In press.

Gibbs, J. and DiMarzio, E. (1958). Nature of the glass transition and the glassy state. *Journal of Chemical Physics*, 28, 373.

Goff, H., Caldwell, K., Stanley, D., and Maurice, T. (1993). The influence of polysaccharides on the glass transition in frozen sucrose solutions and ice cream. *Journal of Dairy Science*, 76, 1268.

Gordon, M. and Taylor, J. (1952). Ideal copolymers and the second order transitions of synthetic rubbers. I. Non-crystalline copolymers. *Journal of Applied Chemistry*, 2, 493.

Karel, M. (1975). Water activity and food preservation. In: *Principles of Food Science*, Part II, Fennema, O. (Ed.), Marcel Dekker, New York, p. 237.

Kerr, W., Lim, M., Reid, D., and Chen, H. (1993). Chemical reaction kinetics in relation to glass transition temperatures in frozen food polymer solutions. *Journal of Science of Food and Agriculture*, 61, 51.

Kokini, J., Cocero, A., Madeka, H., and de Graaf, E. (1994). The development of state diagrams for cereal proteins. *Trends in Food Science and Technology*, 5, 281.

Levine, H. and Slade, L. (1989). A food polymer science approach to the practice of cryostabilization technology. *Comments on Agricultural and Food Chemistry*, 1, 316.

Luyben, W. (1990). *Process Modelling, Simulation and Control for Chemical Engineers*. McGraw-Hill, New York, p. 415

MacInnes, W. (1993). Dynamic mechanical thermal analysis of sucrose solutions. In: *The Glassy State in Foods*, Blanshard, J. and Lillford, P. (Eds.), Nottingham University Press, Nottingham, U.K., p. 223.

Mansfield, M. (1993). An overview of theories of the glass transition. In: *The Glassy State in Foods*, Blanshard, J. and Lillford, P. (Eds.), Nottingham University Press, Nottingham, U.K., p. 103.

Mishima, D., Calvert, L.D., and Whalley, E. (1984). Melting ice I at 77 K and 10 kbar: a new method of making amorphous solids. *Nature*, 310, 393.

Moynihan, C.T., Balitaetac, N., Boone, L., and Litovitz, T.A. (1971). Comparison of shear and conductivity relaxation times for concentrated lithium chloride solutions. *Journal of Chemical Physics*, 55, 3013.

Nelson, K. (1993). *Reaction kinetics in food stability: comparison of glass transiton and classical models for temperature and moisture dependence.* PhD thesis, U. Minnesota.

Peleg, M. (1992). On the use of the WLF model in polymers and foods. *Critical Reviews in Food Science and Nutrition*, 32, 59

Richardson, P. (1995). Assessment of the quality of thawed fruits frozen by different freezing technologies. In: *Proceedings of the First Main Meeting*, Project Process Optimisation and Minimal Processing of Foods, Vol. 2 — Freezing, Published by Escola Superior de Biotecnologia, Porto, Portugal

Roos, Y. (1995). *Phase Transitions in Foods*. Academic Press, London, UK

Roos, Y. and Karel, M. (1991). Amorphous state and delayed ice formation in sucrose solutions. *International Journal of Food Science and Technology,* 26, 553.

Sceats, M. and Rice, S.A. (1982). Amorphous solid water and its relationship to liquid water: a random network model for water. In: *Water: a Comprehensive Treatise*, Vol. 6, Franks, F. (Ed.), Plenum, New York, p. 115.

Slade, L. and Levine, H. (1986). A polymer physico-chemical approach to the study of commercial starch hydrolysis products. *Carbohydrate Polymers*, 6, 213.

Slade, L. and Levine, H. (1991). Beyond water activity: recent advances based on an alternative approach to the assessment of food quality and safety. *Critical Reviews of Food Science and Nutrition*, 30, 115.

Slade, L., Levine, H., and Finley, J. (1988). Protein-water interactions: water as a plasticizer of gluten and other protein polymers." In: *Protein Quality and the Effects of Processing*, Philips, D. and Finlay, J. (Eds.), Marcel Dekker, New York, p. 9.

White, G.W. and Cakebread, S.H. (1966). The glassy state in certain sugar containing food products. *Journal of Food Technology,* 1, 73.

Williams, M., Landel, R., and Ferry, J. (1955). The temperature dependence of relaxation mechanisms in amorphous polymers and other glass-forming liquids. *Journal of the American Chemical Society,* 77, 3701.

chapter eight

The influence of freezing and frozen storage time on structural and other changes in plant tissue

Tamás Sáray, Krisztina Horti, Erna Zackel,
and Ágota Koncz

Contents

Summary 131
1 Introduction 132
2 Changes of texture and other alterations in some fruits caused
   by freezing and frozen storage 133
   2.1 Raspberries 133
   2.2 Strawberries 134
   2.3 Some other fruits 136
3 Effects of freezing and frozen storage on texture and other
   properties of some vegetables 137
   3.1 Vegetable legumes 138
   3.2 Umbellifers 138
   3.3 Some other vegetables 142
Acknowledgments 142
References 143

Summary

A review is provided on the character, causes, and consequences of the primarily structural changes that occur during freezing and the frozen storage of fruits and vegetables. Based on research results and practical experience, the optimum conditions for preserving textural qualities of some products of industrial importance are detailed. The evolution of cellular structure and loss of quality throughout a one-year frozen storage period is monitored for strawberries and celery by microscopic analysis.

1   Introduction

Freezing has long been established as an excellent method for preserving high quality in food products, including vegetables and fruits. Generally, freezing preserves the taste, texture, and nutritional value of foods better than any other preservation method; as a result, ever-increasing quantities of food are being frozen throughout the world. Frozen food markets have experienced enormous growth between 1991 and 1996.

The behavior of foods during processing, whether they are unprocessed, semiprocessed, or fully processed products, is determined mainly by their structure. Food quality and the mechanization or automation of food production are strongly dependent on food structure. The concept of structure relates to the organization of a number of similar or dissimilar elements, their binding into a unity, and the interrelationships between the individual elements or their groupings.

This structuring is necessary in order to develop textural characteristics that are acceptable to consumers. Texture, often the limiting factor in food acceptability, becomes an attribute of the resulting product. It is important to understand the relationship between structure and texture and how these are altered by freezing.

Plant parts used as food are composed of tissues that in turn are composed of many different types of cells, each of which may be peculiar to a specific tissue. Because texture concerns the manner in which the parts of a substance comprise the whole, the origin of textural qualities in fresh and processed fruits and vegetables is essentially histological. The degree of maturity also influences the texture of the preserved product.

The structural analysis of foods has advanced rapidly during the past 20 to 25 years (Monzini et al., 1975; Aguilera and Stanley, 1990). Microscopy (optical or light, electron, and atomic microscopy) and other imaging techniques (e.g., magnetic resonance) generate data in the form of an image. They are an extension of the visual examination of foods that has been practiced by consumers and food processors alike. The visualization of true food structure is extremely difficult; each step of the preparation of a specimen for microscopy alters the food sample to some extent. The power of imaging techniques is in their use as part of an integrated system, where changes observed at various levels of resolution are systematically evaluated (Kaláb et al., 1995).

The parts of fruits and vegetables suitable for consumption consist mainly of parenchyma cells. The cells are interconnected by an amorphous layer, the middle lamella, located outside the cell wall. From the point of view of the proper texture of a product the most important factors are the change in the primary cell wall, the degree of polymerization, estherification of the polygalacturonic acid chain consisting of the middle lamella, and the extent of salt formation. In raw fresh vegetables and fruits of good quality tissues others than parenchyma are present only to a small extent. The plant physiological processes are accompanied by the characteristic turgor pressure, which, however, ceases to exist at freezing together with the life function of cells.

Freezing damage is a consequence of many separate processes. It is not always clear whether the damaging step occurs during freezing or during thawing (Reid, 1995).

Due to the use of the unavoidable supply chain from the factory to the consumer the effect of frozen storage is also part of the processes described so far.

2 Changes of texture and other alterations in some fruits caused by freezing and frozen storage

Frozen fruits constitute a large and important fruit group in modern society. The quality demanded of frozen fruits or frozen fruit products depends upon the intended use of the product. If the fruit will be eaten without any further processing after thawing, texture characteristics are more important than if the fruit will be made into juice (Skrede, 1996). Most fruits are and can be frozen in ripe condition, in a progressive ageing stage when the biochemical degradation has already been started. From the point of view of preserving the original texture the critical factor is the usually inhomogenous ripeness state of the product, and in conformity with this the optimal selection of the freezing rate. In the preservation of the sensitive fruit tissue freeze-cracking damage is well known and occurs many times (Hung and Kim, 1996). This can be minimized through intensive precooling, by ensuring a suitably low center temperature at the end of the quick freezing process, or by the use of cryoprotective materials using the so-called thermal equilibration freezing method.

2.1 Raspberries

Crumbliness in red-ripe stage frozen raspberries is an excellent example of a morphologic cause of poor textural appearance. To reduce structural damage important factors could be, in addition to the general requirements, the size, the variety (apart from the blackberry–raspberry hybrids) of the raw material, gentle and quick transportation, and storage below –20°C without any fluctuations in temperature.

According to variety trials, the texture of raspberries quick-frozen at –30°C in an air blast system with continuous double-belt was damaged regardless of the variety (Kramer Shear press connected to an Instron Universal testing machine). The reduction of the quantity of water-soluble pectins amounted to 27 to 32% in 10 months. The softening process is unavoidable, although the storage stability of coloring and flavoring components is very good (Dégen and Bocsor, 1987). Frozen raspberries, like strawberries, are consumed most of the time prior to complete thawing due to the more favorable feel of texture.

2.2 Strawberries

The largest number of research works investigating the effect of freezing on the structure of fruits from the aspect of hystology are available for strawberries.

The effect of freezing and frozen-storage in strawberries was also investigated by the authors in lengthy trials. The typical conditions were: Gorella cultivar; ripe for consumption; precooling to 5°C; freezing at –30°C in an air blast system with continuous belt up to a center temperature of –15°C; freezing rate 1°C/min; storage at –25°C for 217 days.

The raw material consists of tissue parts with heterogeneous structure and of cells with various size and cell-wall thickness. The state of ripeness is heterogeneous even within a single fruit. Quick freezing has the least effect on the transport tissue elements and the tissue part below the epidermis (see Figure 1).

In other tissue areas the elongation of the cells and the deformation of the cell walls will begin. During the first 3 months of the storage period the structure changes at a relatively moderate rate. Then the intracellular damages accelerate: from many parenchyma cells the plasma is missing, remnants of disintegrated membranes can be observed, and despite the local thickenings of the cell walls, their volume decreases.

Texture of strawberries was measured by back-extrusion with an Instron Universal Testing Machine. The texture properties indicating changes in macrostructure are shown in Figure 2.

The hardness of the fruit that is characteristic of its original condition is damaged most significantly by freezing and this is primarily due to the loss of turgor pressure. No improvement of this situation can be achieved even by using a freezing rate greater than 1°C/min at similar storage temperature. During frozen storage the extrusion values change only to a minimal extent. A significant reduction of the texture properties occurs after 107 days of storage. Our statistical analyses showed that after freezing, the length of the storage period itself has no close relationship with strawberry texture damage measured by instruments.

During the assay of pectins it emerged that parallel to the damage to cell walls and the softening of frozen tissues, the water-soluble pectin content decreases, but this relationship develops mainly in the second half of the storage period. Compositional studies indicated that the changes were associated with the pectin rhamnogalacturonic backbone. After a 7-month storage period the extent of pectin content reduction is 46% compared to the original, and the total drip loss is 27%. The cell structure damage that can

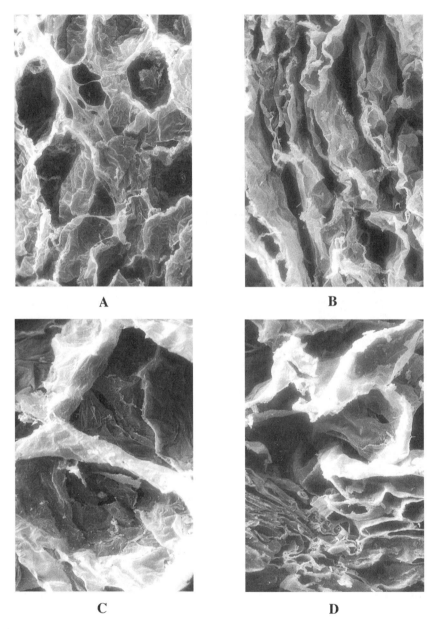

Figure 1 Conventional scanning electron micrographs showing parenchyma of fresh and processed strawberry. A: fresh, before freezing; B: 1 day after freezing; C: 100 days of frozen storage; D: 217 days of frozen storage (Magnification: 500 ×).

be observed on storage days 107 and 217 is demonstrated by the increased reduction of pectin fractions and an increment of drip loss by 5 to 6% compared to earlier measurements.

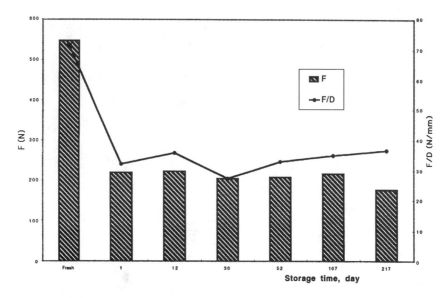

Figure 2 Texture characteristics of fresh and frozen strawberry tested by Instron with back-extrusion cell (F [N]: extrusion force; F/D [N/mm]: coefficient of extrusion; 200 g sample thawed at room temperature/test; crosshead speed: 200 mm/min; n = 6).

In line with consumption requirements, we considered it important to test the stability of the color characteristics of strawberries, and this was carried out by using a Momcolor-D tristimulus colorimeter (Figure 3).

Freezing caused the increase of color saturation of fruits (C*), while the lightness factor (L*) remained unchanged. In the red and yellow hue the decrease in value was observed after 52 days, but later on, compared to the original (fresh) condition, this is insignificant and represents no displacement in either direction.

2.3   Some other fruits

Following the quick freezing of blue plum (cv.: Besztercei) not fully ripened for this purpose, most of the time no changes in structure and texture can be found, but after a 5- to 6-month storage at –20°C the epidermic cell layer damage becomes significant. The tendency to drip increases quickly and the color saturation of the fruit becomes weaker. The additional reduction of storage temperature is definitely beneficial (Zackel, 1989).

As far as sour cherries are concerned, which can be frozen and stored successfully, Sebõk and Schlotter (1984) deal with the conditions (variety, manual and machine picking, stem removal, etc.) that have different impact on processing technology. To preserve the texture characteristics of the fresh fruit they recommend including a 2- to 3-hour-long precooling at 1°C prior to freezing.

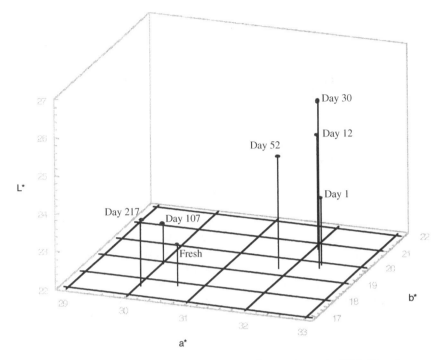

Figure 3   Color points of fresh and frozen stored strawberry in CIELAB color space (samples thawed at room temperature, then sieved, n = 10).

3   Effects of freezing and frozen storage on texture and other properties of some vegetables

The vegetables differ greatly from one another as far as their anatomic structure and the reciprocal arrangement and connection of the different tissues are concerned. The single types of tissues belonging to the different organs and plants may assume various functions and therefore possess very different histologic and cytologic properties.

During ripening and ageing the texture of vegetables, as compared to fruits, shows a definite hardening (starch synthesis, increase of fiber content, etc.). The considerable starch content, through the direct and indirect protection of water retention ability, contributes to the avoidance of structural damage of the frozen, stored, and thawed plant tissue and to the preservation of a favorable texture (Mohr, 1974).

A rapid freezing rate is quite beneficial in preserving the texture of vegetables (Fennema, 1973). The effect of the freezing rate is, however, effective mainly in the case of small geometric sizes. There is no optimal freezing rate specific to vegetable species.

Vegetables must often be blanched before freezing for prolonged shelf life, to inactivate enzymes, remove intracellular gases, reduce bacterial load, etc. The preservation of the structural and other kinds of useful product properties of preheated, frozen, stored, and thawed vegetables and the reduction of irreversible damages assume gentle and concerted technologic treatments.

The Ca^{2+} ion treatment before freezing generally provides more efficient texture improvement in vegetables than in fruits, since the majority of vegetables have a pH value where the free carboxyl groups of pectins are in a dissociated state so the interaction with positive ions is self-evident.

3.1 Vegetable legumes

A good example for the close relationship between the changes observed on the ultrastructural level and the changes in the texture of products is frozen and stored green peas and green beans (Brown, 1977). The structure of the parenchyma of green peas and the texture of tissues hardly change during the process. On the ultrastructural level the reason for this is that most of the cell is filled up by voluminous protoplasmatic material containing relatively much starch and the cells contain small vacuolar spaces. It is probable that the adaptability of these cells against physical and chemical stress generated by heat extraction is considerably better than that of other vegetables. During frozen storage the pectin and alcohol-insoluble solids do not change significantly (Mohr, 1974; Forni et al., 1991).

In green beans large intercellular cavities that are rich in air can be found in the parenchyma. During blanching the air is replaced by liquid. The starch granules pick up water and the volume of the cells shows a relative increase. Some cells are damaged, but the large damage of tissues is prevented by the relatively high freezing rate used also in the industry.

3.2 Umbellifers

The cell structure of carrots is characterized by the serial link-up of cells, thin cell wall, and the differentiation of the transport tissue structure. During freezing and thawing the epidermic and subepidermic cell layer has a high susceptibility to damage. For carrot disks (diameter: 15 mm, thickness: 11 mm) Fuchigami et al. (1994) regard a freezing rate of –5°C/min as optimal for the minimization of the structural and textural damages.

Blanching prior to the freezing and storage of carrots is not used in most cases in order to preserve sensorial quality. Otherwise, a gentle preheating treatment at low temperature (60 to 70°C) is recommended. Its subsequent benefit that might also appear after freezing is the following: by activating the pectin methylesterase (PME), controlling the reduction in methyl ester content, and restraining the probability of splitting of the pectin chain, the texture of the heat-treated product can be stabilized well prior to freezing (Fuchigami et al., 1995a).

The effect and role of freezing and frozen-storage has been investigated by the authors in trials lasting several years on celery roots. The characteristic trial conditions were — cultivar: Imperator; maturity: ripe for consumption; preparation: celery prisms (oblong shape) of industrial origin; blanching in water at 90 to 92°C for 4 minutes; freezing in fluidized bed freezer at –28°C up to a center temperature of –18°C; freezing rate: 1°C/min; frozen storage at –20°C for 1 year (Sáray et al., 1996).

The light and electron microscope tests show that prior to the processing operation fresh, raw celery is characterized by thin-walled cells in which tissue sections with loose and thick texture alternate with older large cells and younger small cells. Blanching, which cannot be avoided from the technologic point of view, has smaller effect than expected. The denaturing of the cytoplasm, the damage of the thin-walled cells, is in fact minimal (Figure 4).

During quick freezing (or rather one day later) in the heterogeneous tissue, cavitation occurs due to the separation of cells. The micelles consisting of cellulose molecules are squashed and some loosening takes place in the cell wall. That part of the product, however, which is made up of supporting or mechanical tissue (relationship of the prepared celery sample to the original root) changes its structure in a smaller degree than could be expected. During storage the damage to the cell wall and the increase of the size of intercellular compartments continues moderately, but the change of cell shape, e.g., shrinkage, hardly occurs. The above-mentioned changes are more moderate along the transport vascular bundles. In spite of these alterations the histological structure of the celery can be satisfactorily preserved during storage.

Freezing in bulk is regarded by the industry and trade as a quality-deteriorating factor, causes some tissue damage, but it is likely that the danger to quality does not lie here but in the local desiccation occurring in the tissue structure of product areas that are in contact with one another and in the resulting effect.

For the objective testing of the product texture, an Instron Universal Testing Machine (type 4302) was used. The size of the celery prisms was $8 \times 10 \times 10$ mm. The measured rheologic properties were: compression (up to a compression of 75%) and relaxation (pressure until the emergence of a counterpressure of 2 N, with a holding time of 3 minutes). The most important results are illustrated in Figures 5 and 6.

The values of the rheologic properties tested by the compression method (F_b, F/D, F_{max}) decrease after blanching by 60 to 82% compared to fresh celery and this process remains characteristic even during freezing. In frozen celery the F_b peak disappears and the product's internal structure has no characteristic destruction point. In addition to the blanching process, therefore, the freezing process contributes also to the softening of products. In the later stage of frozen storage a relative stability can be observed on the macrostructural level. Similar tendency is indicated by the changes in the relaxation elasticity factor of celery prisms. In Figures 5 and 6 typical curves are presented (selected from ten measurements per treatment). Relaxation time

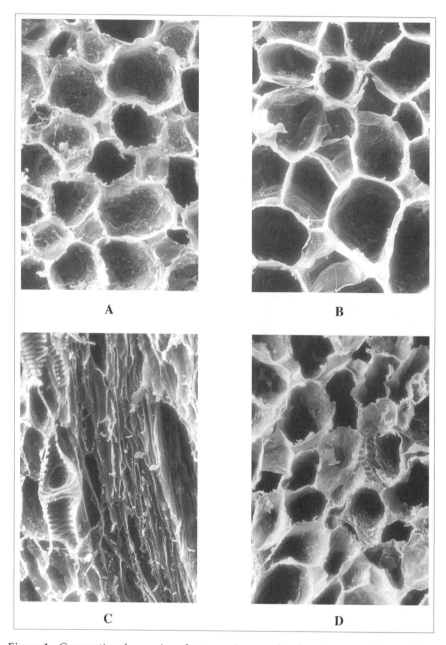

Figure 4   Conventional scanning electron micrographs showing parenchyma of raw and processed celery root. A: raw celery; B: 1 day after freezing (blanched for 4 min); C: 195th days of frozen storage; D: 365 days of frozen storage (magnification: 1500 ×).

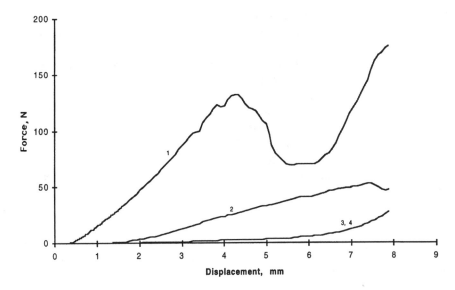

Figure 5 Typical compression curves of raw and processed celery. 1: raw; 2: blanched; 3: frozen storage, 1st day; 4: frozen storage, 365th day.

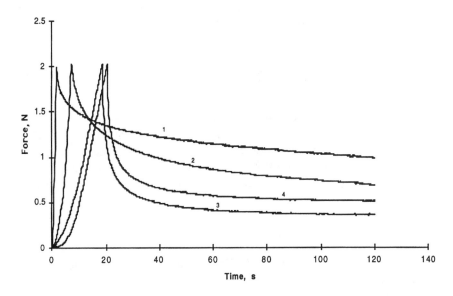

Figure 6 Typical relaxation curves of raw, blanched, and frozen celery. 1: raw; 2: blanched; 3: frozen storage, 1st day; 4: frozen storage, 365th day.

greatly decreases as a function of technologic treatment before storage. This reaction can be explained by the modified water movement caused by the injury of cell walls and by the change of state of the space between cells. During storage no quality change takes place that could be statistically proven and termed as long lasting.

Total pectin content of the celery prisms at the start of processing is 0.48 g/100 g fresh product. After the freezing process the loss, including blanching, is 10%. During the 1-year storage the degradation of pectin is relatively gradual and moderate. Total loss is 23%, of which 13% is due to the storage period. Pectin content in the second period of storage is significantly lower compared to the value measured directly after freezing, and its effect is manifested in cell wall damage.

Drip loss related to the necessary thawing shows a slow increase but stays at an acceptable level during the 1-year period (0.2 to 4.5%) considering both industrial and consumer's demands, proving that the histologic and cytologic stability and the satisfactory texture are coupled with good water retention ability during storage.

3.3   Some other vegetables

The quick freezing and storage of asparagus having the right ripeness level results in no structural changes affecting its quality if the correct technology is used, but with conventional air-blast methods, freezing damage is extensive in the parenchyma tissue surrounding the vascular bundles.

A freezing rate of at least –5°C/min is considered optimal for Chinese cabbage by Fuchigami et al. (1995b) Changes in the structure and composition of leaves and midribs differ during cryogenic freezing. The drip from leaves was less than from midribs.

Despite the genetically engineered hard-skin varieties, the freezing and storage of tomatoes (in whole or sliced) even with the optimal size and ripeness (pink-red) causes problems from the point of view of texture. Freezing in good quality is possible only with a cryogenic medium and cryogenic techniques (George, 1993). Very different research results have been published on the time and effect of storability and on the conditions of thawing.

Acknowledgments

The authors would like to thank A. Csillag and B. Nagy for their valuable contribution.

Our research work on raspberries and celery was supported by the Hungarian Scientific Research Fund (OTKA) under the grant No. 349 (I.3.).

References

Aguilera, J. M. and Stanley, D. W. (1990). *Microstructural principles of food processing and engineering,* Elsevier Applied Science, London, p. 343.

Brown, M. S. (1977). Texture of frozen fruits and vegetables, *Journal of Texture Studies,* 7, 391.

Dégen, Gy. and Bocsor, J. (1987). Quick-freezing of raspberries in a fluidized bed freezer, *Hûtöipar (Refrigerating Industry, Hungary),* 33, 89.

Fennema, O.R. (1973). Nature of freezing process. In: *Low temperature preservation of foods and living matter,* Fennema, O.R., Powrie, W.D., and Marth, E.H. (Eds.), Marcel Dekker, New York, Chap. 3, 4.

Forni, E., Crivelli, G., Polesello, A., and Ghezzi, M. (1991). Changes in peas due to freezing and storage, *Journal of Food Processing and Preservation,* 15, 379.

Fuchigami, M., Hyakumoto, N., Miyazaki, K., Nomura, T., and Sasaki, J. (1994). Texture and histological structure of carrots frozen at a programmes rate and thawed in an electrostatic field, *Journal of Food Science,* 59, 1162.

Fuchigami, M., Miyazaki K., and Hyakumoto, N. (1995a) Frozen carrots texture and pectic components as affected by low-temperature-blanching and quick freezing, *Journal of Food Science,* 60, 132.

Fuchigami, M., Miyazaki, K., Hyakumoto, N., Nomura, T., and Sasaki, J. (1995b). Chinese cabbage midribs and leaves physical changes as related to freeze-processing, *Journal of Food Science,* 60, 1260.

George, R. M. (1993). Freezing processes used in the food industry, *Trends in Food Science and Technology,* 4, 134.

Hung, Y. C. and Kim, N. K. (1996). Fundamental aspects of freeze-cracking, *Food Technology,* 50, 59.

Kaláb, M., Allan-Wojtas, P., and Miller, S.S. (1995). Microscopy and other imaging techniques in food structure analysis, *Trends in Food Science and Technology,* 6, 177.

Mohr, W. P. (1974). Freeze-thaw (and blanch) damage to vegetable ultrastructure, *Journal of Texture Studies,* 5, 13.

Monzini, A., Crivelli, G., Bassi, M., and Buonocore, C. (1975). *Structure of vegetables and modifications due to freezing,* Instituto Sperimentale per la Valorizzazione Technologica (I.V.T.P.A.), Milan, Italy, p. 267.

Reid, D.S. (1995). Basic physical phenomena in the freezing and thawing of plant and animal tissues. In: *Frozen food technology,* Mallett, C.P. (Ed.), Chapman and Hall, Glasgow, Scotland, Chap. 1.

Sáray, T., Zackel, E., Csillag, A., and Horti, K. (1996). Structural and other alterations in plant tissues caused by quick-freezing and storage time, *Acta Alimentaria,* 25, 343.

Sebők, A. and Schlotter, Gy. (1984). The connection between the microstructure and the consistence of the vegetables and fruits, *Élelmezési Ipar (Food Industry, Hungary),* 38, 471.

Skrede, G. (1996). Fruits. In: *Freezing effects on food quality,* Jeremiah, L.E.(Ed.), Marcel Dekker, New York, Chap. 6.

Zackel, E. (1989). Histological changes in deep-frozen plums as a function of storage time, *Hûtöipar (Refrigerating Industry, Hungary),* 35, 87.

chapter nine

Intensification of freezing

Stefan Ditchev and Philip Richardson

Contents

Summary 145
1 Introduction 146
2 Intensification of freezing using a dynamic dispersion medium
   (DDM) 146
   2.1 Development and evaluation of air-blast freezing 146
   2.2 The dynamic dispersion medium — a method for
        intensification of the freezing process 149
3 Modeling approaches to the prediction of freezing times 154
   3.1 Factors affecting validation of predictive models for
        freezing time 155
   3.2 Results 157
4 Comparison of the sensory quality of a range of foods frozen
   by air-blast and cryogenic processes 157
5 Conclusion 160
References 160

Summary

This first part of this text describes the concept and applicability of the dynamic dispersion medium (DDM) method as an effective way to optimize freezing times (DDM is a term introduced by the author, that describes a bicomponent medium whose components are in relative motion). The second part details a comprehensive analysis of the prediction and experimental measurement of freezing times. Processes that led to significantly different freezing times, namely still air, air blast, and cryogenic freezing, were analyzed for their ability to produce different quality products, as perceived sensorially by a trained panel. However, after 1 month and 3 months of storage at –18°C and subsequent slow thawing, no significant differences were found between the samples originally produced by the different freezing methods. This was attributed to the ice crystals' growth during thawing or in the early stages of storage at –18°C.

1. Introduction

The intensification of the freezing process is the main factor for improving the quality of foods preserved in this way. It leads to an optimization focused on the minimization of processing and maximization of product quality.

Food freezing is a complex physical processing technology. From a physical point of view, the heat and mass transfer processes are the key factors, but the microbiologic and biochemical reactions cannot be neglected. The understanding of these parallel and linked processes is necessary for the control and optimization of the freezing process.

One of the main factors for intensification of freezing is the heat transfer coefficient **h**, W/(m²·K), between the product being frozen and the cooling medium.

The prediction of freezing times is of key importance to the design of food freezing operations since the freezing time of a food material determines the residence time required within a freezer and consequently the throughput of a freezing plant. Predictive models also allow establishing the effect of changes in freezing conditions on the freezing time, e.g., effects of product size distribution and effects of initial product temperature. Also, the effects of process factors can be ascertained and related to characteristics of the product, e.g., increasing the air velocity in an air-blast freezer in an attempt to increase the surface heat transfer coefficient will not reduce freezing times significantly if most of the resistance to heat transfer is an internal, conductive effect. Predictive models can indicate such effects without the need for expensive plant trials.

2 Intensification of freezing using a dynamic dispersion medium (DDM)

2.1 Development and evaluation of air-blast freezing

The historical developments in air-blast freezing methods are presented in Figure 1.

Air-blast methods have been developed from a free to a forced air convection and turbulization of boundary air layer with the purpose of increasing the heat exchange between the product and the air stream. The transition from free to forced convection is directly connected to an increase in energy consumption. The optimization of the freezing process essentially means heat exchange optimization and minimum energy consumption. The optimal solution of this problem is shown in point **O** of Figure 2, i.e., the cross point of the lines of the heat transfer coefficient **h**, W/(m²·K) and the specific energy consumption **e**, kWh/kg as a function of air stream velocity w_a, m/s.

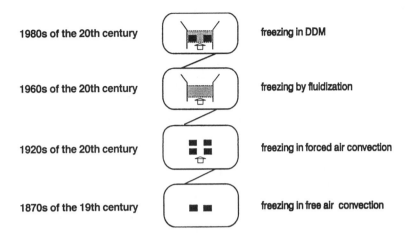

1980s of the 20th century		**freezing in DDM**
1960s of the 20th century		**freezing by fluidization**
1920s of the 20th century		**freezing in forced air convection**
1870s of the 19th century		**freezing in free air convection**

Figure 1 Spiral of development of air-blast freezing.

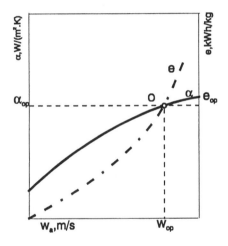

Figure 2 Optimization of freezing from the point of view of process intensity and energy consumption.

Another very important and fundamental index of freezing is the weight loss of the product during the freezing process, caused by the external mass transfer between the product and the air stream. The weight loss ΔG, kg, related to the initial product weight **G**, kg, represents the relative weight loss and is usually expressed in %, i.e., $\Delta G/G \cdot 100 = \Delta g$ %.

The external heat- and mass-transfer processes in freezing are carried out in parallel and are intimately linked (Figure 3).

The heat stream **Q**, W, and the moisture separated ΔG, kg, from the product to the cooling medium are defined by:

$$Q = h_{eff} \cdot F \cdot \overline{\Delta t} = h_{eff} \cdot F \cdot \left(t_s - t_m\right) \tag{1}$$

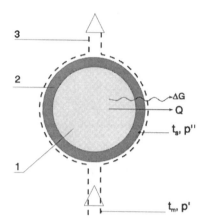

Figure 3 Model of the heat- and mass-transfer processes in freezing: 1 — product core, yet unfrozen; 2 — surface frozen layer; 3 — air stream (cooling medium); Q, ΔG — respectively, heat and moisture stream; t_s, t_m — respectively, temperature on the surface of the product and of the cooling medium ; p′, p″ - respectively, the partial pressure of water vapor on the surface of the product and of the air stream.

$$\Delta G = \beta \cdot F \cdot \overline{\Delta p} \cdot \tau = \beta \cdot F \cdot (p'' - p') \cdot \tau_f \qquad (2)$$

where:

h_{eff} = effective heat transfer coefficient, W/(m²·K);

F   = heat-exchange at the product surface, m²;

t_s, t_m = temperature on the surface of the product and of the cooling medium, respectively, °C;

$\overline{\Delta t}$ = mean integral temperature difference during the whole process duration, K;

β   = coefficient of water evaporation (coefficient of ice sublimation), kg/(m²·h·Pa);

$\overline{\Delta p}$ = mean integral difference in partial pressure of water vapor Pa;

p″, p′ = partial pressure of saturated water vapor on the surface of the product and in the air stream, respectively, Pa;

τ_f = freezing time h.

   The complexity of the mass-transfer processes during cooling and freezing of foodstuffs implied that many investigators used experimental and semiempirical methods in their work (Levy, 1974; Chizov, 1979; Ditchev, 1985).

   The determination of the total mass losses and the dynamics of the mass-transfer process are of great significance in the design and control of freezing processes.

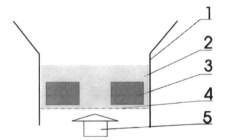

Figure 4 Physical model of a DDM system: 1 — fluidization chamber; 2 — solid particles; 3 — foodstuff; 4 — conveyor (grate); 5 — air stream.

Together with the intensification of the heat exchange, mass losses during freezing decrease because the frozen surface layer of the product is formed quicker and this influences the diffusion resistance for moisture leaving the product.

The relative importance of the various air-blast freezing methods is presented in Table 1.

2.2 The dynamic dispersion medium — a method for intensification of the freezing process

In the methods and equipment for foodstuffs currently known, gaseous, liquid, and solid cooling media have been used.

The utilization of a dynamic dispersion medium (DDM) as cooling medium has been proposed (Ditchev, 1985) as a way for intensifying air-blast freezing of foodstuffs. Dynamic dispersion medium (DDM) is a term introduced by Ditchev (1984) that determines the structure and condition of a bicomponent medium of the type "gas–solid particles," whose components move simultaneously and permanently in upward or downward cocurrent or countercurrent motion. This definition of DDM includes in itself also the terms fluidized bed and dropping bed (layer). The gaseous or liquid component is continuous and the solid component is discrete.

The physical model of a DDM system is shown in Figure 4, where the product being frozen is dipped in DDM.

The intensification of the heat-exchange is due to the intensive movement of the solid particles and their contact with the surface of the product. The turbulent boundary layer on the surface of the product and the conductive component of the heat exchange are the factors that promote the high values of the heat-transfer coefficient **h** (see Table 1).

When freezing in DDM, the external mass transfer decreases and therefore the mass losses of the product diminish. This is due to the intensive surface freezing and to the decrease of the moving factor strength $\overline{\Delta p} = (p'' - p')$ (see Equation 2) if a suitable material for the DDM solid component is chosen. As an example, when using granulated water ice as a DDM solid component,

Table 1 Evaluation of Air-Blast Freezing Methods

N	Freezing method	Relative level			Values of characteristic parameters	
		Heat-transfer intensity	Mass losses	Energy consumption	h, W/(m²·K)	Δg, %
1	Freezing in free or weak air convection	I level	IV level	I level	8–15	3–4.5
2	Freezing in forced air convection (w_a = 2–5 m/s)	II level	III level	II level	10–50	1.5–3.5
3	Freezing by fluidization (w_a = 2–5 m/s)	III level	II level	IV level	60–120	0.5–1.5
4	Freezing in dynamic dispersion medium (w_a = 0.5–2 m/s)	IV level	I level	III level	180–420	0.1–0.35

Table 2 Factors Characterizing and Determining the Selection of the Material (solid component) for DDM

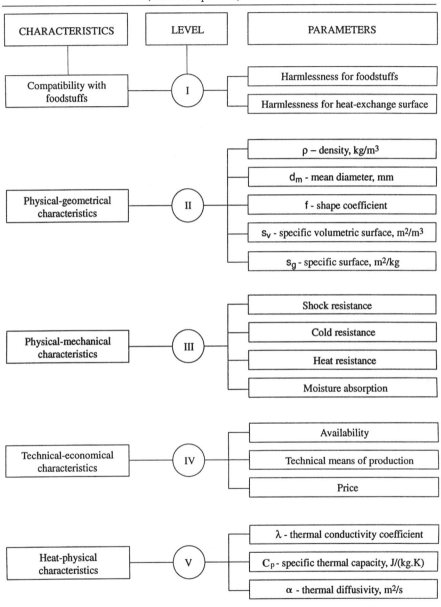

the partial pressure of the water vapor **p′** in the air stream increases, and thus $\overline{\Delta p}$ decreases. This leads to a considerable decrease of mass losses (see Table 1).

There are many factors characterizing and determining the selection of the material (solid particles) for DDM (Table 2).

The practical application and utilization of fluidization and DDM for intensification of various technological processes has been presented by other workers (Zenz, 1949; Ciborowski, 1957; Leva, 1959; Davidson and Harrison, 1963; Sjiromjatnikov et al., 1967; Aerov, 1968; Baskakov, 1968; Zavrodskij, 1968; Romankov and Rashkovskaja, 1968; Gorbie, 1970; Botterill, 1975; Chizhov, 1979; Ditchev, 1984, 1985). The investigations and the utilization of DDM for intensification of refrigeration processes date back to 1982 (Ditchev, 1985). The fields of application of DDM in refrigeration technology are given in Table 3.

Our experimental investigation for meat packages freezing in DDM has been carried out using two materials as DDM. The first variant consisted of a granulated material, polyethylene, with equivalent granule diameter d_v = 3.77 mm and a second variant of DDM involved granulated water ice with d_v = 2 mm.

The results for heat and mass transfer processes can be described with the following equations:

1. The dimensionless freezing time (expressed by the Fourier number):

$$Fo = 1,38\theta^{0,78} \cdot Bi^{-0,72} \cdot Re^{0,21} \qquad (3)$$

applicable in the range: θ = 2.73–3.4; Bi = 1.6–7.5; Re = 50–230.

where: Fo, Bi, Re, θ are, respectively, the numbers of Fourier, Biot, Reynolds and the dimensionless temperature difference.

$$Fo = \frac{\alpha_\omega \cdot \tau_f}{1^2}; Bi = \frac{h_{eff} \cdot 1}{\lambda_f}; Re = \frac{w_f \cdot d_v}{v_a}; \theta = \frac{t_s - t_a}{t_s - t_a}$$

τ_f	= total duration of the freezing process, h:
α_ω	= thermal diffusivity, m²/h;
1	= half thickness of the product (1 = 0,5 δ), m;
h_{eff}	= effective heat transfer coefficient, W/(m².K);
λ_f	= thermal conductivity coefficient of the frozen product, W/(m.K);
w_f	= air stream velocity of fluidization, m/s;
d_v	= equivalent volumetric diameter of the solid particles of the DDM, m;
v_a	= kinematic viscosity of air, m²/s;
t_p^i, t_p^f, t_a	= initial and final temperature in the center of the product and temperature of the air stream, respectively, °C.

2. The dimensionless effective heat transfer coefficient (expressed by the Nusselt number):

$$Nu_{eff} = 1,46 \cdot 108 \cdot Re^{0,077} \cdot Ar_{ld}^{-1,35} \qquad (4)$$

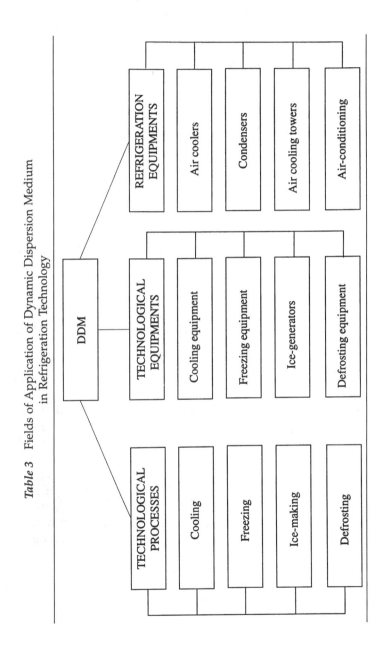

Table 3 Fields of Application of Dynamic Dispersion Medium in Refrigeration Technology

applicable in the range Re = 50–230; Ar_{ld} = (0.2–0.3)·10^6

where:

Nu_{eff}, Re, Ar_{ld} = respectively, the numbers of Nusselt, Reynolds, and Archimedes (for loose density)

λ_a = thermal conductivity coefficient of air, W/(m·K);

g = acceleration of gravity (9.81 m/s²);

ρ_{ld}, ρ_a = respectively, loose density of the granulated material of the DDM and of air, kg/m³.

3. The relative weight loss of the product during freezing, Δg,%:

$$\Delta g = 0.47 + 1,97 \cdot S_g + 0.011 \cdot t_a - 0,005 \cdot \overline{\Delta t} \qquad (5)$$

where:

S_g = specific surface of the frozen product, m²/kg (S_g = 0.079 $\sqrt{G}$ for meat packages; 0.038 $\sqrt{G}$ for fruits);

$\underline{G}$ = single product weight, kg;

$\overline{\Delta t}$ = mean integral temperature difference (for the whole process) between the surface of the product and the air stream, K;

t_a = temperature of air, °C.

3 Modeling approaches to the prediction of freezing times

In order to design and optimize food freezing processes it is necessary to have an effective method for the prediction of freezing times and of the influence of the various parameters on the likely freezing time of specific products.

Generally, there are two approaches to the prediction of freezing time. The first uses analytical or empirical equations to calculate freezing times, utilizing heat-transfer theory to calculate the rate at which freezing should occur under idealized conditions and, depending upon the model, applying empirical corrections to correlate more closely with certain experimental conditions (Ramaswamy and Tung, 1983). Such methods are limited in accuracy by the underlying assumptions and are limited in application to simple situations.

The second approach attempts to solve the basic heat-transfer equations using numerical methods. This approach usually takes the form of finite difference or finite element schemes, which are performed on a computer. The advantage of such an approach is that the schemes can accommodate irregular geometry, time-dependent boundary conditions, and temperature-dependent food thermal properties. The disadvantage of such methods is that substantial computing power is needed, often beyond the resource of food process engineers. Consequently, much attention has been paid in recent years to further developing the analytical methods (George, 1993).

Both approaches require the knowledge of the geometry and physical properties of the food material, along with the physical conditions within the food freezer. Often, it is the inadequate knowledge of these parameters that limits the agreement between predicted and measured freezing times, rather than inherent limitations of the prediction methods. This study has investigated the range of prediction models available, identified the most critical experimental parameters and provided comparisons with experimental freezing times for real food products in freezing situations.

The majority of analytical and empirical models for predicting freezing time have been based upon Plank's model (Plank, 1941), with modifications developed for specific situations. Plank's model considered a slab of immobilized water cooled from both sides and the progression of a "freezing front" toward the center. Plank equated the rate freezing front. In its general form the equation for three-dimension heat transfer is:

$$\frac{\rho L}{t_f - t_a}\left[\frac{1}{h} + \frac{1}{2\lambda_f}\right]\frac{1}{\Phi} \tag{6}$$

where:

ρ = the density of the material, kg/m^3.
L = the latent heat of freezing, kJ/kg.
t_f and t_a = freezing point and temperature of the refrigeration medium, respectively, °C;
h = the heat transfer coefficient, W/(m^2·K);
λ_f = the thermal conductivity of the frozen product, W/(m·K);
l = half thickness of the food, m;
Φ = number depending on the shape (1, 2, and 3 for slab, cylinder, and sphere, respectively).

It is generally accepted that Plank's equation tends to underestimate freezing times as it does not take into account sensible heat loads and also because it assumes that the water is initially at the freezing temperature and remains so throughout the process. Many of the modifications to Plank's equation take into account the additional times associated with precooling and tempering.

Modifications to Plank's equation which take into account sensible heat loads include: Nagaoka et al. (1955), Levy (1958), Earle and Fleming (1967), IIR (1972), Mellor (1976), Ramaswamy and Tung (1983), Cleland and Earle (1984), Pham (1986).

Modifications to extend Plank's equation to other shapes include: Mott (1964), Cowell (1967), Cleland and Earle (1977), Pham (1986).

3.1   Factors affecting validation of predictive models for freezing time

One of the key steps in the modeling of freezing times is the validation of the model. The variables that need evaluation for this purpose can be divided into two categories: those that involve the physical properties of the material (thermal conductivity, latent heat, density, etc.) and those that relate to the system in which they are being frozen (temperatures and heat-transfer coefficient). In terms of the physical properties of the material, many methods have been tried and tested for their evaluation and it is generally accepted that these properties can be measured to a reasonable degree of accuracy over the range of temperatures and phases required.

The evaluation of the heat-transfer coefficient, however, is more difficult as this requires the monitoring of air velocity and the measurement of appropriate temperatures within both the freezer environment and the product. Comparison of the predicted freezing times by the alternative methods allows an assessment of the performance of the various predictive models published in the literature.

Following some preliminary work (Richardson, 1995), a methodology was developed for the measurement of freezing times in both air-blast and cryogenic freezing systems. This allowed an accurate assessment of the prevailing heat transfer coefficient, which is one key datapoint when using the predictive models. Type K thermocouples were used to reduce the influence of conduction errors on the measurement and all thermocouples were located securely at the center of the product samples with the gap around the thermocouple wire carefully sealed to prevent ingress of the freezing medium during the freezing process.

Influence of type and thickness of thermocouple.   Our experiments show that conduction effects contribute to the temperatures measured by the type T thermocouples due to the greater thermal conductivity of copper/constantan over nickel-chromium/nickel-aluminium; also the thermocouples of greater thickness showed increased thermal conduction effects. These results confirm similar observations on thermocouple selection for thermal process measurements (Withers et al., 1993). For subsequent trials, thin, type K thermocouples were used for temperature measurements during the freezing process.

Influence of method of securing thermocouple in food material.   It is imperative to locate the thermocouple securely within products during the trials. Investigations using a PTFE block as a food simulate showed that when no thermal couplant of screw fitting was used, the drop in temperature was most rapid, as cold air was able to penetrate directly to the thermocouple sensing tip. This problem was overcome by sealing the thermocouple and using a thermal couplant and screw fitting. This approach was used for all subsequent freezing trials.

The effects of thermocouple positioning were investigated using a simulation performed on a computer temperature prediction program (PCTemp). For a spherical particle of 35 mm radius, the greatest difference in time taken to respond to a step change in temperature between the radial nodes was 1 minute; therefore, it was not considered of vital importance that the thermocouples were located precisely at the geometric center. A comparison

Table 4 Comparison of Predicted and Measured Freezing Times
for Strawberries

Freezing Conditions:Process Temperature,°C	−20	−20	−30	−30
Air velocity	Low	High	Low	High
Measured freezing time (minutes)	50	36	30	30
Predicted freezing time (minutes):				
Plank's model	57	42	37	27
Cowell's method	56	41	36	27
Nagaoka modification	75	56	48	36
Mott's procedure	81	63	52	40
Levy modification	76	56	48	36
Cleland & Earle	82	61	53	40
Ramaswamy & Tung	76	56	49	36
Mellor modification	64	47	41	30
Pham	76	65	55	41
Cleland & Earle EHTD	83	60	60	43

of experimental and predicted freezing times was made on samples of strawberries frozen in temperatures of −20°C and −30°C, at two air velocities.

The heat transfer coefficients under the two freezing conditions were
measured over four replicates, described by Charm (1971).

3.2 Results

The heat-transfer coefficient measured in the pilot scale air-blast system used
for the experimental program ranged between 26.0 and 38.9 W/(m².K). This
value was higher than anticipated but was confirmed by replication of the
results.

The freezing time was measured and predicted for samples of strawberries with dimensions of 15 to 20 mm being subjected to air-blast freezing
under a range of conditions. The results are summarized in Table 4. The
freezing time is taken as being the time for the strawberries to reach a
temperature of −18°C.

In general the results show that the models tend to overestimate the
freezing time but that they all provide a reasonable estimate of the freezing
time required. The comparison between measured freezing times for differing methods of freezing are shown in Figure 5.

4 Comparison of the sensory quality of a range of foods frozen by air-blast and cryogenic processes

Cryogenic freezing has been reported to offer the food processor the opportunity to produce a higher quality of frozen product as compared with air-
blast techniques. This is hypothesized to be because of the faster freezing
rates that can be obtained with cryogenic freezing units as compared to air-
blast systems.

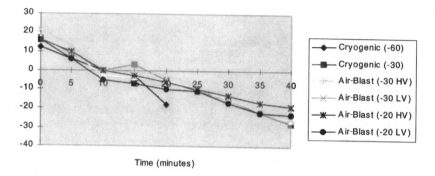

Time (minutes)

Figure 5 Comparison of freezing times with freezing process for strawberries.

The work reported here was undertaken to assess the level of these quality improvements using a trained taste panel.

A range of products were frozen by both air blast freezing (–20°C to –30°C and low air velocities) and cryogenic freezing (–30°C and –60°C) (Richardson, 1995).

The products chosen for this study were peas, Brussels sprouts, strawberries, and a breaded chicken product.

The time–temperature profile for all products was recorded during freezing to –18°C and all products were then stored at –18°C and sampled for sensory tests after 1 month and 3 months.

Key attributes for each product were defined for use in the sensory assessment of the product from these trials. The chosen attributes for each product are listed in Table 5.

The mean panel results are presented together with the least significant difference value (LSD), which indicates when samples are significantly different in any particular attribute. The data show that although there were differences, these were frequently only between the highest and lowest values in a range and may have been by chance. On close inspection, and after multivariate analysis (not presented in the report), it was seen that there was no clear trend to indicate that there were any perceivable differences in the products after processing by either air blast, cryogenic, or indeed slow freezing in a domestic freezer and storage for one month. A few differences were observed: for peas, those processed at –30°C/high air velocity and at –60°C in the cryogenic system were more khaki, grey, and less bright than the others. Brussels sprouts processed in the air-blast system were more yellow and less green than the others; this difference is attributed to the mechanical differences between the processes and the tendency for the Brussels sprouts to rub against each other in the air blast system and so dislodge some of the outer leaves.

Considering the strawberries, some differences in appearance of the product were detected in terms of color. It was, however, unclear as to whether this was due to variation in the raw material, which, although of

Table 5 Product Attributes Used in the Sensory Assessment

Peas	Brussels Sprouts	Strawberries	Chicken
Green	Green	Color	Color
Yellow	Yellow	Brightness	Brightness
Khaki	Khaki	Uniformity	Evenness of surface
Grey	Brightness	Brown	Internal appearance
Brightness	Uniformity	Blue	Salty
Uniformity	Firmness	Grey	Crunchy
Skin firmness	Fibrouseness	Shape retention	Dry
Flesh firmness	Strength of sprout	Sweet	
Mealyness	Sweet	Acid	
Strength of flavor	Bitter	Watery	
Sweet	Earthy	Perfumed/soapy	
Harsh	Nutty	Metallic	
	Stale	Bitter aftertaste	
	Hot	Strength of flavor	
	Cabbage	Stale	
	Sickly/rotten	Earthy	
		Firmness	
		Fibrous/wooly	
		Gritty/seedy	

the same variety, came from different sources. It was noted that in general the fruit used for air-blast freezing was smaller but more equally developed than the fruit used in the cryogenic trials, which had underdeveloped tips. Consideration of the strawberry results suggests that there is as much variation within replicates as between the treatments, allowing a conclusion to be drawn that no significant sensory differences could be found between product frozen by the two methods. This trend was also observed when further samples were evaluated after 6 months storage.

Considering the chicken product, of the 21 attributes of appearance (external and internal), flavor and texture, which were rated for intensity, significant differences between any samples were only found in the following eight: external appearance — depth of color, brightness, evenness of surface; internal appearance — pastiness (wetness) of coating crumb; flavor — salty, toasted; and texture — crunchy, dry.

However, the differences between mean values (for chicken) for individual samples was frequently greater than the differences between treatments. Moreover, there was no consistent trend that any treatment affected quality more than another. The results suggest that the method of freezing had no consistent effect on the coated chicken samples.

Since there were no measurable differences in the sensory quality of the strawberries frozen by the different techniques, microscopic examination of the strawberries was undertaken to try and identify any structural differences that might be due to the different processing techniques. Scanning electron microscopy (SEM) did not show any differences because of high levels of frosting of the samples caused by the high humidity at the time of processing. Light stereomicroscopy confirmed that the raw material was of the same variety but differed in maturity. It was impossible to determine if this in itself had masked any sensory differences in the products frozen by the different techniques.

Damage to products during freezing is largely due to the growth of ice crystals. These also grow during thawing and so it is possible that the slow thawing process undertaken prior to sensory analysis could have negated any benefits that may have been present because of the fast freezing. It is however believed that any benefits that were there by virtue of the faster freezing rate achieved with cryogenic freezing were lost during the early stages of storage at $-18°C$.

It was concluded that further microscopy work was needed to establish the differences between freezing techniques, but it is important to note that no significant differences were observed by the trained sensory panel.

5 Conclusion

DDM can be used as a method for intensifying refrigeration processes (cooling and freezing) of foodstuffs. By means of the selection of the DDM, the heat- and mass-exchange from the product to the cooling medium can be influenced and controlled. Consequently, DDM used as a cooling medium can ensure minimum processing and optimal quality preservation of the processed foodstuffs.

The key factors that influence the measurement of temperature during the freezing process were investigated and a temperature measurement protocol was established utilizing type K thermocouples, which were required to be firmly located and sealed into products if accurate data was to be collected.

A range of mathematical models were applied to the data for the prediction of freezing time and found to agree with experimental data within 10%. Best results were obtained by using a number of the published methods and averaging the result.

No relationship was found between the mathematical prediction of freezing time and the sensory quality, since no significant differences were found in the sensory data for any product.

No statistically significant differences were measured by the sensory panel in a range of products frozen by air-blast or cryogenic techniques. This work was carried out in both pilot plant and production scale equipment.

References

Aerov, M.E. and Todes, O.M. (1968). Hydraulic and Thermal Fundamentals of Apparatuses Operation with a Stationary and Granulated Fluidized Bed, *Chimija*, Leningrad (St. Petersburg), Russia. In Russian.

Baskakov, A.P. (1968). Processes of Heat- and Mass Transfer in a Fluidized Bed. *Metalurgizdat*, Moscow, Russia. In Russian.

Botterill, J.S. and Fluid, M. (1975). *Bed Heat Transfer,* Academic Press, London.

Charm, S.E. (1971). *Fundamentals of Food Engineering.* AVI Publishing, Westport, CT.

Chizhov, G. B. (1979). Thermal and Physical Processes in Refrigeration Technologies of Food Products, *Food Industry,* Moscow, Russia. In Russian.

Ciborowski, J. (1957). Fluidyzacia. *PWT,* Warsaw, Poland. In Polish.

Cleland, A.C. and Earle, R.L. (1977). A comparison of analytical and numerical methods of predicting freezing times of foods. *Journal of Food Science,* 42 (5), 1390.

Cleland, A.C. and Earle, R.L. (1984). Freezing time prediction for different final product temperatures. *Journal of Food Science,* 49, 1230.

Cowell (1967). The calculation of freezing time. In: *Proceedings of the XIII International Congress of Refrigeration,* 2, 667.

Davidson, J. F. and Harrison, D. (1963). *Fluidized Particles,* University of Cambridge, U.K.

Ditchev, S. (1984). Intensification of the Processes. In: *Refrigeration Equipment by Using Dynamic Disperse Medium,* D.Sc. Thesis HIFFI-Plovdiv, Bulgaria. In Bulgarian.

Ditchev, S. (1985). Intensification of the refrigeration process by using dynamic disperse medium, *Proceedings of the ICEF 4 Conference,* Edmonton, Alberta, Canada.

Earle, R.L. and Fleming, A.K. (1967). Cooling and freezing of lamb and mutton carcasses. 1. Cooling and freezing rates in legs. *Food Technology,* 21, 79.

George, R.M. (1993). A review of methods for predicting the freezing times of foods. In: *Technical Bulletin No. 94,* Campden and Chorleywood Food Research Association, Chipping Campden, U.K.

Gorbie, Z.P. (1970). Heat-exchange and Hydromechanics of Dispersion Open Streams, *Energija,* Moscow, Russia. In Russian.

IIR (1972). *Recommendations for the Processing and Handling of Frozen Foods,* 2nd Ed., International Institute of Refrigeration, Paris.

Leva, M. (1959). *Fluidization,* Marcel Dekker, New York.

Levy, F. (1958). Calculating freezing time of fish in air blast freezers. *Journal of Refrigeration,* 1, 55

Levy, F. (1974). Meat — toward better understanding of the mechanics of the weight loss. *Annexe au Bulletin de l'IIF,* N 3.

Mellor (1976). Cited by Cleland, A.C. and Earle, R.L. (1976). A comparison of methods for predicting the freezing times of cylindrical and spherical foodstuffs. *Journal of Food Science,* 44, 958.

Mott, L.F. (1964). The prediction of product freezing time. *Australian Refrigeration and Air Conditioning Heat,* 18 (12), 16.

Nagaoka, J., Tajaki, S., and Hotami, S. (1955). Experiments on the freezing of fish in an air blast freezer. *Journal of Tokyo University of Fisheries,* 42 (1), 65.

Pham, Q.J. (1986). A simplified equation for predicting freezing times. *Journal of Food Technology,* 21, 209–219.

Plank, R. (1941). Bietrafe zur berechnung und bewertung der gerfrierfesch. In German.

Ramaswamy, H.S. and Tung, M.A. (1983). A review on predicting freezing times of foods. *Journal of Food Process Engineering,* 7, 169.

Richardson, P. (1995). Assessment of the quality of thawed fruits frozen by different freezing technologies. In: *Proceedings of the First Main Meeting,* Project Process Optimisation and Minimal Processing of Foods, Vol. 2 — Freezing, Published by Escola Superior de Biotecnologia, Porto, Portugal.

Romankov, P. G. and Rashkovskaja, N.V. (1954). Drying by Fluidization. *Leningradskojl otdelenije.* In Russian.

Sjiromjatnikov, N.I., Vasanova, L.K., and Shumanskij, N. (1967). Heat- and Mass-Exchange in Fluidized Bed, *Chimija*, Moscow, Russia. In Russian.

Withers, P., Tucker, G.S., and Holdsworth, M. (1993). A comparison of conduction errors in type T and type K thermocouples for food applications. In: *Technical Memorandum No. 680*. Campden and Chorleywood Food Research Association, Chipping Campden, U.K.

Zabrodskij, C. (1968), Problems About Heat-Exchange in Fluidization Systems. *Collection of Heat and Mass Transfer*, 5, Minsk, Belarus. In Russian.

Zenz, F. A. (1949). Two- phase fluid solid flow. *Industrial and Engineering Chemistry*, 41.

chapter ten

Developments in fish freezing in Europe with emphasis on cryoprotectants

T. Ronan Gormley

Contents

Summary 163
1 Introduction 164
2 Fresh fish history.. 164
3 Freezing conditions/equipment.. 165
4 Storage conditions.. 166
5 Thawing conditions 167
6 Multiple freezing.. 167
7 Cryoprotectants 167
    7.1 Dairy ingredients as cryoprotectants............................. 168
    7.2 Inclusion level and freezing rate 169
    7.3 Sensory properties 170
    7.4 Tests in underutilized species 170
8 Conclusions.. 172
Acknowledgment.. 172
References.. 172

Summary

Selected aspects of recent fish freezing R & D in Europe are discussed under the headings fresh fish history, freezing conditions/equipment, storage conditions, thawing conditions, multiple freezing, and cryoprotectants. It is concluded that studying the interactions between these is the most fruitful direction for future R & D.

0-8493-7905-9/99/$0.00+$.50
© 1999 by CRC Press LLC

Dairy ingredients act as cryoprotectants in a wide range of fish species by enhancing the water-holding capacity of thawed tissue. Caseinates and whey protein concentrates showed the extremes of effect in terms of water-holding capacity and gel properties of the minced fish. Whey protein concentrate also enhanced the sensory properties of breaded cod nuggets.

1   Introduction

Freezing is an important technology for fish preservation, and though most aspects have been fairly extensively researched the overall picture regarding fish freezing and storage is still incomplete. This may be due to the fact that various components of the frozen fish chain have been studied individually, rather than collectively, and important interactive effects and synergy have been missed and/or neglected. The freezing and storage of fish and shellfish have been reviewed recently by Morrison (1994). It is clear from the review that continued research on the long-established PPP (product-process-package) and TTT (time–temperature-tolerance) factors (Jul, 1984), in combination with hazard analysis of critical control point (EU, 1994), and quality systems (Jakobsen and Lillie, 1992), are the key to ensuring high-quality frozen seafood products for the consumer. The findings from a recent survey (Peavey et al., 1995) in the U.S. on consumer attitudes toward fresh and frozen fish showed that there is no room for complacency. Many of the consumers held a persistent belief that frozen fish is substantially inferior to fresh, i.e., less nutritious, more bones, and using only "offcuts." This highlights the need for concerted efforts to rectify possible misunderstandings about frozen fish/shellfish products, including packing them in see-through packs so that consumers can actually see the fish they are buying.

This chapter addresses some of the developments in fish freezing, and in related areas, occurring in Europe in the last decade under six headings; fresh fish history; freezing, storing, and thawing conditions; multiple freezing; cryoprotectants. Most attention is focused on cryoprotectants.

2   Fresh fish history

High-quality fresh fish is a prerequisite for a high-quality frozen product. A recent survey (Gormley, 1995) on the R & D requirements of 819 food SMEs (small to medium sized enterprises) indicated that research on raw material quality was their top priority, i.e., virtually all of them were experiencing problems with their raw material. Tests in Norway (Magnussen and Johansen, 1995) have indicated that more attention should be paid to freshness than to freezing rate and that fish frozen in the prerigor state gave by far the best yield and quality. Rigor status of fish at time of freezing has major implications for gaping. Rigor mortis is a state of muscular contraction that occurs between 1 to 12 hr after death and may last for 24 hr or longer. Delays in icing cause a rapid onset of rigor, which means there may be insufficient time for gutting. Attempts to straighten "curved" fish for freezing will cause the muscle segments to separate, thus leading to "gaping," i.e., the surface develops cracks (Aitken and Mackie, 1980). Hence, the requirement for immediate icing, and for filleting and freezing prerigor.

Procedures for assessing fish freshness have been reviewed by Whittle et al., (1990), while Southey et al. (1995) found large variations in drip loss, thiobarbituric acid substances (TBARS), protein extractability, and hypoxanthine, both within and between three batches of good-quality frozen cod fillets. However, differences between good- and bad-quality fish (as determined by sensory evaluation) were large enough to warrant industrial use of some of the physical and chemical quality indices.

The upgraded quality-index method (Nielsen and Jessen, 1997) is used as a sensory system for determining the freshness of thawed whole cod and is particularly relevant as much raw material (for further processing) is received as frozen fish. The method is based on the significant sensory parameters for whole fish, where each weighted parameter is scored from 0 to 1 or 2 or 3 (depending on the weighting) with 0 the best quality. By assessing many parameters no single parameter can unduly imbalance the score. The scores for all the characteristics are then added to give an overall sensory score, i.e., the so-called quality index.

The multilingual guide to EC freshness grades for fishery products (Howgate et al., 1992) uses four freshness grades [E, A, B, C (reject)], and 33 fish species are currently provided for in the regulations. Classification is based on descriptors relating to the skin, outer slime, eyes, gills, peritoneum (in gutted fish), internal (body cavity) odor, and gill odor.

3   Freezing conditions/equipment

Freezing and freezing methods have a major effect on water relations in fish and the consensus is that fast freezing (1°C/min) produces the best quality in that it induces small ice crystals, evenly distributed in the muscle. Slow freezing (5°C/h) produces large, damaging ice crystals (Nielson and Jessen, 1997). The effect of the type of material being frozen also has a major bearing on frozen/thawed quality as indicated by centrifugal drip loss values of 278, 267, and 345 g/kg for fillets from frozen fish, block frozen fillets, and block frozen mince respectively (Gormley et al., 1993), i.e., mincing prior to freezing had a particularly adverse effect on the water-holding capacity of these samples of silver smelt (*Argentinus silus*). There is also considerable interest in two-stage freezing with the second stage being termed superfreezing (Magnussen, 1992; Magnussen and Johansen, 1995) where the product is held at −50 to −60°C in order to give an extended shelf life. In tests, the first freezing (salmon fillets) was in a tunnel at −28°C to a core temperature of −22°C. The product was then transferred to a regional superfreezing low-temperature cold store where the temperature was lowered to −50 to −60°C for long-term storage. The results showed that superfreezing minimized quality loss in the frozen salmon as indicated by sensory evaluation and color measurement. Research on superfreezing at –60°C is ongoing at The National Food Center, Dublin (Gormley, 1997) on salmon and smoked mackerel, and will be extended soon to a wider product range.

Recent developments in freezing equipment include the AIR Products Cryo-Dip, and the Dybvad Stal Industri (DSI) range of horizontal plate freezers. The former has a capacity of up to 5 ton/h, depending on the product, and allows companies easy entry to IQF (liquid nitrogen) products because of its relatively low capital cost. The DSI plate freezers are easily cleaned and are convertible with short change-over which benefit freezer trawlers and land-based plants switching between herring and mackerel seasons (Anon., 1995). Developments in cryogenic freezing of food, including fish, have been discussed by Tomlin (1995), where FGP rotary freezing equipment was used (Ionov and Akimova, 1994) for low-temperature (–55°C) freezing of fish onboard ship. The results showed that FGP freezing had beneficial effects on fish quality in comparison with conventional freezing at –25°C. FGP freezing in combination with microwave thawing reduced protein degradation and weight loss, and improved the water-holding capacity of the fish tissue.

4 Storage conditions

There are relatively few data in the literature on the effects of storage temperatures below –30°C on fish quality, although there is increasing evidence (Magnussen and Johansen, 1995; Scudder, 1995) that temperatures of –50 to –60°C are beneficial in terms of product quality, especially for fatty fish such as mackerel and herring. This relates to the two stage or superfreezing outlined above. Scudder (1995) claims that the reason –18°C is used as a storage temperature is that it corresponds to 0°F. However, he stresses that the breakdown of trimethylamine oxide (TMAO) to dimethylamine (DMA) and formaldehyde occurs readily at –18 to –20°C. The level of breakdown is greatly reduced below –24°C and becomes lower still below –30°C. The breakdown of TMAO to DMA and formaldehyde is of major significance in that the latter bonds with the fish proteins to cause toughening. These findings are supported by those of Howell (1995), who showed that toughening due to protein aggregation increased more rapidly during storage at –20°C than at –30°C for cod, and especially for hake. Fluctuating cold-storage temperatures below the freezing point also adversely affect the quality of frozen fish (Nilsson, 1997) due to recrystallization and other effects. This topic is being studied in the ongoing EU Concerted Action on "The Preservation of Frozen Food Quality and Safety throughout the Distribution Chain" (FAIR Program). Tests are also ongoing at The National Food Center, Dublin on the effects of fluctuating temperatures (three cycles of –28°C to –10°C followed by storage at –28°C) on the quality of frozen salmon and smoked mackerel.

5   Thawing conditions

A thermal arrest (freezing plateau) occurs between 0 and –5°C, even in the fast freezing of fish, and 5 to 10% of the muscle water has not formed ice crystals even at –20°C. This water is bound in the protein membrane structures and can still be active in chemical reactions (Nilsson, 1997). The glass transition state, where the whole fish forms a crystalline inert state, occurs when the temperature falls to –50°C or below. During thawing the opposite situation arises. The ice crystals turn to water, but with a longer thermal arrest as the heat transfer is faster in ice compared to water, and during thawing heat is transferred through a growing layer of water. The thawing process influences fish muscle structure and biochemistry (Nilsson and Ekstrand, 1995) almost as much as freezing. Fast thawing does not allow ice crystal growth or recrystallization to the same extent as slow thawing, and gives a more intact appearance to the muscle structure, and also a lower drip loss. Water heated to 40 to 60°C is usually used for defrosting, although forced humid air is also used, but to a lesser extent. Microwave and dielectric thawing systems are also under consideration.

6   Multiple freezing

Logistics often demand that fish are frozen "round," and then thawed, filleted, and refrozen. Many processors also use fish from frozen fillet blocks that are thawed and are used for consumer fish products, which are then refrozen, i.e., double freezing is commonplace in today's fish processing industry. A second freezing results in new ice crystal formation and a "repeat" of the chemical changes occurring during the first freezing (Nilsson, 1997). Hurling and McArthur (1996) found that refreezing of cod fillet blocks did not necessarily result in greater deterioration (relative to a "once-frozen" control) of cooked fish sensory attributes after nine months at –22°C. However, thawed and refrozen fish muscle displayed a faster decline in myofibril protein solubility than once-frozen controls, and had reduced water-holding capacity. Tests in Sweden (Nilsson and Ekstrand, 1994) indicated that refreezing after glazing affected cod (fillets) and rainbow trout (cutlets) muscle tissue. Fast refreezing after glazing resulted in less drip after thawing, both spontaneous or forced, in both species. Leakage of intracellular marker enzyme was highest in slowly refrozen trout samples.

7   Cryoprotectants

Extension of shelf life of fish during frozen storage can be achieved by the incorporation of ingredients (e.g., cryoprotectants) that are able to prevent ice crystal growth and the migration of water molecules from the protein, thus stabilizing the protein in its native form during frozen storage (Fennema et al.,1973). Some cryoprotectants such as mono- and dissaccharides, glycerol, sorbitol, some salts, ascorbic acid, citric acid, carboxymethyl cellulose, gums or their combinations, are satisfactorily used to freeze-preserve fish, and also fruit and poultry (Fennema et al., 1973). The cryoprotective effect of some nonfish proteins, for example milk protein derivatives and egg white, in surimi and fish-based products has been reported by several authors (Nielson et al., 1985; Bugarella et al., 1985a; Lanier, 1986; Chang-Lee et al., 1990; Chung and Lee, 1990; Ryder, 1990). Although the study (Chung and Lee, 1990) on the functional and physicochemical properties of nonfish proteins proved useful in some cases for predicting gel-forming ability and other functional properties of fish-based products, a good cryoprotective effect can be obtained only when a synergism between nonfish and fish proteins has been established (Bugarella et al., 1985b; Lanier, 1986).

Four projects have been carried out recently at The National Food Center, Dublin on the use of dairy ingredients in fish mince: (1) their potential as cryoprotectants; (2) the effect of inclusion level and freezing rate on cod mince quality and on the properties of full-strength and added-water cod-mince gels; (3) their effect on the sensory properties of added-water cod nuggets; (4) their use in the mince of 12 underutilized fish species.

7.1　Dairy ingredients as cryoprotectants

The potential cryoprotective effect of nine dairy powders [lactose (LAC), skim milk isolate (SKI), 90% demineralized whey (DEM), milk protein isolate (MPI), whey (WHE), whey protein concentrate (35%) (W35) and (80%) (W80), spray-dried calcium caseinate (CCA), and sodium caseinate (NCA)] added at 80 g/kg of fish mince, was investigated in the minces of cod (*Gadus morhua*), haddock (*Gadus aeglefinus*), salmon, and spent salmon (*Salmo salar*). Cod and haddock were chosen as low-fat white fish, salmon as an oily fish, and spent salmon as a fish with a lower fat and a higher moisture content than ordinary salmon. The minced, treated fish was subjected to three freeze-thaw cycles (each cycle embraced blast freezing at −35°C for 2 h followed by thawing at 3°C overnight). The thawed mince was evaluated for water-holding capacity (WHC), gel properties (compression, elasticity, penetration) and color (Hunter L). Experimental details and results were published in 1995 (Anese and Gormley) and 1996 (Anese and Gormley), and the main findings are summarized here.

Each of the nine dairy ingredients exerted a cryoprotective effect (i.e., a raised WHC in the thawed fish tissue) in salmon and spent salmon. In frozen cod and haddock mince only NCA, CCA, and MPI increased the WHC, while W35 and W80 gave a reduced WHC in thawed fish in comparison with controls.

Whey protein concentrate gave the firmest gels and the caseinates the softest. There was generally an inverse relationship between the WHC of the thawed fish tissue and the compressive strength of the fish gels. The reason may be that ingredients giving a low fish WHC could facilitate extensive cross linking of proteins, thereby leading to a firmer gel (Johnson and

Table 1   Percentage of Unfrozen[a] Water and of Water
Available for the Hydration of Frozen Cod
Mince Samples with Added Dairy Ingredients

Ingredient	Unfrozen water (%)	Hydration water (%)
Control[b]	8.9	40.1
Skim milk powder	12.8	46.1
Demineralized whey	15.1	54.2
Milk protein isolate	14.1	50.9
Whey protein concentrate (80%)	12.6	45.3
Calcium caseinate	21.5	77.5
Sodium caseinate	20.1	72.6

[a] Calorimetric tests (carried out at the University of Udine).

[b] no dairy ingredients added.

Zabik, 1981; Chung and Lee, 1990). Conversely, proteins, i.e., caseinates, with a very good WHC when incorporated into fish, may distribute so finely throughout the matrix that they cause a dilution of the minced fish, resulting in a weakening of the gel (Bugarella et al., 1985b; Lanier, 1986; Chung and Lee, 1990). The dilution effect of some of the dairy ingredients studied can be attributed to a strong protein-water interaction. Table 1 shows the percentage of unfrozen water surrounding the proteins, and the percentage of water available for hydration of fish proteins of ingredient-incorporated minced cod. The caseinates gave a much higher hydration value than did MPI, SKI, DEM, or W80, which in turn had higher hydration values than the control sample. Moreover, it must be pointed out that among the dairy ingredients studied W80 had the lowest hydration value (as well as the lowest WHC value and the highest gel compression value).

Although more research is needed, a possible application of those dairy ingredients that are effective in preventing protein denaturation could be their addition to formulated fish-based products as improvers of the quality of the frozen fish.

7.2   Inclusion level and freezing rate

The cryoprotectants showing the extremes of performance in the previous experiment (see 7.1.), i.e., W80 and NCA, were included in cod mince at levels of 20, 40, and 80 g/kg of fish mince. The samples were frozen either by fast (air blast, −35°C) or slow (cabinet, −20°C) methods and were subjected to three freeze-thaw cycles (Glancy et al., 1995). The thawed mince was tested for WHC and gel properties as described previously (Anese and Gormley, 1996).

Fast freezing of cod mince resulted in a higher WHC and stronger gels than slow freezing. Different inclusions (20, 40, 80 g/kg of mince) of W80 did not affect the WHC values, whereas NCA at the 80 g/kg inclusion level gave a much higher WHC value than the 20 or 40 g/kg inclusions. These data agree with those of Anese and Gormley (1995) (see 7.1.). For example,

W80 had a slight negative effect on cod WHC and so it could be expected that different levels of W80 would not have much effect on WHC values. In contrast, NCA enhanced the WHC of cod mince (Anese and Gormley, 1996), and so inclusion level would be likely to increase cod WHC values, which did in fact take place (Gormley et al., 1995).

The W80 or NCA did not affect gel compression values for gels made from fast frozen mince. However, for samples of slow frozen mince, increasing inclusions gave stronger gels in the case of NCA, and weaker gels in the case of W80. The addition of water greatly reduced gel strength, and neither W80 nor NCA increased added-water gel strength values over those of the control.

7.3 Sensory properties

Taste panels were carried out to investigate the acceptability of deep fried (previously frozen) cod nuggets containing dairy ingredients. Six panels were conducted (see Tables 2 and 3) on nuggets with and without added water, and the dairy ingredients were included at 80 g/kg of fish mince. Test 1 (panels 1–3) involved the comparison of W35, W80, CCA, NCA, and a control (no dairy ingredient), while test 2 (panels 4–6) evaluated the effect of added water, i.e., 1:0, 2:1, 1.5:1 (fish to water).

The overall taste panel data for test 1 (Σ column, Table 2) show that panelists preferred cod nuggets containing W35 followed by those containing W80. The inclusion level of 80 g/kg of minced fish enhanced sensory quality rather than detracting from it. This agrees with the data of Ryder (1990) who found that inclusions of dairy ingredients at 60 g/kg of fish mince had no negative impact on flavor. The nuggets containing the caseinates received the lowest rating while the control was third among the five samples. However, there was an interaction for the level of added water; this occurred for W80 (ranked 5th in panel 1 and 2nd in panels 2 and 3) and for the control (ranked 5th in panel 2, and 2nd and 3rd in panels 1 and 3). Data for test 2 (Σ column, Table 3) showed a clearcut preference for fish nuggets without added water; the sample with the most added water was the least preferred. These effects were consistent from panel to panel and suggest that the dairy ingredients are not able to "carry" the extra water in added-water fish nuggets; this is a similar effect to that found for the gels.

7.4 Tests in underutilized species

The performance of W35 and NCA as cryoprotectants (inclusion level 80 g/kg of minced fish) was tested in the mince (from frozen fillets) of 12 underutilized fish species obtained from trawlers fishing out of Loughinver in Scotland. The species were roundnose grenadier (*Coryphaenoides rupestris*), roughead grenadier (*Macrourus berglax*), orange roughy (*hoplostethus atlanticus*), blue ling (*Molva dypterygia*), Portuguese shark (*Centroscymnus coelolepis*), dogfish (*Scyliorhinus stellaris*), tusk (*Brosme brosme*), forkbeard (*Phycis blennoides*),

Table 2 Taste Panel Preference Scores (rank sums)[a] for Fish Nuggets with Added Dairy Ingredients (different water levels) (Test 1)

Dairy ingredient[b]	Ratio (fish mince: added water)			
	Panel 1 1:0	Panel 2 2:1	Panel 3 1.5:1	Σ
Whey protein conc. (35%)	35	29	31	95
Whey protein conc. (80%)	55	31	36	122
Calcium caseinate	50	53	51	154
Sodium caseinate	49	52	58	159
None	36	60	49	145

[a] For 15 tasters; lowest sum = preferred sample
  range for statistical significance (15 tasters × 5 samples) = 32–58* P < 0.05
  30–60** P < 0.01
[b] 80 g/kg of mince

Table 3 Taste Panel Preference Scores (rank sums)[a] for Fish Nuggets with Added Water (different dairy ingredients) (Test 2)

Ratio fish/water	Dairy ingredient[b]			
	W35[c] Panel 4	NCA[c] Panel 5	None Panel 6	Σ
1:0	25	30	28	83
2:1	43	36	35	114
1.5:1	52	54	57	163

[a] For 20 tasters: lowest sum = preferred sample
  range for statistical significance
  (20 tasters x 3 samples) = 32–48* P < 0.05
  30–50** P < 0.01
[b] 80 g/kg of mince
[c] See Table 2

bluemouth rockfish (*Helicolenus dactylopterus*), cardinal fish (*Epigonus telescopus*), black scabbard (*Aphanopus carbo*) and leafscale gulper shark (*Centrophorus squamosus*). The samples were thawed at 4°C, filleted and were refrozen (air blast –35°C for 2 h) and stored (–28°C) until required, i.e., they were twice-frozen fish. A sample of locally caught cod (*Gadus morhua*) was twice frozen and was used as a control. The samples were prepared and the tests were carried out as described by Maier et al. (1997).

Cardinal fish, Portuguese shark, and dogfish had the most favorable WHC and gel compression values (full-strength gels), and roundnose grenadier, forkbeard, and leafscale gulper shark the least favorable. Both W35 and NCA acted as cryoprotectants in most of the species in that they raised the WHC of the fish after freezing/thawing. The effect was greatest for mince with NCA, which is in agreement with the data of Anese and Gormley (1996) for cod, salmon and haddock.

Added water (2 parts mince:1 part water) gels were prepared for four of the species, i.e., cardinal fish, dogfish, blue ling, and cod (as control). The inclusion of W35 increased gel compression values for three of the species but not for blue ling. Sodium caseinate (NCA) greatly increased the gel compression values of cardinal fish and dogfish. This contrasts with previous findings for cod and haddock gels (Anese and Gormley, 1996) (see Section 7.1) where caseinates gave softer gels than whey proteins. This change in performance of NCA cannot be explained without further research but could be due to the significant fat content of cardinal fish and dogfish.

Full-strength gels of roundnose grenadier, orange roughy, and black scabbard were the whitest and those of dogfish, forkbeard, and blue ling the least white. Whiteness is desirable in the manufacture of whitefish products.

8 Conclusions

Fish freezing, frozen storage, and thawing are influenced by many factors, a number of which interact. It is in the study of the interactions that the greatest progress will be made in fish freezing research in the future.

Dairy ingredients act as cryoprotectants in a wide range of fish species. Overall, caseinates and whey proteins showed the extremes of effect in terms of water-holding capacity and gel strength. The inclusion of whey proteins also enhanced the sensory properties of deep fried cod nuggets.

The research items presented here are only a small selection of recent R & D activity on fish freezing in Europe.

Acknowledgment

Thanks are extended to Kerry Ingredients, Listowel, Co. Kerry, Ireland.

References

Anon. (1995). Denmark lead the way with hi-tech equipment, *Seafood International*, 10 (8), 41.

Aitken A. and Mackie, I.M. (1980). *Fish Handling and Processing*, 2nd Edition, HMSO, London.

Anese, M. and Gormley,T. R. (1995). Maintaining quality in frozen fish mince, *Farm and Food*, 5 (2), 13.

Anese, M. and Gormley, T. R. (1996). Effect of dairy ingredients on some chemical, physico-chemical and functional properties of minced fish during freezing and frozen storage, *Food Science and Technology* (LWT), 29, 151.

Bugarella, J. C., Lanier, T. C., and Hamann, D. D. (1985a). Effects of added egg white or whey protein concentrates on thermal transitions in rigidity of croaker surimi, *Journal of Food Science*, 50, 1588.

Bugarella, J. C., Lanier, T. C., Hamann, D. D., and Wu, M. C. (1985b). Gel strength development during heating of surimi in combination with egg white or whey protein concentrate, *Journal of Food Science*, 50, 1595.

Chang-Lee, M. V., Lampila, L. E., and Crawford, D. L. (1990). Yield and composition of surimi from Pacific Whiting (*Merluccius productus*) and the effect of various protein additives on gel strength. *Journal of Food Science,* 5(1), 83.

Chung, J.H. and Lee, C.M. (1990). Relationships between physicochemical properties of nonfish proteins and textural properties of protein-incorporated surimi gels. *Journal of Food Science,* 55(4), 972.

EU FLAIR Concerted Action N. 7 (1994). *HACCP User Guide,* European Commisssion, Brussels, Belgium.

Fennema, O. R., Powrie, W. D., and Marth, E. H. (1973). Freezing injury and cryoprotectants. In: *Low Temperature Preservation of Foods and Living Matter,* 1st Edition, Fennema, O.R., Powrie, W.D., and Marth, E. (Eds.), Marcel Dekker, New York, p. 476.

Glancy, C., Anese, M., and Gormley, T. R. (1995). The effect of dairy ingredient inclusion level and rate of freezing on the water-holding capacity and gel properties of frozen cod mince, *Irish Journal of Agricultural and Food Research,* 34 (2), 212.

Gormley, T. R. (1995). RTD needs and opinions of European food SMEs, *Farm and Food,* 5 (2), 27.

Gormley, T. R. (1997). Upgrading the cold chain, *Research Report,* Teagasc, The National Food Centre, Dublin, Ireland.

Gormley, T. R., Ward, P., and Somers, J. (1993). A note on the effect of long-term frozen storage on some quality parameters of silver smelt (*Argentinus silus*), *Irish Journal of Agricultural and Food Research,* 32, 201.

Gormley, T. R., Glancy, C., and Anese, M. (1995). Dairy ingredients as cryoprotectants in frozen fish mince. In: *Proceedings of the First Main Meeting,* Project Process optimization and minimal processing of foods, Vol. 2 — Freezing, Oliveira, J. and Oliveira, F. (Eds.), Published by Escola Superior de Biotecnologia, Porto, Portugal, p. 38.

Howell, N. K. (1995). Elucidation of aggregation mechanisms of proteins in frozen fish, *Final Report to EU FAR Programme, UP.3.647,* University of Surrey, U. K.

Howgate, P., Johnston A., and Whittle K. J. (1992). *Multilingual Guide to EC Freshness Grades for Fishery Products,* Torry Research Station, Aberdeen, Scotland (published in collaboration with The West European Fish Technologists Association — WEFTA).

Hurling, R. and McArthur, H. (1996). Thawing, re-freezing and frozen storage effects on muscle functionality and sensory attributes of frozen cod (*Gadus morhua*), *Journal of Food Science,* 61(6), 1289.

Ionov, A. G. and Akimova, E. Y. (1994). Quality of fish frozen in an FGP apparatus, *Rybnoe-Khozyaistvo,* 3, 50.

Jakobsen, M. and Lillie, A. (1992). Quality systems for the fish industry. In: *Quality Assurance in the Fish Industry* (Quality Assurance Symposium), Ministry of Fisheries, Copenhagen, Denmark, 515.

Johnson, T.M. and Zabik, M.E. (1981). Gelatinisation properties of albumen protein, singly and in combination, *Poultry Science,* 60, 2071.

Jul, M. (1984). *The Quality of Frozen Foods,* 1st Edition, Academic Press, London.

Lanier, T. C. (1986). Functional properties of surimi, *Food Technology,* 3, 107.

Magnussen, O. M. (1992). Superfreezing low temperature technology, *Scan-Ref,* 21 (4), 34.

Magnussen, O. M. and Johansen, J. (1995). Effects of freezing and storage temperatures at −25°C to −60°C on fat fish quality. In: *Proceedings of the 19th International Congress of Refrigeration,* 11, 249.

Maier, K., Gormley, T. R., Connolly, P. L., and Auty, M. (1997). Assessment of underutilised fish species, *Farm and Food*, 7 (2), 30.

Morrison, C.R. (1994). Fish and shellfish. In: *Frozen Food Technology*, 1st Edition, Blackie Academic & Professional, London, p. 196.

Nielsen, J. and Jessen, K. (1997). New developments in sensory analysis for fish and fishery products. In: *Seafood from Producer to Consumer, Integrated Approach to Quality*, 1st Edition, Luten, J.B., Børresen, T. and Oehlenschläger, J. (Eds.), Elsevier, Amsterdam, Netherlands, p. 537.

Nielsen, J., Braun, A., and Madsen, A. (1985). Milk powders as cryoprotectants in frozen minced fish, IIF-IIR *Commissions C_2, D_3*, Aberdeen, Scotland, p. 169.

Nilsson, K. (1997). Effect of freezing and thawing on water relations in fish. Paper presented at *WEFTA Industry Forum 97 on Water Management in Fish and Fishery Products*, April 14–15, Brussels, Belgium.

Nilsson, K. and Ekstrand, B. (1994). Refreezing rate after glazing affects cod and rainbow trout muscle tissue, *Journal of Food Science*, 59 (4), 797.

Nilsson, K. and Ekstrand, B. (1995). Frozen storage and thawing methods affect biochemical attributes of rainbow trout, *Journal of Food Science*, 60 (3), 627.

Peavey, S., Work, T., and Riley, J. (1995). Consumer attitudes toward fresh and frozen fish, *Journal of Aquatic Food Product Technology*, 3 (2), 71.

Ryder, D.N. (1990). Study on the use of milk proteins in fish gel products. IIF-IIR Commission C_2, *International Institute of Freezing*, Aberdeen, Scotland, p. 107.

Scudder, B. (1995). Icelandic research warning on frozen fish, *Seafood International*, 10 (6), 51.

Southey, A., O'Neill, E., and O'Connor, T. (1995). Assessment of the variation in quality indices in frozen cod. In: *Proceedings of the First Main Meeting*, Project Process optimization and minimal processing of foods, Vol. 2 — Freezing, Oliveira, J. and Oliveira, F. (Eds.), Published by Escola Superior de Biotecnologia, Porto, Portugal, p. 49.

Tomlin, R. (1995). Cryogenic freezing and chilling of food. *Food Technology International*, Europe, 145.

Whittle, K. J., Hardy, R., and Hobbs, G. (1990). Chilled fish and fishery products. In: *Chilled Foods — The State of the Art*, 1st Edition, Gormley, T.R. (Ed.), Elsevier Applied Science, London, p. 87.

chapter eleven

Advances in osmotic dehydration

Harris N. Lazarides, Pedro Fito, Amparo Chiralt,
Vassilis Gekas, and Andrzej Lenart

Contents

Summary 176
1 Introduction 176
2 Process variables 178
   2.1 Product identity.. 178
      2.1.1 Species, variety, maturity level 178
      2.1.2 Tissue location 178
      2.1.3 Size and shape.. 179
   2.2 Product pretreatment 179
   2.3 Osmotic solution 180
      2.3.1 Osmotic solution composition 180
      2.3.2 Osmotic solution concentration.. 181
      2.3.3 Osmotic solution characterization.. 181
   2.4 Phase contacting.. 181
   2.5 Process temperature 182
   2.6 Process pressure 183
   2.7 Process duration.. 183
3 Process analysis and modeling.. 183
   3.1 Scope and objectives of process modeling 183
   3.2 System description.. 184
      3.2.1 Driving forces 184
      3.2.2 Mass transfer mechanisms 185
   3.3 Equilibrium stages and controlling mechanisms 186
   3.4 Volume and compositional changes 188
4 Osmotic preconcentration as pretreatment in convective drying 190
   4.1 General considerations... 190
   4.2 Impact of osmotic dehydration on convective dehydration
      kinetics 191

0-8493-7905-9/99/$0.00+$.50
© 1999 by CRC Press LLC

175

4.3   Impact of osmotic dehydration on final product
      properties...192
4.4   Impact of osmotic dehydration on water sorption
      kinetics 193
5 Research and development needs...194
6 Final remarks 195
References..196

Summary

The pioneering work of Ponting's group on osmotic dehydration of fruits in the mid 1960s triggered an ever increasing interest of food scientists and engineers in osmotic processing as a basic unit operation for food processing. Osmotic dehydration is already considered a valuable tool in minimal processing of food. It can be applied either as an autonomous process or as a processing step in alternative processing schemes leading to a variety of end products. Despite its well-recognized advantages and the large amount of research work that has been published in this area, large-scale industrial applications of osmotic processing are hindered by fundamental issues that have to be resolved. Such issues include satisfactory process control, environment-friendly solution management, and microbiologic validation of the process. This chapter presents recent developments regarding proper control of process variables, process analysis and modeling, research needs, and current efforts to overcome hurdles that delay a plethora of food processing applications.

1   Introduction

Osmotic dehydration (OD) is the partial removal of water by direct contact of plant or animal tissue with a suitable hypertonic solution, i.e., highly concentrated solutions of sugars, salts, sugar/salt mixtures, etc. During the last years there has been a strong research and industrial interest in the OD process. An overwhelming number of papers have been published and several reviews have been written on the subject (Ponting, 1973; Le Maguer, 1989; Raoult-Wack et al., 1992; Torreggiani, 1993; Fito et al., 1994; Lazarides, 1994, 1995; Raoult-Wack, 1994; Dalla Rosa et al., 1995; Lenart, 1995; Fito et al., 1998).

During OD two major countercurrent flows take place simultaneously. Under the chemical potential gradients existing across the product–medium interface, water flows from the product into the osmotic medium, while osmotic solute is transferred from the medium into the product. A third transfer process, leaching of product solutes (sugars, acids, minerals, vitamins) into the medium, although quantitatively negligible, is recognized as affecting the sensorial, nutritional, and functional characteristics of the final product.

Solute uptake (during OD) modifies the composition of the final product and provides a means of direct product formulation, i.e., controlled introduction of selected solutes to enhance nutrition, taste, color, texture, stability, safety, etc.

Due to continuous immersion in the osmotic medium, OD is an oxygen-free process. Therefore, there is no need to use sulphur dioxide and/or blanching for protection against oxidative and enzymatic discoloration.

Besides, mild heat treatment at relatively low process temperatures (<50°C) favors color and flavor retention, often resulting in products with superior sensorial characteristics (Ponting, 1973).

Finally, OD is an effective way to reduce overall energy requirements in dehydration and dehydro-freezing processes.

The above advantages make OD a versatile unit operation and an intermediate process-step in several alternative process schemes, where product safety and stability is assured through complimentary processing, i.e., pasteurization, canning, freezing, moderate or complete drying (air-, vacuum-, freeze-drying), and various possible hurdle combinations, i.e., impregnation with preservatives or antimicrobials, dehydro-salting-smoking, acidification, edible coating, etc.

Despite the large amount of research work that has been published in the area of osmotic processing, industrial scale applications are held back by fundamental issues which have to be faced. Such issues include efficient, environment-friendly solution management, microbiologic validation, and satisfactory process control.

Satisfactory process control, however, requires adequate predictive models based on detailed process analysis; and such models are still missing. This is mainly due to poor understanding of fundamentals of mass transfer in biological cellular tissues. Therefore, emphasis of this presentation is placed on process analysis and modeling. Since drying is a major (predominant) area of application of OD, the impact of OD on complimentary air drying is also offered special attention.

Plant and animal tissues present substantial differences in terms of structure and behavior. Therefore, a common treatment of both would introduce unnecessary complexity in an already difficult to solve problem. Thus, the discussion of this chapter focuses on OD of plant (fruit and vegetable) tissues.

With respect to OD applications on animal products (i.e., meat and fish), the reader can find valuable information in other review and research papers (Collignan and Raoult-Wack, 1992, 1994; Lucas and Raoult-Wack, 1996; Deumier et al., 1996, 1997; Collignan and Montet, 1998). These papers present rather interesting approaches to animal product processing and stabilization, including immersion chilling/freezing of fish aboard, dehydro-salting-liquid-smoking of salmon, squid tenderization with enzyme impregnation, meat curing, etc.

2   Process Variables

Before we study the fundamentals of mass transfer, we need to examine the relative importance of product characteristics and identify universal properties that characterize the entire system, including product (raw, intermediate, final), osmotic solution, and process set-up (equipment). A short presentation on important process variables will be valuable in understanding process analysis and modeling. A more detailed discussion on the role of process variables in mass exchange can be found in a previous review (Lazarides, 1994).

2.1   Product identity

2.1.1   Species, variety, maturity level

Species, variety, and maturity level all have a significant effect on natural tissue structure in terms of cell membrane structure, protopectin to soluble pectin ratio, amount of insoluble solids, intercellular spaces, tissue compactness, entrapped air, etc. In turn, these structural differences substantially affect diffusional mass exchange between product and osmotic medium.

Work at Thessaloniki laboratory revealed large differences in apparent water and solute diffusivities for apple and potato tissues during OD in a 38 dextrose equivalent (DE) corn syrup solids (CSS) solution (Lazarides et al., 1997). The differences were explained by drastically (order of magnitude) higher intercellular porosity in apple (25% porosity) compared to potato (*ca.* 2.5% porosity). Under identical process conditions, different potato varieties gave substantially different water loss (Hartal, 1967). Finally, firmness of raw fruit was found to affect quality characteristics of osmo-dehydrated kiwi fruit slices (Torreggiani et al., 1991).

2.1.2   Tissue location

Tissue location within the same, homogeneous-looking, fruit was also found to affect mass transport rates during osmotic drying. Figure 1 shows the variation of dehydration efficiency index (water loss/sugar gain-WL/SG ratio) for inner and outer parenchyma in apple tissue (Granny Smith variety) during osmotic preconcentration at two temperatures. At both experimental temperatures, the dehydration efficiency index was higher for outer parenchyma tissue throughout the entire duration of the process (3 hours).

This was a finding of Lund laboratory based on apple-structure observations made initially by other workers (Vincent, 1989). Since the two kinds of tissue show different porosity and pore interconnectivity, it was postulated that the two tissues could show different transport rates, on the ground that intercellular space structure details determine pathways of transport, especially when cellular membranes are intact. The validity of this assumption was confirmed for two apple varieties (Kim and Granny Smith) and for process temperatures between 5 and 40°C.

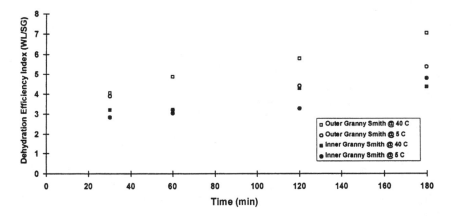

Figure 1 Effect of raw tissue location/structure (apple, Granny Smith variety) on the dehydration efficiency index (WL/SG) in 50% sucrose solution. Filled and open symbols indicate samples of inner and outer parenchyma, respectively.

2.1.3 Size and shape

Size and shape also play a significant role in mass exchange, since they result in different specific surface areas or surface to thickness ratios.

Up to a certain total surface/half thickness (A/L) ratio, samples of higher specific surface (i.e., rings) gave higher water loss (WL) and sugar gain (SG) values compared to lower specific surface shapes (i.e., slice and stick) (Lerici et al., 1985). Past this A/L limit, however, higher specific surface samples (i.e., cubes) favored SG at the expense of lower WL, resulting in lower weight reduction (WR). The lowest WL associated with the highest A/L ratio was explained as the result of reduced water diffusivity due to high sugar uptake.

2.2 Product pretreatment

Product pretreatments and process conditions affecting the integrity of natural tissue have a severe effect on mass exchange. Disruption of structural barriers improves water and solute diffusivities within the product, resulting in faster equilibration in favor of higher solute uptake. Blanching, freeze/thawing, sulfating, acidification and high process temperatures all favor solids uptake yielding lower WL/SG ratios (Ponting, 1973; Lerici et al., 1988; Biswal and Le Maguer, 1989; Lazarides and Mavroudis, 1995).

The Thessaloniki laboratory investigated the effect of freeze/thawing after partial OD (i.e., for 30 to 90 min) on mass transfer rates in freeze/thawn fruit tissue during additional osmotic processing (Lazarides and Mavroudis, 1995). Freeze/thawing did not seem to affect the rate of water removal during complimentary OD, but it resulted in a tremendous increase of solute uptake, compared to continuous (uninterrupted) OD (Figure 2).

Structural changes seem to result in decreased tortuosity of diffusion paths, favoring solids transfer (Oliveira and Silva, 1992).

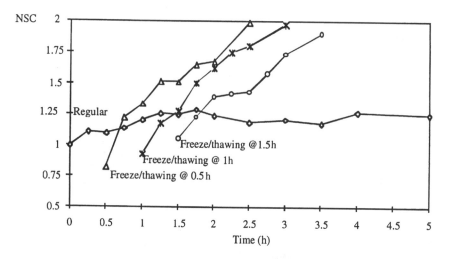

Figure 2 Impact of a freeze/thaw step on solute uptake during osmotic dehydration of apple slices in corn syrup solids. Experimental conditions: 7 mm apple slice, CSC 38 DE 55%, 50°C (Lazarides and Mavroudis, 1995). Printed with kind permission from the Institute of Food Technologists.

2.3 Osmotic Solution

Composition and concentration of the osmotic solution predetermine its chemical potential, which is the driving force for mass exchange. Both composition and concentration are gradually changing during solution recirculation and reuse. They therefore need continuous control and adjustment at the desired levels. Eventually, the osmotic solution must be disposed of in environmentally acceptable, economically efficient ways. Solution management is a major issue that has to be resolved before OD can be exploited industrially. A concerted action research cooperation on osmotic processing is under way within the 4th EU-Framework program. Among other objectives, this project is also meant to face this specific issue in the most efficient manner.

2.3.1 Osmotic solution composition

Several solutes, alone or in combinations, have been used in hypertonic solutions for osmotic dehydration. Based on effectiveness, convenience, and flavor, salt and sugar solutions proved to be the best choices.

There is a marked difference in the way that sugar and salt penetrate the processed tissue. While sucrose accumulates in a thin subsurface layer, resulting in surface tissue compacting (an extra mass transport barrier), salt was found to penetrate the osmosed tissue to a much greater depth (Lenart and Flink, 1984a). Presence of salt in the osmotic solution can hinder the formation of the compacted surface layer, allowing higher rates of water loss and solids gain. Besides, increasing salt concentration in sugar/salt solutions of constant weight concentration leads to lower water activity solutions with respectively increased osmotic capacity.

Acidification of concentrated sucrose solutions had a favorable effect on water loss (Moy et al., 1978). Pectin hydrolysis and depolymerization is probably responsible for the improved dehydration efficiency of acidified solutions, especially at higher process temperatures.

Finally, there is good evidence that the molecular size of the osmotic solute has a significant effect on water loss to solids gain ratio. The smaller the solute, the larger the depth and the extent of solute penetration (Hawkes and Flink, 1978; Bolin et al., 1983; Lenart and Lewicki, 1987, 1989; Lerici et al., 1988; Lenart, 1992; Lazarides et al., 1994). The Thessaloniki laboratory found a strong linear relationship between solute size and solids uptake. Large dextrose equivalent (DE) corn syrup solids (CSS) favored sugar uptake, resulting in lower WL/SG ratios and vice versa. In fact, lower DE (large size) CSS gave negative net solid gain values, indicating that solute uptake was inferior to leaching (loss) of natural tissue solids (Lazarides et al., 1994).

2.3.2 Osmotic solution concentration

Increased solute concentrations resulted in increased WL and SG. The observed increase in WL, however, was much higher than the increase in SG. As a result, increased concentrations lead to increased weight reductions (Hawkes and Flink, 1978, Lenart, 1992, Conway et al., 1983).

How far can we increase concentration to affect mass exchange? It seems that there is a maximum concentration above which water loss cannot be improved through increased solution concentration. At high sugar concentrations, additional increase in concentration (above 65%) did not promote faster weight loss (Ponting et al., 1966; Contreras and Smyrl, 1981).

2.3.3 Osmotic solution characterization

It is well known that concentrated solutions used in OD processes are real (not ideal) solutions, which may deviate strongly from ideal situations. Therefore, activities rather than concentrations should be used to provide a sound theoretical basis for characterization of osmotic solutions and/or interpretation of OD data. Unfortunately, there is a lack of solute activity or osmotic pressure data. In an effort to face this problem, Lund laboratory initiated a project on characterizing osmotic solutions through proper modeling and measuring water and solute activity data.

2.4 Phase contacting

Phase contacting is mainly determined by two factors:

1. Product/solution ratio, and
2. Solution movement relative to product surface (solution Reynolds number).

A small product/solution ratio (<1/30) is usually used to avoid the dilution effect caused by the water removal. On the other hand, large Reynolds numbers are needed to safely assume negligible mass transfer resistance on the product surface.

While the first requirement is easy to meet, the second one strongly depends on experimental set-up (i.e., equipment specifications, operation conditions) and is not always easy to calculate and control. Due to this problem, a great deal of published research data are not directly comparable and cannot be easily replicated or complemented by additional work. To overcome this obstacle, within the above-mentioned European project, an effort is being made to specify a prototype experimental set-up and a standard experimental protocol to be used by all laboratories participating in the project.

2.5 Process temperature

It is well recognized that diffusion is a temperature-dependent phenomenon. Higher process temperatures seem to promote faster water loss through swelling and plasticizing of cell membranes, faster water diffusion within the product, and better mass (water) transfer characteristics on the product surface, due to lower viscosity of the osmotic medium.

At the same time, solids diffusion within the product is also promoted by higher temperatures, only at different rates, mainly dictated by the size and concentration of the osmotic solute (Lazarides et al., 1994).

It is interesting to mention that potato and apple tissues seem to respond in a different manner to limiting process temperatures. During 3 h of OD at 50°C in the same medium (CSS) under the same concentration and process conditions, apple slices gave a small (*ca.* 8%) net solid loss, while potato sticks of the same thickness gave a large increase (by 85%) of normalized solids content (Lazarides et al., 1994; Lazarides and Mavroudis, 1996). Besides tissue differences in structure, compactness, intercellular spaces and soluble solids content, the larger specific surface of potato sticks (compared to apple slices) also promoted solute uptake, but this was not the controlling factor. OD of potato sticks at temperatures up to 45°C gave negative to zero net solute gain, which indicates that extensive solute uptake at 50°C is the result of heat-induced damage of the potato tissue at temperatures >45°C. Apple did not show a similar behavior.

In the case of sucrose solutions, temperature affected water loss without affecting sucrose penetration (Lenart and Flink, 1984b).

Therefore, whenever it is desirable to achieve higher water removal and lower solids gain, it pays to use a higher process temperature within the applicable range. In fruits and vegetables this range cannot go beyond 50°C, if we want to keep cell membranes intact and semipermeable. Above this temperature there is also a negative impact on final product quality due to softening, browning, flavor loss, etc. (Ponting et al., 1966). Minimal processing of fruits usually refers to temperatures up to 40°C.

2.6 Process pressure

The effect of pressure changes on osmotic dehydration has been thoroughly analyzed by the research group of Valencia (Fito, 1994; Fito and Pastor, 1994).

When a porous solid is immersed in a liquid, it is affected by changes in pressure or temperature. Internal pore gas could expand or get compressed, depending on pressure gradients in the food/solution system. This phenomenon was called the *hydrodynamic mechanism* (HDM). The model of this mechanism predicts the entry of external liquid in a porous product, when any overpressure is applied to the system; nevertheless liquid will be released when product returns to normal pressure.

Pressure gradients provoke a coupling of the HDM with deformation-relaxation phenomena in the solid (Fito and Chiralt, 1995; Salvatori et al., 1998a). Changes in product volume and composition have been modeled (Fito and Chiralt, 1995) and the impact on OD has been evaluated (Fito et al., 1996). Application of vacuum impregnation (VI) for a specified time before OD has been named *pulsed vacuum osmotic dehydration* (PVOD) (Shi et al., 1995; Barat et al., 1997).

2.7 Process duration

While true equilibrium (characterized by an equality of water chemical potential in product and solution) is approached after a rather long period of time, it was found that mass exchange is not significantly changed after a period of 4 to 5 hours (Lenart and Flink, 1984b; Lazarides et al., 1994). For the particular experimental set-ups this can be considered as the *practical end-point* of the process.

In every case, the practical end-point is far before true equilibration and is actually specified by the observed relative changes of mass transfer data in combination with the desired final product characteristics (Table 1).

3 Process analysis and modeling

3.1 Scope and objectives of process modeling

During the last years, several attempts have been made to model the operation of osmotic dehydration of foods. The Valencia laboratory proposed the general layout of a model aiming at the following substantial targets (Fito and Chiralt, 1996; Barat, 1998; Fito et al., 1998):

- to evaluate water and solute fluxes
- to calculate weight losses suffered by the product
- to predict changes in product composition

Obviously, the ultimate objective of such a model is to predict quality, stability, and shelf life of the final product and evaluate the economical viability of the specific process.

Table 1 Criteria for Practical Definition of Equilibrium Stages
in Osmotic Processes of Different Length

Process length	Time scale	Operation	Controlling mechanism	Equilibrium definition
VSTP	Minutes	Vacuum impregnation	HDM	$\Delta p_I = 0$
STP	Hours	Osmotic preconcentration (minimally processed food)	Pseudo-diffusion and Matrix deformation.	$\Delta a_w = 0$
				$z_w = y_w$
LTP	Hours/days	Extensive osmotic dehydration. (Dehydrated fruits)	Matrix relaxation	$\Delta V = 0$ $\Delta p_{DRP} \sim 0$
VLTP	Days/weeks	Candy fruits production.	Gas release and pore filling.	$\Delta M = 0$ $\Delta p_{DRP} = 0$

VST: very short time process. STP: short time process. LTP: long time process. VLTP: very long time process. (The actual length of each process depends on process temperature, osmotic solution concentration, and other operation variables, Barat, 1998.)

The model has been applied to atmospheric osmotic dehydration (OD) and to pulsed vacuum osmotic dehydration (PVOD), where the samples were vacuum-impregnated (VI) with the osmotic solution prior to OD. This section deals with the last improvements that were included in the proposed model (Barat, 1998), and some others that can be introduced for its further upgrading.

3.2 System description

From an engineering point of view, food can be defined as composed of two separate phases (Barat, 1998):

- The food solid matrix (FSM) consisting of all insoluble compounds (mostly cellular walls), and
- The food liquid phase (FLP) consisting of water and water-soluble solutes.

The ratio of the two phases (FLP/FSM) is strongly related to an important quality factor, namely the water-retention capability of the final product.

The two product phases are interacting with the third phase of the system, which is the osmotic solution (OS).

3.2.1 Driving forces

The basis for this analysis is the theoretical fact that the property that most properly characterizes an osmotic solute, from the mass transfer point of view, is its water activity lowering capacity. Indepedent of their composition, two solutions of the same water activity (or osmotic pressure) will be equivalent in their osmotic drying effect, since the driving force for water transport is not concentration but chemical potential.

The Lund research group has tested the above idea with a "clear cut" equilibrium experiment for osmotic dehydration of apple tissue. They used two "iso-osmotic" (isotonic) solutions, that is, solutions having the same water activity (aw = 0.884) but different water concentrations. The first solution was a binary one, at 60% sucrose concentration. The second was a multicomponent mixture containing sucrose, NaCl, fructose, and glucose. The mixture composition was set to prevent solute losses from the tissue to the osmotic solution. In fact, the concentration of each solute in the osmotic solution was higher (i.e., twice as high) than the one found in fruit tissue, in order to compensate for possible solute activity effects in the concentrated solution. As a first approximation, the assumption was made that no coupling effects are present, in the spirit of the Onsager approach of irreversible thermodynamics. Of course, for a more thorough study of possible coupling effects, more research is needed. The water loss observed in the two cases was statistically the same for the two solutions, whereas the solid uptake was not (Gonzalez, 1996). The time for reaching a pseudo-equilibrium state was 24 hours for the binary solution and 36 hours for the multicomponent one, which indicates that, at least for a given time scale, states of quasi-equilibrium or quasi-steady-state do occur due to minimization of one of the driving forces present in the system. This is correct according to Prigogine ideas for equilibrium states of biological systems (Prigogine, 1968).

Another finding of this study was the impressive increase of sucrose activity with increased concentrations of the osmotic solution, when the Gibbs-Duhem theorem was applied. Thus, for a 60% sucrose solution, the solute activity coefficient was at relatively low levels (i.e., below 10), while at 70%, the solute activity coefficient reached the tremendous value of 24. These values were based on molality concentrations.

3.2.2 Mass transfer mechanisms

Mass transfer kinetics have been modeled in the first hours of treatment (Fito and Chiralt, 1996) by assuming that mass transfer mechanisms may be classified in two groups as follows:

1. *Hydrodynamic mechanisms (HDM):* They are pressure gradient dependent and develop at the very beginning of the process (t~0). These mechanisms include capillary and external pressure effects. In the PVOD process, action of the HDM creates fast and significant changes in product weight and soluble solids concentration. These changes are evaluated in the model as taking place at time t = 0.

2. *Pseudo-diffusion mechanisms (PDM):* They are water and solute activity dependent and they apply at time t>0. The effect of these mechanisms may be modeled by using Fick's equations on the basis of the following assumptions:

- Equilibrium in this scale of time may be defined by equal water and solute activities (or concentration in some cases) in FLP and the osmotic solution;
- Mass transfer driving forces (in terms of FLP concentration) may be defined by using the FLP solute concentration, and can be described by the following equation:

$$\frac{dz^j}{dt} = D_e \nabla^2 z \tag{1}$$

where: z^j is the concentration of component (j) in FLP and D_e is the effective diffusion coefficient

- Weight loss kinetics may be modeled by an exponential equation.

The model has been applied to several products with quite satisfactory results (Barat, 1998).

An interesting approach to understanding transfer mechanisms taking place in apple tissue at near room temperatures has led to a generalized equation predicting concentration profile changes during the OD process (Salvatori, 1997). The equation was based on the concept of *advancing disturbance front* (Salvatori et al., 1998a). Simultaneous experimental analysis of structure and composition profiles confirmed the usefulness of this approach in developing a generalized equation.

3.3   Equilibrium stages and controlling mechanisms

In such a complex system, the evolution to equilibrium is rather complex as well. The group of Valencia (Barat, 1998) studied equilibration behavior of cylinders of apple (Granny Smith) in water–sucrose solutions (0% to 65% w/w) at 30, 40, and 50°C for periods up to 14 weeks, in OD and PVOD experiments. Experimental data showed that true thermodynamic equilibrium is achieved within some months (Barat, 1998, Barat et al., 1998), after passing through several pseudo-equilibrium steps, as is usual in such complex systems (Nicolis and Prigogine, 1977).

A general overview of changes in total mass and total volume of apple tissue (cylindrical samples) submerged in sucrose solution at different concentrations and temperatures is given in Figure 3. With OD experiments, the sample-solution system was maintained under atmospheric pressure (and controlled temperature) at all times, while in the PVOD experiments, samples were vacuum impregnated with the osmotic (sucrose) solution prior to equilibration. As shown in this figure, OD samples suffered a strong loss of weight within a time period of 0.5 to 2 days (from 0 to t_2), depending on temperature and osmotic solution concentration. The minimum weight at t_2 corresponds to a minimum volume and to an equal concentration and a_w in the FLP and the osmotic solution.

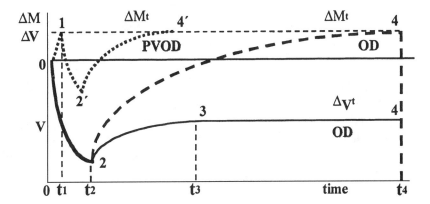

Figure 3 Changes of total mass (ΔM) and total volume (ΔV) during equilibration of apple tissue in sucrose solutions. (The PVOD samples were vacuum impregnated with the sucrose solution (step 0–1), prior to equilibration) (Barat, 1998).

On the other hand, PVOD samples increased their mass in a first step (t_1 = 5–15 min) as a result of VI, but quick water loss due to the development of osmotic gradients caused a total mass decrease thereafter, following the pattern of water loss in OD samples. Nevertheless, minimum weight was substantially higher and it was achieved earlier in this case, compared to OD (point 2′ of Figure 3). The important water loss experienced during this period resulted in great deformations of the FSM (Barat et al., 1998). As a consequence, strong tensions accumulated in the structure.

For $t>t_2$ stresses in structure relax, deformations recover, and product volume is increased again with simultaneous suction of external solution. As a result, total mass and volume increase without any change in the FLP composition. At the microstructure level, Cryo-SEM observations (Salvatori, 1997; Salvatori et al., 1998b) indicate that, in a first step (from t_2 to t_3), matrix relaxation yields an increase in overall volume of cells, causing external liquid to flow into the empty spaces generated between cell walls and vacuoles. At the same time, the porous structure remains partially occupied by gas. When cell wall relaxation is practically complete (point 3), gas inside pores is slowly released and is replaced by external liquid, due to the action of capillary forces. When no capillary or stress forces are present in the product, and a_w in external and internal liquid phase is the same, the chemical potential for any component in all phases of the system is the same, and true equilibrium is reached (point 4).

Based on these observations, the following equation has been proposed by Barat (1998) to calculate changes in water chemical potential ($\Delta\mu^w$) in an osmotically treated product:

$$\Delta\mu^w = RT \cdot \ln a_w + \overline{V} \cdot \Delta P_I + \overline{V} \cdot \Delta P_{DRP} + \overline{V} \cdot \Psi \tag{2}$$

where:

the first term ($RT \ln a_w$) evaluates the contribution of water-solute interactions;

the second, the effect of hydrostatic pressures (turgor, capillary and external pressures);

the third would be the contribution of relaxation of accumulated stresses in the matrix (such relaxation induces pressure gradients in the liquid and gas phases, with corresponding bulk flows); and

the last term represents interactions between water and solid matrix, and plays a less important role in foods, where water concentration is high.

The different steps leading to final equilibration reflect the changing impact of each contribution (each term of Equation (2)) in the development of specific chemical potential gradients. In fact, step 0–1 in PVOD experiments (Figure 3) is a very short time process (VSTP) of vacuum impregnation (VI) lasting a few minutes, with the second term of Equation (2) (hydrostatic pressure) as prevailing driving force. Step 1–2 in PVOD (and step 0–2 in OD) is a short time process (STP) lasting a few hours and it reflects equilibration of the first term, due to the action of a_w gradients. Step 2–4, both in OD and PVOD, reflects the action of pressure gradients (due to matrix relaxation) as driving forces (i.e., the third term of Equation (2) has a prevailing impact). This is a long or very long time process (LTP or VLTP) lasting some hours or days in the case of extensive osmotic dehydration, and some days or even weeks in the case of candy fruit production. In the last case, gas release and pore filling play a significant role as driving mechanisms.

A list of equilibrium stages and practical definitions of equilibration criteria are presented in Table 1 (Barat et al., 1998). With respect to osmotic dehydration, a_w equilibration is a good criterion. In very long time treatments, however, (i.e., candy fruits) slow equilibration processes controlled by matrix relaxation, gas release, and pore filling play an important role in the final FLP/FSM ratio.

3.4   Volume and compositional changes

Shrinkage studies were undertaken by the laboratories of Lund and Thessaloniki in collaboration. The geometric shrinkage behavior of apple and potato tissue subjected to osmotic drying was found to be anisotropic.

Volume changes were linearly related to water content. In fact, this correlation was found to hold for both apple varieties (Granny Smith and Kim) and all temperatures studied (Figure 4). Interestingly, in the 40°C trials of Kim variety, the slope was less steep and its intercept was higher compared to Granny Smith variety, implying a heat-induced better volume retention, which has to be analyzed and explained.

On a practical point of view, changes in composition can be described in a three-component diagram, as in usual food engineering solid–liquid

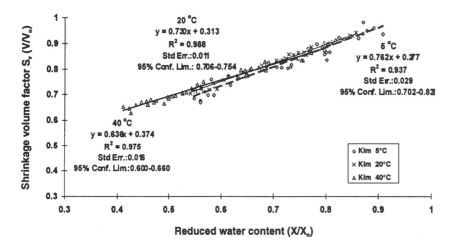

Figure 4 Tissue shrinkage as a function of water content at three different temperatures (apple tissue, Kim variety).

operations (Barat et al., 1998). In Figure 5, the axis shows water and solute concentrations. The inert phase concentration is the distance of any point to the hypotenuse. Point 1 represents the raw material and point 2 the first pseudo-equilibrium step at $t = t_2$, when equality of concentration in FLP and osmotic solution (OS) is achieved. Any point between 1 and 2 may be the final situation in an atmospheric osmotic dehydration (OD) operation. In the case of candy fruit processes, fruits arrive to point 2 and continue increasing their FLP/FSP ratio, moving from 2 to 3 (the line from origin to OS represents samples with same FLP composition). Point 5 represents concentration changes produced during vacuum impregnation (VI) of sample 1 with OS. Point 6 would be the result of OD after application of VI. Lines 1–5 and 5–6 show compositional changes of a sample during (1–5) and after PVOD (5–6).

The Warsaw laboratory found that the kind of osmotic solute had a remarkable influence on soluble solids diffusivity in apple, carrot, and pumpkin (Lenart and Iwaniuk, 1993). As an example, the use of starch syrup instead of sucrose caused a significant decrease in soluble solids diffusivity during osmotic processing of pumpkin when the water content in the sample was about 3 g/g dry matter, soluble solids diffusivity was about threefold lower for dewatering in starch syrup than in sucrose solution.

Osmotic preconcentration (OP) of apple slices in sugar solutions was followed by massive exchange of sugars between product and the osmotic medium (Lazarides et al., 1995). This exchange was strongly related to composition of the osmotic solution and was faster within the first hour of OP. OP in solution of CSS resulted in massive leaching (loss) of sucrose and limited losses of fructose and glucose, providing a final product with half its initial sugar content. OP in sucrose solution, however, provided a more stable total sugar balance, due to extensive uptake of sucrose counteracting most losses in fructose and glucose. Early interruption of OP had much

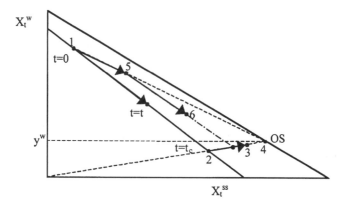

Figure 5 Composition changes in different osmotic processes (Barat, 1998). (x) are the water and solute concentrations in the food system. The concentration of the osmotic solution is (y). The point OS represents the osmotic solution concentration, because the concentration of the inert phase is zero.

greater impact on solids exchange after use of CSS than after use of sucrose. OP in CSS favored retention of fructose and glucose and diminished loss of sucrose and total solids.

Solute exchange during osmotic dehydration modifies the composition and taste of the final product. This so-called "candying" (or "salting") effect is sometimes desirable, as it tends to improve the taste and acceptability of the final product. In many cases, however, extensive solute uptake is undesirable, because of its negative impact on taste and the nutritional profile of the product, which can no longer be marketed as "natural." In every case, solute uptake results in the development of a concentrated solids layer under the surface of the product, upsetting the osmotic pressure gradient across the tissue–medium interface and decreasing the driving force for water flow (Hawkes and Flink, 1978).

4   Osmotic preconcentration as pretreatment in convective drying

4.1   General considerations

Depending on raw material properties, osmotic solution characteristics, process parameters, and process duration, up to 70% of the initial water can be removed from a plant tissue by osmotic dehydration. The amount of water remaining in the material, however, does not ensure its stability, as water activity is generally higher than 0.9. When shelf stability is an ultimate process objective, other, complementary methods of water removal, such as convective drying, are suggested (Lenart and Cerkowniak, 1996).

It is important to notice that osmotic preconcentration is an oxygen-free (anaerobic) process; and as a pretreatment to convective drying, it offers certain advantages over conventional enzyme inactivation pretreatments, e.g., blanching or sulfating. Blanching of the raw material destroys the cell membrane and damages tissue texture, causing a greater shrinkage of the dried material. On the other hand, sulfating does not cause such changes, but it has a negative nutritional impact, due to the well-recognized toxicity of sulphur compounds (Lenart, 1994).

Considering the industrial interest for combining osmotic preconcentration with vacuum drying, convective drying and freezing, it is important to study the impact of osmotic treatment on physicochemical characteristics of intermediate products.

4.2 Impact of osmotic dehydration on convective dehydration kinetics

Besides its negative effect on the rate of water loss during osmotic preconcentration, solute uptake "blocks" the surface layers of the product, posing an additional resistance to mass exchange and lowering the rates of complementary (vacuum-, convection-, freeze-) dehydration (Lenart and Cerkowniak, 1996). Even a simple immersion of a raw material into an osmotic solution, caused a substantial decrease of water removal rates in convective drying (Lenart, 1994; Lenart and Cerkowniak, 1996). On the other hand, partial dehydration and solute uptake protect the product against structural collapse during complementary dehydration and against structural disorganization and exudation (loss of juices and texture) upon freeze/thawing.

According to experiments run at the Warsaw laboratory, in the course of convective drying of osmotically preconcentrated apple tissue, there appears no period of constant drying rate (Figure 6). Osmotic pretreatment for 3 hours in sucrose solution at temperatures between 20 and 40°C caused a decrease of complementary, convective drying rates (Lenart, 1994). In the course of convective drying of strawberry, the greatest changes in drying rate were caused by osmotic pretreatment in sucrose solution.

Drying rates of apple dehydrated in starch syrup were higher than those of apple dehydrated in sucrose solution. This is probably due to different extents of sample impregnation with osmotic solutes of different molecular size (Lazarides et al., 1995).

In the case of osmotic pretreatment of apple and strawberry in sugar solutions of various molecular sizes, there was no substantial influence of osmotic dehydration time on the kinetics of convective dehydration. This was particularly evident when starch syrup, which contains a considerable amount of dextrines, was used (Lenart, 1994; Lenart and Cerkowniak, 1996). These results are due to rapid impregnation of a subsurface tissue layer right at the beginning of the osmotic process. Experiments on the influence of sample rinsing after osmotic pretreatment on the kinetics of complementary, convective drying showed that rinsing causes a 20 to 30% increase in the drying rate of apples, which were osmotically dehydrated in either sucrose or starch syrup solution.

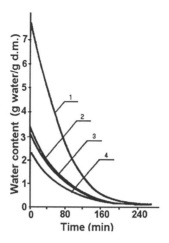

Figure 6 Effect of osmotic preconcentration on the kinetics of convection drying of apple tissue 1: Convective drying of blank (raw apple-no osmotic pretreatment); 2, 3, 4: Convective drying of samples after osmotic preconcentration for 3 hours in sucrose solution at 20°C, 30°C, and 40°C, respectively (Lenart and Cerkowniak, 1996).

As osmotic processing lowers the rates of complementary, convective drying, total dehydration time is practically the same as in direct convective drying to water contents below 10 to 15% (a_w = 0.4–0.6). Contrary to this fact, total drying time is spectacularly shortened in the case of dehydration to intermediate water contents (a_w = 0.6–0.9) (Lenart, 1994).

4.3   Impact of osmotic dehydration on final product properties

Solute uptake and leaching of natural acids, color, and flavor compounds out of osmo-dehydrated plant tissue affect its organoleptic properties, since they modify (to a certain extent) its natural composition (Lenart, 1991; Lazarides et al., 1995).

On the other hand, osmotic pretreatment contributes to retention of flavor in convectively dried fruits, making them more acceptable as ready-to-eat snacks compared to totally air-dried products.

Foods preserved by osmo-convective drying have better texture and lower shrinkage compared to traditionally dried products. The former appear to have a porous structure similar to that of lyophilized products.

While composition changes may have a positive or negative impact on final product acceptance, they certainly affect product rehydration properties in a rather negative manner. Rehydration of osmotically dried fruit is lower than the untreated, in both rate and extent (Lenart, 1991; Lenart and Iwaniuk, 1993). Even a very short contact of apple with sucrose solution prior to convective drying resulted in a two and a half-fold decrease of water adsorbed within 1 hour of rehydration, compared to apple dried without osmotic pretreatment (Lenart and Iwaniuk, 1993). This is due to rapid impregnation of a subsurface tissue layer with sugar and the lower rehydration of sugar in the product, compared to the natural tissue itself. It has also been shown that the longer the osmosis time, the lower the rehydration rate and extent of osmo-convection in dried carrots (Lenart, 1991).

Water diffusivity during rehydration of osmotically pretreated samples seems to depend on the kind of osmotic substance. Starch syrup solutions gave a lower water diffusivity compared to sucrose solutions when osmotically dried apple and carrot were rehydrated (Lenart, 1991). Pumpkin did not exhibit a similar response. Water adsorption rate in dried plant tissue also depends on the content of osmotic substance and its spatial distribution as well as on the structure of the dried product. In the case of comminuted dried products, the water adsorption rate mainly depends on chemical composition of the dried product, decreasing significantly with increasing sucrose content (Lenart, 1991).

Due to low hygroscopicity, products that were dried after osmotic pretreatment can be kept in a quite humid atmosphere without the risk of becoming sticky (no caking).

4.4 Impact of osmotic dehydration on water sorption kinetics

Plant tissues are heterogeneous materials with regard to both chemical composition and structure. This heterogeneity significantly affects their water sorption kinetics. The effect of osmotic pretreatment on sorption properties of dried plant tissue is clearly expressed in the shape of sorption isotherms.

Changes in soluble solids composition and the prevailing presence of osmotic solutes have a significant effect on water sorption behavior. Osmotic pretreatment of fruits in sucrose solution was found to flatten the shape of the isotherm, making the product less hygroscopic (Lenart, 1991; Lazarides et al., 1995). Sugar exchange during OP of apple slices in sucrose and corn syrup solids solutions resulted in drastically modified sugar profiles of vacuum-dried products, causing an isotherm shift of the sorption isotherm to the right (Lazarides et al., 1995). The shift was more pronounced in the product treated in sucrose solutions, and it was explained by differences in concentration and water-binding properties of the three main sugars, namely fructose, glucose, and sucrose. The practical consequence was a different product response to dehydration and rehydration processes.

As expected, the longer the osmotic treatment the larger is the impact on the qualitative and quantitative characteristics of water sorption isotherms (Lenart, 1991).

The above water sorption changes do not depend only on sugar composition, but they are also affected by the form in which sugars are found in the product. Depending on water content, sugars may occur in three forms: crystalline, amorphous, and in solution. Immediately after drying sucrose in dried plant tissue occurs in an amorphous form. During storage, however, it gradually turns into a stabler crystalline form, leading to a significant further lowering (flattening) of the water sorption isotherm. Presence of simple sugars and other soluble cell constituents leads to an increase of equilibrium water content, which makes the dry product more hygroscopic (Lenart, 1991).

Adsorption properties of dried plant tissue also depend on their inner structure. The osmotic pretreatment leads to less shrinkage during drying, causing at the same time an increase of the dried product density (Sitkiewicz et al., 1996). As a result, meso- and macropores take a higher share of the total volume of pores occurring in the material.

5   Research and development needs

The above discussion reveals the potential of osmotic dehydration as a fundamental process step in efficient, minimal processing of high-moisture agricultural products. At the same time it sketches the needs and shortages that need to be faced before we can proceed with industrial applications.

As with every food-processing operation, most research efforts are eventually directed toward process control and process/product optimization. In addition to the usual concerns of many chemical engineering processes, food processing design has to consider nutrition, sensorial characteristics, product stability, and consumer safety. Economics and environmental impact are also crucial issues that must be taken into consideration. Overall, the OD process (alone or in combination with other treatments) must lead to a safe and stable product, with satisfactory nutritional and organoleptic characteristics, at an affordable price. Without a satisfactory level of these properties, a product cannot satisfy the demands of a modern consumer; therefore it will not survive in a competitive market.

Under the light of these goals and limitations, in food engineering terms, the following research and development work needs to be undertaken:

1. Objective characterization of raw and processed materials with a minimal number of universal properties (i.e., hardness, elasticity, porosity, density, insoluble/soluble solids, etc.).

    Such characterization is essential for establishing universal models that will adequately predict the behavior of products of different origin and/or properties (i.e., species, variety, maturity level, tissue location, and other major sources of product variation). Besides, this is absolutely necessary in establishing universal process control criteria and designing suitable equipment and process automation.

2. Development of adequate models to describe/monitor the process and predict product behavior and final product properties.

    Models should be based on a minimal number of universal process/product parameters and should be evaluated on several markedly different applications, looking for *"adequate for engineering purposes performance."*

This major task will be accomplished through further work on the following subtasks:

a. delineate mass transfer mechanisms at microscopic (cell) and macroscopic level,
b. study the kinetics of mass exchange and deformation/relaxation-coupled phenomena,
c. study short- and long-term pseudo-equilibrium stages and define practical equilibrium criteria.

3. Characterization and microbiologic evaluation of osmotic solutions in order to define:
   a. suitability for a particular process application,
   b. circulation and adjustment characteristics,
   c. reuse cycles and rejection criteria,
   d. economically acceptable management, without (or with minimal) negative environmental impact.
4. Microbiologic validation of the process and evaluation of storage stability/safety of final product. Formulation of possibilities could offer efficient solutions to both product stability and safety (i.e., controlled impregnation with preservatives and/or antimicrobials).
5. Study of possible process applications and appropriate process combinations leading to well-recognized final products with specified shelf life and handling requirements (i.e., packaging and storage specifications). Special attention should be paid to innovative applications that help to solve specific processing problems, i.e., chilling/formulation of microbiologically sensitive material (e.g., fish) immediately after harvesting.

In trying to realize the above tasks, it is essential to utilize all written research experience, which means that we need to collect and evaluate all published data, establishing a database.

The current European project on *"Improvement of overall food quality by application of osmotic treatments to conventional and new processes"* is expected to help achieving the above tasks through integration of ongoing research and efficient exchange of information among universities, research centers and industrial partners throughout Europe.

6 Final remarks

Most of the work done so far refers to osmotic processing of fruits and vegetables with emphasis on the manipulation of process parameters to control product dehydration or impregnation. Recent recognition of direct formulation possibilities has opened a whole new world of potential applications in conventional and new processes for plant and animal products. The strong, renewed interest has brought about coordinated efforts to study the fundamentals of osmotic processing with a new, global approach.

The osmotic treatment is now viewed as a complex solid–liquid system that has to be analyzed and modeled as a basic unit operation. In this context, fundamental findings such as properties and behavior of biological materials, mass transfer mechanisms, and process modeling, will be directly applicable to every solid/liquid process.

System complexity mainly comes from structural and organizational variations of biological materials. Therefore, transfer mechanisms have to be studied at cell level, taking into account newly acquired knowledge on physical, physicochemical, and biological properties of biomaterials.

Extensive industrial applications are awaiting a successful address of fundamental issues, such as adequate process modeling, microbiologic process validation, and efficient, environment-friendly osmotic solution management.

References

Barat, J. M. (1998). *Desarrollo de un modelo de la deshidratacion osmotica como operacion basica*, Ph.D. Thesis, Universidad Politecnica de Valencia, Spain. In Spanish.

Barat, J. M., Alvarruiz, A., Chiralt, A., and Fito, P. (1997). A mass transfer modelling in osmotic dehydration. In: *Engineering and Food at ICEF 7*, Jowitt, R. (Ed.), Sheffield Academic Press, p.G-81.

Barat, J. M., Albors, A., Chiralt, A., and Fito, P. (1998). Structural changes throughout equilibration of apple in osmotic treatments. In: *Proceedings of the Third Main Meeting*, Project Process optimization and minimal processing of foods, Vol. 3 — Drying, Oliveira, J. and Oliveira, F. (Eds.), Published by Escola Superior de Biotecnologia, Porto, Portugal.

Biswal, R. N. and Le Maguer, M. (1989) Mass transfer in plant materials in aqueous solutions of ethanol and sodium chloride: equilibrium data. *Journal of Food Process Engineering*, 11(3), 159.

Bolin, H. R., Huxsol, C. C., Jackson, R., and Ng, K. C. (1983). Effect of osmotic agents and concentration on fruit quality, *Journal of Food Science*, 48 (1), 202.

Collignan, A. and Montet, D. (1998). Tenderizing squid by treatment with endocellular enzymes. *LWT-Food Science and Technology*. In press.

Collignan, A. and Raoult-Wack, A. L. (1992). Dewatering through immersion in sugar/salt concentrated solutions at low temperature. An interesting alternative for animal foodstuffs stabilization. In: *Drying '92*, Mujumdar, A.S. (Ed.), Elsevier Science Pub., Amsterdam, Netherlands, p. B-1887.

Collignan, A. and Raoult-Wack, A. L. (1994). Dewatering and salting of cod by immersion in concentrated sugar-salt solutions. *Lebensmittel Wissenchaft und Technologie*, 27, 259.

Contreras, J. E. and Smyrl, T. G. (1981). An evaluation of osmotic concentration of apple rings using corn syrup solids solutions, *Canadian Institute of Food Science and Technology Journal*, 14 (4), 310.

Conway, J., Castaigne, F., Picard, G., and Vovan, X. (1983). Mass transfer considerations in the osmotic dehydration of apples, *Canadian Institute of Food Science and Technology Journal*, 16 (1), 25.

Dalla Rosa, M. Bressa, F., Mastrocola, D., and Pittia, P. (1995). Use of osmotic treatments to improve the quality of high moisture-minimally processed fruits. *Proceedings of the Second International Symposium on Osmotic Preconcentration of Fruits and Vegetables*, April 18–19, Warsaw, Poland, p. 69.

Deumier, F., Zakhia, N., and Collignan, A. (1996). Formulation of a cured meat product by the dewatering-impregnation soaking (DIS) process: Mass study and assessment of product quality. *Meat Science*, 44 (4), 293.

Deumier, F. Mens, F., Heriard, B., and Collignan, A. (1997). Control of immersion processes: A novel system for monitoring mass transfer tested with Herring brining. *Journal of Food Engineering*, 32 (3), 293.

Fito, P. (1994). Modelling of vacuum osmotic dehydration of food, *Journal of Food Engineering*, 22, 313.

Fito, P. and Pastor, R. (1994). On some non-diffusional mechanism occurring during vacuum osmotic dehydration, *Journal of Food Enginering*, 21, 513.

Fito, P. and Chiralt, A. (1995). An update on vacuum osmotic dehydration. In: *Food Preservation by Moisture Control: Fundamentals and Applications*, Barbosa-Cánovas, G.V. and Welti-Chaves, J. (Eds.), Technomic Pub. Co., Lancaster, U.K., p. 351.

Fito, P. and Chiralt, A. (1996). Osmotic dehydration: An approach to the modelling of solid food-liquid operations. In: *Food Engineering 2000*, Fito, P., Ortega-Rodriguez, E., and Barbosa-Canovas, G. (Eds.), Chapman & Hall, New York, p. 231.

Fito, P., Andres, A., Pastor, R., and Chiralt, A. (1994). Vacuum osmotic dehydration of fruits. In: *Minimal Processing and Process Optimization of Foods: An Interface*, Singh, R.P. and Oliveira, F. (Eds.), CRC Press, Boca Raton, FL, p. 107.

Fito, P., Andrés, A., Chiralt, A., and Pardo, P. (1996). Coupling of hydrodynamic mechanism and deformation-relaxation phenomena during vacuum treatments in solid porous food-liquid systems, *Journal of Food Engineering*, 27, 229.

Fito, P., Chiralt, A., Barat, J., Salvatori, D., and Andres, A. (1998). Some advances in osmotic dehydration of fruits, *Food Science and Technology International*, In press.

Gonzalez, Ch. (1996). *Osmotic dehydration of apples — Some equilibrium studies.* Diploma Thesis, Lund University (Sweden) and Valencia Polytechnic University (Spain).

Hartal, D. (1967). *Osmotic dehydration with sodium chloride and other agents,* Ph.D. Thesis, University of Illinois, Urbana.

Hawkes, J. and Flink, J. M. (1978). Osmotic concentration of fruit slices prior to freeze dehydration, *Journal of Food Processing and Preservation*, 2, 265.

Lazarides, H.N. (1994). Osmotic dehydration: developments and prospects. In: *Minimal Processing and Process Optimization of Foods: An Interface*, Singh, R.P. and Oliveira, F. (Eds.), CRC Press, Boca Raton, FL, p. 73.

Lazarides, H.N. (1995). Osmotic preconcentration as a tool in freeze preservation of fruits and vegetables. In: *Proceedings of the Second International Symposium on Osmotic Preconcentration of Fruits and Vegetables*, April 18–19, Warsaw, Poland, p. 88.

Lazarides, H. N. and Mavroudis, N. (1995). Freeze/thaw effect on mass transfer rates during osmotic dehydration. *Journal of Food Science*, 60 (4), 826.

Lazarides, H. N. and Mavroudis, N. (1996). Kinetics of osmotic dehydration of a highly shrinking vegetable tissue in a salt-free medium. *Journal of Food Engineering*, 30, 61.

Lazarides, H.N., Katsanidis, E., and Nickolaidis, A. (1994). Mass transfer kinetics during osmotic preconcentration aiming at minimal solid uptake. *Journal of Food Engineering*. 25 (2), 151.

Lazarides, H.N., Nicolaidis, A., and Katsanidis, E. (1995). Sorption behavior changes induced by osmotic preconcentration of apple slices in different osmotic media. *Journal of Food Science*, 60 (2), 348–350.

Lazarides, H.N., Gekas, V., and Mavroudis, N. (1997). Apparent mass diffusivities in fruit and vegetable tissues undergoing osmotic processing. *Journal of Food Engineering*. 31, 315.

Le Maguer, M. (1989). Osmotic dehydration: Review and future directions. In: *Proceedings of the International Symposium on Progress in Food Preservation Processes*, Vol. 1, CERIA, Brussels, Belgium, p. 283.

Lenart, A (1991). Sorption properties of apples and carrot preserved by osmo-convection method. *Food Technology and Nutrition*, 19, 27.

Lenart, A. (1992). Mathematical modelling of osmotic dehydration of apple and carrot, *Polish Journal of Food Nutrition and Science*, 1/42 (1), 33.

Lenart, A. (1994). Osmotic dehydration of fruits before drying. In: *Minimal Processing and Process Optimization of Foods: An Interface*, Singh, R.P. and Oliveira, F. (Eds.), CRC Press, Boca Raton, FL, p. 87.

Lenart, A. (1995). Osmotic dehydration of foods before convection drying. In: *Proceedings of the Second International Symposium on Osmotic Preconcentration of Fruits and Vegetables*, April 18–19, Warsaw, Poland, p. 99.

Lenart, A. and Cerkowniak, M. (1996). Kinetics of convection drying of osmodehydrated apples. *Polish Journal of Food Nutrition and Science*, 5/46 (2), 73.

Lenart, A. and Flink, J. M. (1984a). Osmotic concentration of potato. I. Criteria for the end-point of the osmosis process, *Journal of Food Technology*, 19 (1), 45.

Lenart, A. and Flink, J. M. (1984b). Osmotic concentration of potato. II. Spatial distribution of the osmotic effect, *Journal of Food Technology*, 19, 65.

Lenart, A. and Iwaniuk, B. (1993). Mass transfer during rehydration of dewatered apple, pumpkin and carrot. *Polish Journal of Food Nutrition and Science*, 2/43 (4), 69.

Lenart, A. and Lewicki, P.P. (1987). Kinetics of osmotic dehydration of the plant tissue. In: *Drying '87*, Mujumdar, A.S. (Ed.), Hemisphere Publ., New York, p. 239.

Lenart, A. and Lewicki, P.P. (1989). Osmotic dehydration of apples at high temperature. In: *Drying '89*, Mujumdar, A.S. and Roques, M. (Eds.), Hemisphere Publ., New York, p. 501.

Lerici, C. R., Pinnavaia, G., Dalla Rosa, M., and Bartolucci, L. (1985). Osmotic dehydration of fruit: Influence of osmotic agents on drying behaviour and product quality, *Journal of Food Science*, 50, 1217.

Lerici, C. R., Mastrocola, D., Sensidoni, A., and Dalla Rosa, M. (1988). Osmotic concentration in food processing. In: *Proceedings of the International Symposium on Preconcentration and Drying of Foods*, Bruin, S. (Ed.), Elsevier, Amsterdam, Netherlands, p. 123.

Lucas, T. and Raoult-Wack, A.-L. (1996). Immersion chilling and freezing: phase change and mass transfer in model food. *Journal of Food Science*, 61 (1), 127.

Moy, J. H., Lau, N. B. H., and Dollar, A. M. (1978). Effects of sucrose and acids on osmo-vacuum dehydration of tropical fruits, *Journal of Food Processing and Preservation*, 2, 131.

Nicolis, G. and Prigogine, I. (1977). *Self-organization in nonequilibrium systems: From dissipative structures to order through fluctuations*, John Wiley & Sons, New York.

Oliveira, F.A.R. and Silva, C.L.M. (1992). Freezing influences diffusion of reducing sugars in carrot cortex, *Journal of Food Science*, 57 (4), 932.

Ponting, J.D. (1973). Osmotic dehydration of fruits — Recent modifications and applications. *Process Biochemistry*, 8, 18.

Ponting, J. D., Watters, G. G., Forrey, R. R., Jackson, R., and Stanley, W. L. (1966). Osmotic dehydration of fruits, *Food Technology,* 29 (10), 125.

Prigogine, I. (1968). *Introduction to Thermodynamics of Irreversible Processes,* John Wiley & Sons, New York.

Raoult-Wack, A.L. (1994). Recent advances in the osmotic dehydration of foods *Trends in Food Science and Technology,* 5, 255.

Raoult-Wack, A.L., Guilbert, S., and Lenart, A. (1992). Recent advances in drying through immersion in concentrated solutions. In: *Drying of Solids,* Mujumdar, A.S. (Ed.), Elsevier Science Pub., Amsterdam, Netherlands.

Salvatori, D. (1997). *Deshidratacion osmotica de frutas: cambios composicionales y estructurales a temperaturas moderadas,* Ph.D. Thesis, Universidad Politecnica de Valencia, Spain. In Spanish.

Salvatori, D., Andrés, A., Chiralt, A., and Fito, P. (1998a). The response of some properties of fruits to vacuum impregnation, *Journal of Food Processing Engineering.* In press.

Salvatori, D., Albors, A., Andres, A., Chiralt, A., and Fito, P. (1998b). Analysis of the structural and compositional profiles in osmotically dehydrated apple tissue. *Journal of Food Science.* In press.

Shi, X.Q., Fito, P., and Chiralt, A. (1995). Influence of vacuum treatment on mass transfer during osmotic dehydration of fruits, *Food Research International,* 28, 445.

Sitkiewicz, I., Lenart, A., and Lewicki, P.P. (1996). Mechanical properties of osmotic-convection dried apples. *Polish Journal of Food Nutrition and Science,* 5/46 (4), 105.

Torreggiani, D. (1993). Osmotic dehydration in fruit and vegetable processing. *Food Research International,* 26, 59.

Torreggiani, D., Forni, E., Maestrelli, A., and Quadri, F. (1991). Influence of firmness of raw fruit on quality of dehydrofrozen kiwifruit disks. In: *Abstracts of the Proceedings of 8th world Congress of Food Science and Technology* Toronto, Canada, Sept. 29–Oct. 4.

Vincent, J. F. V. (1989). Relationship between density and stiffness of apple flesh *Journal of the Science of Food and Agriculture,* 47, 443.

chapter twelve

Rehydration of dried plant tissues: basic concepts and mathematical modeling

Fernanda A.R. Oliveira and Leonard Ilincanu

Contents

Summary 202
1   Introduction 202
2   Factors affecting rehydration 204
   2.1   Intrinsic factors..204
      2.1.1   Product chemical composition.............................209
      2.1.2   Predrying treatments...209
      2.1.3   Product formulation 210
      2.1.4   Drying techniques and conditions.......................210
      2.1.5   Postdrying procedures 216
   2.2   Extrinsic Factors 217
      2.2.1   Composition of the immersion media...................217
      2.2.2   Temperature 217
      2.2.3   Hydrodynamic conditions.....................................218
3   Modeling rehydration processes 218
   3.1   Mechanisms of rehydration 218
   3.2   Kinetics modeling 220
   3.3   Structural changes modeling 222
4   Conclusions..223
List of symbols 225
References..225

0-8493-7905-9/99/$0.00+$.50
© 1999 by CRC Press LLC

Summary

Many dried foods are consumed or further used for industrial purposes after rehydration and therefore optimization of drying should also consider its effects on rehydration characteristics: rehydration rate and rehydration capacity. These characteristics depend on the size/geometry, composition, and structure of the food products, as well as on rehydration conditions, such as immersion medium or temperature. Although research in drying is quite extensive, a great need still exists in the field of rehydration. Mathematical modeling is very important in this context, as it may not only be very useful for prediction purposes, but also provide some insight into the underlying physical mechanisms. Mathematical modeling should concern not only the process kinetics but also structural changes, as these highly influence the process. This chapter discusses the most important aspects on rehydration, including the role of intrinsic and extrinsic factors and modeling of process kinetics and structural changes.

1   Introduction

Drying is a conventional and widely used process in food engineering. In recent years research has focused on the improvement of the quality of dried products, either by developing new drying processes, or by optimizing processing conditions of conventional processes. Many dried products are consumed or further industrially used after rehydration (also named as rewetting or soaking). Therefore, concerns on the quality of dried foods should be extended to their characteristics after rehydration. The level of rehydration depends on the application. Most dried foods should undergo a full and fast rehydration in order to improve their characteristics before cooking, but other products, such as dried fruit pieces mixed with breakfast cereal, should remain crispy while immersed in milk for a short period.

Many studies have been published on drying of fruits and vegetables, but literature on rehydration is relatively scarce, except for dry legumes (see Figure 1). Much of the progress in this field has been made by analogy with dehydration processes (Garcia-Reverter et al., 1994), but several aspects of rehydration cannot be satisfactorily addressed by simply "reversing" the phenomena that occur during drying. Rehydration of dried plant tissues is essentially a sorption process. However, because of the nonhomogeneity of the tissues and the complexity of the many physical and biochemical phenomena involved, mathematical modeling of the process kinetics is often based on empirical models.

Rehydration highly depends on microstructure. Microstructural features such as seed coat thickness and porosity, large and open micropyle and hilum, affect the imbibition rates of beans (Deshpande and Cheryan, 1986, Agbo et al., 1987). During dehydration the cells undergo significant changes affecting their shape and volume, these effects being measured in terms of volume shrinkage, changes in porosity and case hardening. The nature and

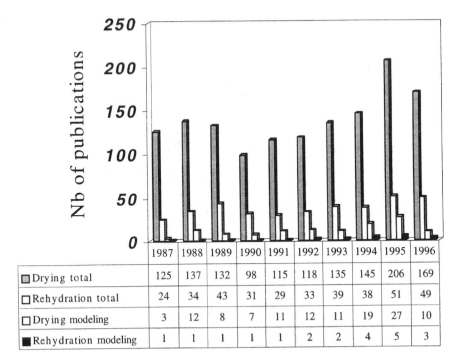

	1987	1988	1989	1990	1991	1992	1993	1994	1995	1996
▨ Drying total	125	137	132	98	115	118	135	145	206	169
☐ Rehydration total	24	34	43	31	29	33	39	38	51	49
☐ Drying modeling	3	12	8	7	11	12	11	19	27	10
■ Rehydration modeling	1	1	1	1	1	2	2	4	5	3

Year

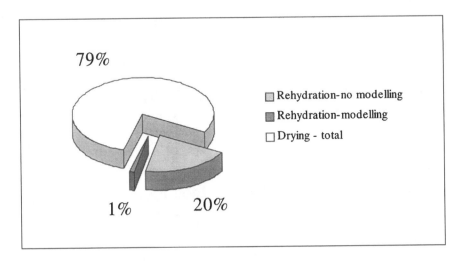

Figure 1 Number of publications on drying and rehydration (cited in FSTA from 1987 to 1996). (a) distribution per year; (b) total number of publications.

extent of these changes greatly depends on the drying process and on the drying operating conditions. Mizza and Le Maguer (1980) discussed the effects of the process of water removal on rehydration characteristics. Several authors developed correlations between processing conditions and porosity, but in general literature is still poor in experimental data concerning structural changes during drying and rehydration and quantitative analysis is lacking.

This chapter presents the basic issues concerning rehydration of plant tissues, in particular dried fruits and vegetables, focusing on the effects of intrinsic and extrinsic factors on rehydration and modeling aspects.

2　Factors affecting rehydration

When assessing rehydration, it is important to consider two aspects: rehydration rate and rehydration capacity. Rehydration rate usually decreases with time of immersion, and its initial value is very important to characterize products that will be subjected to short immersion times, such as dried fruits mixed with breakfast cereal. In these cases a low rehydration rate is required, so that the products will remain crisp before consumption. Rehydration capacity or rehydration ratio refers to the maximum amount of water that the product is able to uptake upon immersion. This concept is important to products that are to be used after full reconstitution and both a high rehydration capacity and rate are desirable. The efficiency of a rehydration process may be calculated as the ratio between rehydration and dehydration ratios. Rehydration capacity should be calculated as the increase in weight or moisture content after equilibrium is reached, although often rehydration capacity is measured after a given immersion time of practical interest and not after full rehydration. Some authors applied a correction for the amount of soluble solids that leaked into the medium from the plant tissue (Shomer et al., 1990) while others included the dehydration ratio to express the water uptake as a function of the product fresh weight (Neumann, 1972; Jayaraman et al., 1990). Table 1 presents some reported values for rehydration capacity of dried fruits and vegetables. These values are difficult to compare as they were obtained under different conditions and some of them do not refer to equilibrium. Furthermore, some authors simply report rehydration ratios without describing how they were measured. The efficiency of the rehydration process was calculated, when data were available, and its values are also included in Table 1. It can be seen that rehydration efficiency highly depends on the food product, and values ranging from 4 to 93% were found.

Rehydration is influenced by several factors depending either on the lattice forces acting inside the plant tissue or on the interactions between the matrix and the immersion medium. The first group may be classified as intrinsic factors and the second as extrinsic factors (see Figure 2).

2.1　Intrinsic factors

In addition to the size and geometry of dried food products that obviously affect rehydration, with smaller particles rehydrating faster (Staal et al., 1990), many other factors affect both rehydration rate and capacity.

Table 1 Rehydration Capacity and Efficiency of Dried Vegetables

Food product	Drying technique	Reference	DR (%)	T_{im} (°C)	t_{im}	RC	η (%)
Cabbage	na	Fellows (1988)	11.5#	na	na	10.5#	91
Carrots	Air drying	Sterling and Shimazu (1961)		20	24 h		
	No pretreatment		91			6.65[b]	61
	Preblanching		91			5.6[b]	51
	Air drying	Eshtiaghi et al. (1994)					
	No pretreatment		86[a]			1.6[c]	26
	Preblanching		89[a]			2.1[c]	26
	Prefreezing		88[a]			3.1[c]	42
	Pre-high-pressure treatment		88[a]	100	20 min	1.5[c]	21
	Preblanching and freezing		88[a]			2.1[c]	29
	Pre-high pressure tr. + freezing		88[a]			2.5[c]	34
	Freeze-drying	Curry et al. (1976)					
	No pretreatment		99.7	24	10 min	5.80[d]	47
	Soaking in distilled water					7.95[d]	69
	Soaking in 0.1M NaCl					8.77[d]	77
	Soaking in 0.2M NaCl					9.06[d]	80
	na	Fellows (1988)	7.5#	na	na	7.0#	93
Cauliflower	Air drying	Jayaraman et al. (1990)					
	No pretreatment		99.7	100	14 min	0.422[f]	4.1
	Preblanching + soaking in 3% salt		99.6		10 min	0.415[f]	4.7
	Preblanching + soaking in 6% sucrose		99.6		7 min	0.625[f]	15
	Preblanching + soaking in 3% salt+6% sucrose		90		7 min	0.690[f]	28

Table 1 (continued)　Rehydration Capacity and Efficiency of Dried Vegetables

Food product	Drying technique	Reference	DR (%)	T_{im} (°C)	t_{im}	RC	η (%)
Celery	Air drying						
	No pretreatment		96[a]			0.21 to 0.32[g]	17 to 29
	Prefreezing	Neubert et al. (1968)	96[a]	4.5	17 h	0.19 to 0.69[g]	15 to 68
	Prefreezing and pressing		9[a]			0.16 to 0.65[g]	15 to 64
	Preblanching, freezing, pressing and puffing		98[a]			0.22 to 0.65[g]	21 to 64
	Prefreezing		96[a]			0.260[h]	21
	Prefreezing and soaking in 15% sucrose		91[a]			0.432[h]	34
	Prefreezing and soaking in 30% sucrose		85[a]			0.612[h]	46
	Prefreezing and soaking in 45% sucrose		79[a]			0.631[h]	42
	Prefreezing and soaking in 60% sucrose		80[a]			0.581[h]	38
	Prefreezing and soaking in 1.5% Na_2CO_3+0.5%$NaHSO_3$+60% sucrose	Neuman (1972)	83[a]	100	15 min	0.710[h]	54
	Prefreezing and soaking in 1.5% Na_2CO_3+60% sucrose		83[a]			0.712[h]	55
	Prefreezing and soaking in 0.5% Na_2CO_3+60% sucrose		84[a]			0.538[h]	38
	Prefreezing and soaking in 1.5% Na_2CO_3+0.5% $NaHSO_3$		97[a]			0.389[h]	36
	Prefreezing and blanching in 65% sucrose		68[a]			0.777[h]	45

Product	Pretreatment	Reference					
	Prefreezing and blanching in 65% sorbitol		64a			0.777h	42
	Prefreezing and blanching in 65% dextrose		66a			0.793h	45
	Prefreezing and blanching in 65% glycerol		68a			0.774h	45
	Prefreezing and blanching in 65% n-propanol		97a			0.307h	27
Green beans	Air drying	Eshtiaghi et al. (1994)					
	No pretreatment		86a			3.5c	57
	Preblanching		89a			4.1c	51
	Prefreezing		88a			4.8c	66
	Pre-high-pressure treatment		86a	100	20 min	3.2c	52
	Preblanching and freezing		90a			5.9c	66
	Pre-high-pressure treatment and freezing		88a			4.8c	66
Green peppers	na	Fellows (1988)	17.0#		na	8.0#	47
Mushrooms	Air drying	Lee et al. (1995)	na	18	25 h	12.4d	na
		Li Shing Tat & Jelen (1987)	89	20	40 min	6.39d	72
	Vacuum drying	Li Shing Tat & Jelen (1987)	na		2 min	7.16d	80
	Freeze drying	Li Shing Tat & Jelen (1987)	na		1.5 min	7.21d	81
		Fang et al. (1971)	na		na	6 to 8e	na
		Le Loch-Bonazi et al.(1992)	99.8	90	na	na	49
		Baek et al. (1989)	7.0#		na	7.16 to 7.77e	na
		Hammami (1998)	87a	5	0.5–1 min*	5.4e	44
		Fellows (1988)	81a		na	5.5#	79
Onions	na						
	Air drying						
	No pretreatment	Schwimmer (1969)	83a	90		0.826g	80
	Pre-soaking in 40%AC+2.3%CL		99.8	100	24 h	0.918g	90
	Pre-soaking in 40% AC		84a	100		0.797g	76
Persil	Freeze drying	Hammami (1998)	81a		1 min*	7.1e	89

Table 1 (continued)   Rehydration Capacity and Efficiency of Dried Vegetables

Food product	Drying technique	Reference	DR (%)	T_{im} (°C)	t_{im}	RC	η (%)
Potatoes	Air drying						
	No pretreatment		83[a]			1.8[c]	34
	Preblanching		81[a]			2.3[c]	54
	Prefreezing	Eshtiaghi et al. (1994)	81[a]	na	20 min	1.5[c]	31
	Pre-high-pressure treatment		82[a]			2.0[c]	47
	Preblanching and freezing		13.0[#]			3.8[c]	89
	Pre-high-pressure treatment and freezing		14.0[#]			2.1[c]	46
Spinach	na	Fellows (1988)			na	5.0[#]	38
Tomato (flakes)	na	Fellows (1988)			na	5.0[#]	36

DR — Drying ratio (% w/w water lost in drying/water in fresh product)

η — rehydration efficiency (% w/w water uptake in rehydration/water lost in drying)

* equilibrium was confirmed

units not available

a % w/w water lost in drying/fresh product weight or water lost in drying/frozen product weight

b g water/g dried product

c ml water uptake/g dried product

d g water/g dry solids

e g rehydrated product/g dried product

f g water/g rehydrated product

g g water/g fresh product

h g rehydrated product/g frozen product

AC — acrylamide; CL — N,N-methylene-bis(acryamide)

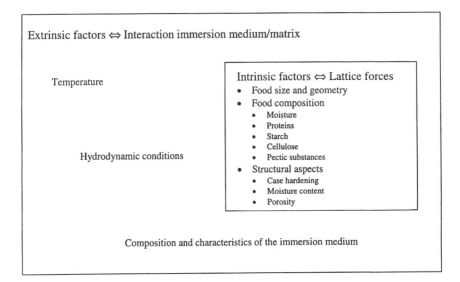

Figure 2 Factors that affect rehydration.

2.1.1 Product chemical composition

The chemical composition of the food product may be expected to affect the rehydration characteristics, as it may control the ability of the material to sorb water. Sopade et al. (1992) reported higher absorption capacities for legumes compared to grains, because of their higher protein content. According to these authors, the main component absorbing water is protein, although other components such as cellulose, starch, and pectin substances may play an important role in the phenomenon. However, Sefa-Dedeh and Stanley (1979) and Hsu et al. (1983) earlier reported contradictory results. Coagulation of proteins during storage of cowpeas was also reported to limit the water diffusion and hydration (Liu et al., 1993).

2.1.2 Predrying treatments

The nature and extent of pretreatments, the type and extent of size reduction and peeling affect the rehydration characteristics (Fellows, 1988). Probably blanching is the most common pretreatment. Careful control of blanching was shown to reduce the extent of changes in the pectic substances and improve water uptake (Plat et al., 1991). Eshtiaghi et al. (1994) studied the effect of a number of pretreatments (blanching, freezing, high pressure, freezing/high pressure, and freezing/blanching) on the rehydration capacity of green beans, carrots, and potatoes (Figure 3). Their results show that in general the pretreatments increase the rehydration capacity, although they are dependent on the food product. The effect of high pressure is negligible, whereas blanching and particularly freezing increase the rehydration capacity. Osmotic dehydration has also been suggested as a predrying treatment, as this is said to contribute to a better texture, retention of aromas, and color

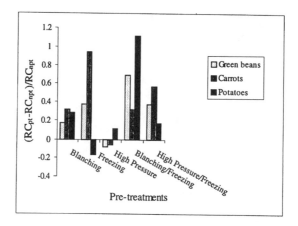

Figure 3 Effect of pretreatments on the rehydration capacity of dried products, expressed in terms of the fractional variation in relation to samples dried without pretreatments (RC_{pt} — rehydration capacity of samples dried after the respective pretreatment; RC_{npt} — rehydration capacity of samples dried without any pretreatment). Values were calculated from data reported by Eshtiaghi et al. (1994).

(Lerici et al., 1998). The sugar concentration was reported by Lerici et al., (1977) to reduce the rehydration ratio, whereas the results of Neumann (1972) and Jayaraman et al. (1990) show an increased rehydration capacity when samples are soaked in sucrose solutions prior to drying. Neumann (1972) suggests this to be the result of shrinkage prevention rather than an effect of sucrose on the water tissues water-holding capacity.

2.1.3 Product formulation

Rehydration characteristics of plant tissue may be improved by incorporation of some compounds in their formulation. Incorporation of NaCl into freeze-dried carrot slices (Curry et al., 1976) and addition of calcium salts have been reported to increase the texture characteristics of the products after rehydration. *In situ* formation of hydrophilic polymers in carrots, onion, and celery were reported to improve the appearance and the textural properties of these products (Schwimmer, 1969). Introduction of hydroxyl groups into the cell wall may minimize hydrogen bonding of polysaccharides that compose it, and contribute to minimum cell deformation, thus improving rehydration. Infiltration of vegetable tissue with edible polyhydroxyl compounds (e.g., sucrose, sorbitol, dextrose, or glycerol) for this purpose has been reported by several authors (Neumann, 1972; Jayaraman et al., 1990).

2.1.4 Drying techniques and conditions

Drying may be accomplished by a variety of techniques such as sun drying, air drying, pneumatic drying, fluidized bed drying, spouted bed drying, explosion puffing, foam drying, microwave, dielectric and radio-frequency drying, and freeze-drying. In general the heat applied during drying reduces the hydration of starch and the elasticity of the cell walls and coagulates proteins to reduce their water holding capacity (Fellows, 1988). Foods that are dried under optimum conditions suffer less damage and rehydrate more rapidly and to a greater extent than poorly dried foods. Rapid drying and high temperatures cause greater changes when compared to moderate rates of drying and lower temperatures (Fellows, 1988).

In spite of the development of new and improved drying techniques, most fruits and vegetables are still air dried on trays or belts, which is the most simple and economical way. Air drying greatly affects the structural rigidity and produces tissue collapse. Shrinkage and case hardening adversely influence rehydration characteristics. Freeze-drying is used to dehydrate high-added-value fruits and vegetables, when visual appearance is an important aspect. Freeze-drying produces highly porous and hygroscopic products that rehydrate almost instantaneously and in general show a high rehydration capacity. Puff explosion can be accomplished by processes such as vacuum drying and also leads to porous products with a non-shrunken structure and thus with good rehydration features. Some new techniques to remove water from foods, such as microwave or radio-frequency drying are still at a developmental stage and improvements are required from a technical point of view.

The physical effects of drying on plant tissues, in what concerns their rehydration characteristics, are mostly linked to shrinkage, case hardening and porosity changes.

Shrinkage. When subjected to drying, plant tissues usually shrink. Shrinkage depends on the type of food and also on the drying technique and conditions. In fibrous products shrinkage is nonisotropic (Madamba et al., 1994). A product that shrinks fast is subjected to great mechanical stresses that produce cracks. A shrivelled product will present a small effective area for mass transfer. Figure 4 shows the dependency of rehydration efficiency on dehydration ratio and shrinkage, based on data collected from Fellows (1988) for a number of foods: cabbage, carrots, onions, peppers, spinach, and tomato. According to these data, low rehydration efficiency is obtained when the product has both intermediate dehydration and shrinkage ratios. Products with a large shrinkage ratio but where that is the result of a high water loss, show good rehydration efficiency. These conclusions are however based on limited data and would need further confirmation.

Case hardening. Case-hardening refers to the formation of a hard impermeable skin at the product surface, due to chemical and physical changes caused by high temperatures. This skin reduces the rate of drying, produces a food with a dry surface and a moist interior (Fellows, 1988) and hinders rehydration. Case-hardening often occurs in air drying. Microwave drying also leads to case-hardening and thus is usually applied in combination

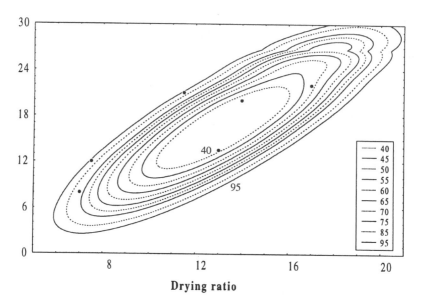

Figure 4 Effect of shrinkage and dehydration ratio on rehydration efficiency (the lines represent isorehydration efficiency, i.e., the pairs of dehydration ratio and shrinkage that lead to the same rehydration efficiency).

with air drying, so that drying rate can be improved but the dried product shows the desired quality. Freeze-drying does not produce case-hardening.

Porosity. Dried plant products are porous solids and porosity may be expected to play an important role on rehydration. The pores within the plant tissue are often interconnected, creating large intracellular networks, and the percentage of the pores connected to the outside is usually referred to as open pore porosity. Porosity changes during dehydration depend on the initial moisture content, composition, size of the material, and also on the type of drying and drying conditions (Saravacos, 1967; Madamba et al., 1994).

According to the studies of Roman et al. (1979) and Lozano et al. (1980, 1983), as air dehydration proceeds, the intracellular water moves toward outer boundaries, the volume of the cells reduces and the pores remain open until a certain level of moisture content is reached. After this stage the pores are split into two categories: open-pores and locked-in pores. This critical moisture content has been reported to be 1.5 $g/g_{dry\ weight}$ for apples (Lozano et al., 1980) and experimental data have confirmed that when this level is reached, a significant change in the three-dimensional arrangement of the cellular tissue occurs. The cellular collapse leads to slow and hindered rehydration.

Different correlations were reported in literature to describe porosity as a function of total volume shrinkage, cell shrinkage, and moisture content (Lozano et al., 1980, 1983; Zogzas et al., 1994) and in general a linear relationship was found between porosity and moisture content: Bakshi and

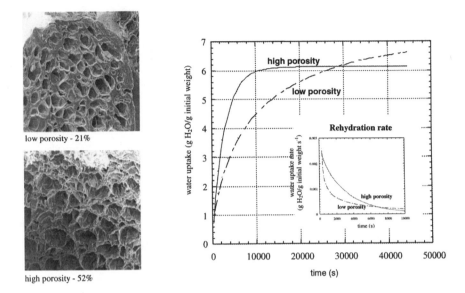

low porosity - 21%

high porosity - 52%

Figure 5   Rehydration of apple samples with different initial porosity.

Singh (1980) reported a linear correlation between moisture content and equivalent spherical radius for rice during parboiling; Thakor et al. (1995) reported a linear relationship between volume and moisture content for canola seeds during hydrothermal processing. Porosity also depends on the type of food. According to Zogzas et al. (1994) potatoes develop an almost negligible porosity during drying (less than 10%), while apples porosity increases significantly, from 20% (for fresh product) to 70% (for dried material).

Figure 5 shows water uptake curves for apples with different initial porosity. Results show that rehydration proceeds at a faster rate in the higher-porosity apples. The rehydration capacity is greater for low-porosity samples, but if one takes into consideration the rehydration ratio, tenderized samples show a better rehydration efficiency (Table 2).

Moisture content.   Water removal induces a number of changes in the plant tissue, particularly at the cell level. The membrane integrity depends on the difference of osmotic pressures across the membrane. Water is drawn due to the chemical potential difference across the plasma membrane. As the moisture content decreases, the removal of water from the cytoplasm induces changes, which are responsible for the failure of cells to uptake water and regain original shape. This change has been explained by Beg et al. (1995) as a result of the fusion of lipid bilayers. These authors suggested that drying also leads to compression of plasma membrane and when loss of moisture exceeds the critical level, the packing ability of the phospholipids and proteins, which are main components of plasma, tonoplast, and membrane, is

Table 2   Rehydration Capacity and Efficiency of Dried Fruits

Food product	Drying technique	Reference	DR (%)	T_{im} (°C)	t_{im}	RC	η (%)
Apple	Air drying						
	No pretreatment	Sinnamon (1968)	85			3[a]	16
	Quick prefreezing		85	20	5 min	3.8[a]	22
	Slow prefreezing		86			4.4[a]	26
	Air drying						
	No pretreatment	Kabbert et al. (1993)	na	na	na	11[b]	na
	Predehydrated with ethanol		na	na	na	30[b]	na
	Air drying						
	High moisture		82	5		5.09[c]	77
			82	20		4.73[c]	71
			82	50		4.66[c]	70
			85	5		5.78[c]	87
	Low moisture	Our data	85	20	5 min to	4.92[c]	72
			85	50	20 h*	5.16[c]	76
	Tenderized		86	5		6.11[c]	92
			86	20		6.10[c]	92
			86	50		6.03[c]	92

	Reference					
Freeze drying	Hammami (1998)	99.98	50	1 to 30 min*	3.5^a	53
Vacuum drying						
No pretreatment	Sinnamon (1968)	86	20	5 min	3.1^a	16
Quick prefreezing		85			4.7^a	31
Slow prefreezing		85			4.1^a	27
Raspberries Freeze drying						
No pretreatment	Medas and Simatos (1971)	na	na	na	4.04 to 4.2^a	na
Presoaking in 0.15% Tween 80					4.5 to 4.6^a	
Strawberries Freeze drying	Hammami (1998)	99.8	50	5 to 20 min*	3.0^a	26
	Von Baumunck & Hondelman (1969)	na	20	0.5 to 5 min	na	35 to 50
	Carballido and Rubio (1970)	na	20	5 to 10 min	5.0^a	na

DR — Drying ratio (% w/w water lost in drying/water in fresh product)

η — rehydration efficiency (% w/w water uptake in rehydration/water lost in drying)

* equilibrium was confirmed

[a] g rehydrated product/g dried product

[b] g water/g dry solids

[c] g water/g dried product

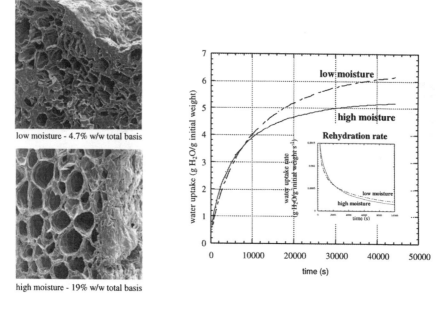

low moisture - 4.7% w/w total basis

high moisture - 19% w/w total basis

Figure 6 Rehydration of apple samples with different initial moisture content

greatly disturbed. However, this critical moisture level where the irreversible wilting occurs depends on the tolerance of the plant tissue to desiccation. Some works cited by Willis and Teixeira (1988) reported that tobacco leaf disks fail to rehydrate when the tissue was reduced to 40% of its fresh weight, while corn roots have lost their viability when 75% of their water has been removed.

Figure 6 shows water uptake curves for apples with different initial moisture content. Rehydration capacity was greater for the low-moisture samples, but if one takes into consideration the dehydration ratio, it can be seem that rehydratation efficiency values are very close (Table 2). Rehydration rates were also similar. Probably, the low-moisture samples were not dried below the critical moisture level where effects on rehydration would become important.

2.1.5 Postdrying procedures

After drying, food products may be subjected to a number of treatments, the most usual being coating, flaking, and tenderizing. Puffing, granulating, and infusing are also common. The general purpose of coating is to avoid oxidation and preserve the appearance of the product. Sometimes, as in the case of freeze-drying, the presence of sugars affects the drying process; therefore coating of the samples is usually carried out after drying. Coating is usually done with hygroscopic substances, such as polysaccharides, which increase the rehydration rate. Flaking has been reported by several authors (Summers, 1994; Roetenberg, 1995) to be an efficient method to increase the rehydration rate and to retain the piece identity. Staal et al. (1990) showed that the rehydration rate of dried potato flakes was greater than that of potato granules; these two products were also found to have a faster rehydration than native or modified starch. Tenderization is a particular case of flaking. In a series of additional processing steps, low-moisture food dices are heated into an elastic form, compressed in a roller mill, and rapidly cooled using CO_2. This rapid cooling results in a flat flake with a crushed cell structure, which has a very high rehydration rate. Tenderized apple pieces are used in refrigerated dough and breakfast cereals (Summers, 1994).

2.2   Extrinsic factors

2.2.1   Composition of the immersion media

Almost all the studies reported in literature used water as the rehydration media; however the characteristics of the immersion media may affect both the rehydration rate and capacity. Some dried fruit are mixed with breakfast cereal and consumed after immersion in milk; thus rehydration in milk is also of interest. Figure 7 shows rehydration curves for apple pieces immersed in water and in whole milk. Rehydration is much slower in milk. In water, hydration of the solid matrix occurs to a significant extent, thus increasing the driving force to mass transfer but decreasing the water movement rate through the fruit piece, as the cellular structure is lost and mass transfer occurs mainly through an enlarged solid matrix. In milk, a layer composed mainly of lipids and micelles surrounds the fruit pieces (Figure 7), creating a major barrier to moisture transfer to the solid matrix, and the fruit piece retains its cellular structure better. Relaxation phenomena of the solid matrix are therefore very important in the moisture uptake process.

Del Valle et al. (1992) found that addition of carbonate salt to the soaking solution reduced water absorption and swelling of beans, probably due to the water binding effect.

2.2.2   Temperature

Temperature increases the rehydration rate but usually does not play a significant effect on rehydration capacity. However, Staal et al. (1990) reported that the rehydration capacity of dried potato granules increased with temperature, whereas for flaked products a maximum was found around 30°C. The effect of temperature on rehydration rate is due both to the decreased viscosity of the immersion medium and to the effects of temperature on the food material structure. Temperature effects are usually described by an Arrhenius-type relationship (Thakor et al., 1995; McLaughlin, 1996; Ilincanu et al., 1997). When a discontinuity is found at a given temperature, a WLF-type equation usually yields good results.

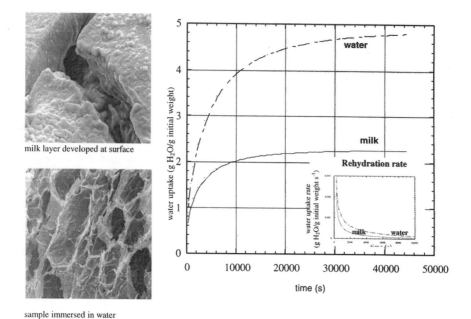

milk layer developed at surface

sample immersed in water

Figure 7   Rehydration of apples immersed in whole milk and in water.

2.2.3   Hydrodynamic conditions

Stirring or agitation of the immersion media improves the rehydration rate. Neubert et al. (1968) reported that shaking or swirling mixtures of dried celery and water during rehydration significantly increases the rate of water absorbed. However, in general the major resistance to water transfer lies within the food material, and thus, except for highly viscous immersion media, agitation is not expected to play a major role.

3   Modelling rehydration processes

3.1   Mechanisms of rehydration

Rehydration is a complex phenomenon that involves different physical mechanisms such as internal diffusion, convection at the surface and within large open pores, and relaxation of the solid matrix (Figure 8). Capillary imbibition is very important at the early stages, leading to an almost instantaneous uptake of water. Tension effects between the liquid and the solid matrix may also be relevant. During rehydration water moves inside the food material and soluble solids are lost and possibly also gained, depending on the immersion medium. Furthermore, not only the food material is nonisotropic, but also its geometry, volume, porosity and structure in general change with moisture uptake. Rigorous mathematical modeling of rehydration is therefore very difficult.

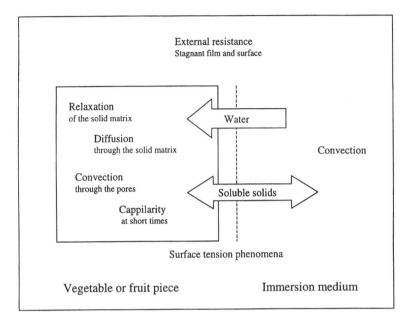

Figure 8   Mechanisms involved in rehydration

Food products may be considered as blends of polymeric materials, and thus many concepts earlier developed for polymer engineering may be adapted to food engineering. Studies developed on polymers, reported by Peppas and Branon-Peppas (1994), have shown that during the water transport at low and medium temperatures, the density of the polymeric network decreases as water penetrates. This effect results in a decrease of the glass transition temperature and also induces an increase of the large molecular chain motion. The structural changes associated to this process include swelling, microcavity formation, and phase transition, all these phenomena being dominated by the structural relaxation phenomena that occur at the solid hydrated matrix. Two limit situations of moisture uptake have been reported by Peppas and Branon-Peppas (1994) and classified as Case I (Fickian) or Case II diffusion, depending on the relative rates of moisture diffusion and of the polymer structure rearrangement to accommodate the water molecules. In Case I, moisture diffusion is much slower, thus controlling rehydration, whereas in Case II, the relaxation at the boundary between the swollen and the unplasticized material becomes the determining factor and in the limit situation where this boundary is very well defined, one has a shrinking core behavior.

From our studies in food products, we have in fact identified two key issues pertaining to rehydration rate: (1) the ability of the solid matrix to hydrate and to go through structural relaxation mechanisms, and (2) the rate of water movements through the pores (that may include convection) and through the hydrated structure (see Section 2.2.1).

3.2   Kinetics modeling

Modeling the kinetics of rehydration of dried fruits and vegetables is indeed scarce. Thus, the discussion presented in this section is based to a large extent on the hydration of dry legumes.

Rigorous mathematical models require a fundamental knowledge of the mechanisms involved in the water and soluble solids transport. They are based on microscopic balances of the material properties, which often lead to differential equations describing their behavior in time and/or space. Fink's 2nd law of diffusion is often applied to model rehydration (wheat — Becker, 1960; Fan et al., 1961; peanuts — Whittier and Young, 1972; sorghum — Fan et al., 1963; soybeans — Hsu et al., 1983; various legumes — Hsu, 1983a, 1983b; corn — Fan et al., 1963, Misra et al., 1981; rice — Banyopadhyay and Roy, 1977, 1978; Bakshi and Singh, 1980, Engels et al., 1986, Hendrickx et al., 1986, 1987; apples — Lomauro et al., 1985; dried potato — McLaughlin, 1996; dried onion — Mizza and Le Maguer, 1980), but because it is hard to believe that the process is controlled by a simple diffusion mechanism, this is to some extent an empirical approach. Additionally, the solutions of Fick's 2nd law used were obtained for a well-defined geometry and specific boundary conditions. Time-dependent conditions might improve the model, but this usually leads to the necessity of numerical solutions, requiring extra computational effort that often is not worthwhile, as the model would still differ from reality. A common approach is to attribute the deviations of the Fickian behavior to a time- or concentration-dependent diffusivity (Hsu, 1983a, 1983b; Zhang et al., 1984; Khan, 1995).

Other models consider that the process at very short times proceeds by capillary action and is then followed by a diffusional mechanism. Because water sorption by capillarity is very fast, this is accounted in the model as an instantaneous gain of water upon immersion, ΔM_0. This model was first applied by Becker (1960) to quantify water uptake during soaking of seeds:

$$\frac{M(t) - M_0}{M_x - M_0} = \Delta M_0 + k\sqrt{t} \tag{1}$$

The model was further used to predict water uptake by corn and sorghum (Fan et al., 1963), by rice (Banyopadhyay and Ghose, 1965; Banyopadhyay and Roy, 1976, 1977, 1978, Lu et al., 1994) and by soybean and pigeonpea grains (Singh and Kulshrestha, 1987).

No studies could be found where relaxation was considered as the limiting step. Increasing interest is being paid to a different approach based on irreversible thermodynamics, but this has not been applied to rehydration so far.

Because of its complexity, rehydration is often described on the basis of empirical models. In spite of their simplicity, empirical models are very useful and their interpretation can provide a valuable insight into the process mechanism. The development of empirical models should proceed according to the following steps (1) collection of experimental data for different operating conditions (e.g., temperature); (2) selection of a mathematical model with an adequate functional form; (3) estimation of the model parameters by fitting the model to the data and statistical analysis (e.g., determination coefficient, normality of residuals, lack of fit, correlation between parameter estimates); (4) analysis of the influence of operating conditions (e.g., temperature) on the model parameters and (5) interpretation of the results of the model.

First-order kinetics, and the Peleg and the Weibull (or Page) models are the most frequent in rehydration modeling. First-order kinetics basically assume that rehydration is controlled by the external surface, what may be the situation in extremely porous products or when case-hardening occurs. This model has been applied for rice (Suzuki et al., 1976; Lu et al., 1994), raw and roasted corn semolina (Bhattacharya, 1995), red kidney beans (Abu-Ghannam and McKenna, 1997a) and freeze-dried blueberries (Lim et al., 1995).

The Peleg model has been proposed by Peleg (1988) to fit experimental data on sorption of water by milk powder and rice. The model has two parameters, k_1 — the rate parameter and k_2 — the equilibrium parameter:

$$M(t) = M_0 + \frac{t}{k_1 + k_2 t} \qquad (2)$$

k_1 is temperature sensitive, and its dependence is often described by an Arrhenius-type relationship. This model has been applied to a large number of foods, such as maize, millet, and sorghum (Sopade et al., 1992), soybean, cowpea, peanuts (Sopade and Obekpa, 1990), red kidney beans (Abu-Ghannam and McKenna, 1997b) and rice (Lu et al., 1994).

The Weibull model describes the process as a sequence of probabilistic events. This model involves two factors: the scale parameter (α) which is related to the process rate, and the shape parameter (β) that depends on the mechanisms controlling the process.

$$\frac{M(t) - M_0}{M_x - M_0} = 1 - e^{-\left(\frac{t}{\alpha}\right)^\beta} \qquad (3)$$

The model is based on a distribution function whose characteristics depend on the shape factor β. This property confers a wide flexibility to the model, allowing it to be applied for a number of situations where the underlying mechanisms are quite different. When the shape parameter equals 1, this model reduces to a first-order kinetics. Cunha et al. (1998) showed that this model could adequately describe rehydration processes controlled by internal diffusion, external convection, and relaxation. The diffusional process was clearly identified when $\beta = 0.6$, whereas processes controlled by an external resistance or relaxation phenomena could not be differentiated from the values of β, that averaged 1 to 1.1 in both cases.

Misra and Brooker (1980) applied the Weibull model to describe rewetting data for shelled yellow corn using experimental data collected from a variety of sources. The shape parameter can be calculated from their data and varies between 0.52 and 0.68, depending on drying conditions, with air humidity decreasing the parameter value. Thakor et al. (1995) used the Weibull model to fit moisture sorption data of canola seeds during soaking and steaming. The equilibrium moisture content attained in steaming was lower compared to the level reached in soaking. The rate parameter was dependent on temperature and followed an Arrhenius-type equation with activation energy of 20 kJ/mol. The shape parameter was in average 0.84 for soaking and 1.17 for steaming, showing that the rehydration mechanisms were different in the two media, with internal mass transfer playing a greater role in soaking. Ilincanu et al. (1997) found that the Weibull model adequately described the experimental data for water uptake and soluble solids losses during rehydration of dried apple pieces. Temperature dependency of the rate parameter was well described by an Arrhenius-type equation with an activation energy of 18.9 and 25.8 kJ/mol, respectively, for samples with an initial moisture content of 19.0 and 4.0% (w/w, total basis). The shape factor for moisture uptake averaged 0.62 and 0.77, respectively for high- and low-moisture samples. Concerning the soluble solids losses, the initial moisture content and temperature did not affect the leaching rates, and the shape factor averaged 0.57. Soluble solids were lost at a higher rate than water was gained. This behavior is probably due to an accumulation of soluble solids in the collapsed cells of surface layers: as water is removed during drying, solutes move from the interior of the food to the surface and evaporation of water causes solute concentration (Fellows, 1988).

3.3 Structural changes modeling

During rehydration, plant tissue undergoes significant structural changes. Qualitative analysis of structural changes is important to study the phenomena involved in the process, but quantitative analysis would bring an interesting potential to model the process kinetics. Earlier attempts to model geometrically the structure of plant tissue were done by Rotstein and Cornish (1978). These authors proposed a simple geometrical model, where cells are cubically truncated spheres occupying the common volume of a cube and a sphere intersecting in such way that they have a common center. However, the Voronoi tessellation method is probably the one that offers the greatest potential. This is a convenient and powerful method to carry out a random discretization of space. Voronoi diagrams generated by a random distribution of points are an important tool of stochastic geometry. According to its definition, a Voronoi diagram for a finite set of data points in a plane is the set of convex polygons, where the sides are the locus of points that are closer to the point enclosed by the polygon than to any other. These polygons, which were first studied by the Russian mathematician G. Voronoi, are irregular and convex. They are also called Dirichlet regions, Thiessen polygons, or Wigner-Seitz cells, but the more descriptive term is "proximal polygons" (Preparata and Shamos, 1985). Although Voronoi diagrams can be defined in three dimensions, a two-dimensional (2-D) approach is usually sufficient. The relevant properties of a 2-D tessellation may be found in Preparata and Shamos (1985).

Voronoi diagrams have been widely used to model structured media: liquids, polycristals, and proteins. They have also been applied to model the fragmentation of the universe. In archeology, Voronoi polygons are used to map the spread of the use of tools in ancient cultures, and in ecology diagrams of field species are used to investigate the effect of overcrowding (Preparata and Shamos, 1985). Literature about the use of Voronoi diagrams to model food microstructure is very scarce (Mattea et al., 1989; Roudot et al., 1990). In the work of Mattea et al. (1989) structural changes during drying of apple tissue were described using the Voronoi tessellation approach. Their model assumes that all the cells behave identically as a function of the moisture content, which is a simplification of reality.

Ilincanu et al. (1998) assessed the applicability of Voronoi diagrams to simulate raw and dried apple structures, using SEM micrographs as the source of information for cell size and cell distribution. The matrix with the coordinates of the centroids was used to generate the Voronoi diagram, rather than having a random distribution. Similarity was in general very good (Figure 9), although two situations were found where the Voronoi diagrams failed to reproduce the original micrographs. The first case occurred when very small cells surrounded very large ones. Because of the characteristics of the Voronoi tessellation algorithm, edges must be equally distanced between two neighboring centroids and this promotes a leveling effect, as shown in Figure 10a. The second case occurred when the cells were concave (see Figure 10b). By definition, the Voronoi diagram is a set of convex polygons; therefore, the situation where the cell is concave (either as a result of processing or during sample preparation for the SEM) cannot be well described.

By changing the location of the vertices of the Voronoi polygons, one can model structural changes at the cellular level. Thus, Voronoi diagrams show an interesting potential to simulate the swelling and relaxation phenomena during the rehydration of dried plant tissue.

4   Conclusions

The effects of drying conditions on rehydration characteristics should be further explored, so that drying–rehydration can be analyzed in an integrated

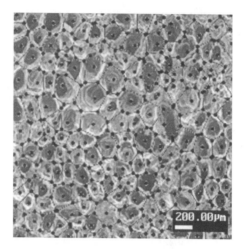

Figure 9 Mathematical modeling of raw apple tissue using the Voronoi tessellation approach.

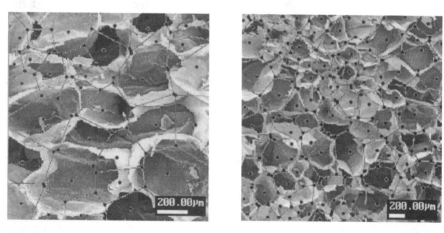

Figure 10 Apple tissue structures that cannot be modeled by the Voronoi tessellation approach: (a) cells with different sizes; (b) concave cells.

way. Mathematical modeling of rehydration kinetics should be improved as not only is this essential for process optimization, but also because mathematical modeling may be a tool for further understanding the phenomena involved. A model considering both the relaxation of the solid matrix, thus including structural changes, and the water movement through the hydrated matrix and through the pores appears to be a suitable alternative to the empirical models currently used. Finally, the effects of drying/rehydration on the quality characteristics of the reconstituted product should be assessed and quantified.

List of symbols

DR   dehydration ratio (% w/w)
K   lumped parameter in Equation (1)
k_1, k_2 parameters in Equation (2)
M(t)   moisture content at time t
M_0 initial moisture content
M_∞ equilibrium moisture content
ΔM_0 instant gain in initial moisture content due to immersion in liquid
RC   rehydration capacity
RC_{npt} rehydration capacity of samples without pretreatment
RC_{pt} rehydration capacity of samples pretreated
t   time
tim   immersion time (min or h)
α rate parameter of the Weibull model (Equation (3))
β shape parameter of the Weibull model (Equation (3))
η efficiency of the rehydration process (% w/w)

References

Abu-Ghannam, N. and McKenna, B. (1997a). *Journal of Food Science*, 62 (3), 520.

Abu-Ghannam, N. and McKenna, B. (1997b). *Journal of Food Engineering*, 32 (4), 391.

Agbo, G.N., Hosfield, G.L., Uebersax, M.A., and Klomparens, K. (1987). *Food Microstructure*, 6, 91.

Baek, H.H., Kim, D.M., and Kim, K.H. (1989). *Korean Journal of Food Science and Technology*, 21, 145.

Bakshi, A.S. and Singh, R.P. (1980). *Journal of Food Science*, 45 (5), 1387.

Banyopadhyay, B. and Ghose, T.K. (1965). *Indian Journal of Technology*, 3, 360.

Banyopadhyay, S. and Roy, N.C. (1976). *Indian Journal of Technology*, 14, 27.

Banyopadhyay, S. and Roy, N.C. (1977). *Journal of Food Science and Technology*, 14 (3), 95.

Banyopadhyay, S. and Roy, N.C. (1978). *Journal of Food Technology*, 13, 91.

Becker, H.A. (1960). *Cereal Chemistry*, 37, 309.

Beg, M.A.A., Nazar, M., Anwar, M., and Ehteshamuddin, A.F.M. (1995). *Pakistan Journal of Scientific and Industrial Research*, 38 (3–4), 159.

Bhattacharya, S. (1995). *Journal of Food Engineering*, 25 (4), 21.

Carballido, A. and Rubio, J.L. (1970). *Anales de Bromatologia*. 22, 229. In Spanish.

Cunha, L.M., Oliveira, F.A.R., Ilincanu, L.A., and Drumond, M.C. (1998). In: *Proceedings of the Third Workshop of the Copernicus project Process Optimisation and Minimal Processing of Foods*, Vol. 3 — Drying, Oliveira, J.C. and Oliveira, F.A.R. (Eds.). Escola Superior de Biotecnologia, Porto, Portugal.

Curry, J.C., Burns, E.E., and Heidelbaugh, N.D. (1976). *Journal of Food Science*, 41 (1), 176.

Del Valle, J.M., Stanley, D.W., and Bourne, M.C. (1992). *Journal of Food Processing and Preservation*, 16, 75.

Deshpande, S.S. and Cheryan, M. (1986). *Journal of Food Science*, 51, 1218.

Engels, C., Hendrickx, M., De Samblanx, S., De Gryze, I., and Tobback, P. (1986). *Journal of Food Engineering*, 5, 55.

Eshtiaghi, M.N., Stute, R., and Knorr, D. (1994). *Journal of Food Science*, 59 (6), 1168.

Fan, L., Chung, D.S., and Shellenberger, J.A. (1961). *Cereal Chemistry*, 38, 540.

Fan, L., Chu, P., and Shellenberger, J.A. (1963). *Cereal Chemistry,* 40, 303.

Fang, T.T., Foottrakul, P., and Luh, B.S. (1971). *Journal of Food Science,* 36, 1044.

Fellows, P. (1988). *Food Processing Technology,* Ellis Horwood, Ltd. Chichester, U.K.

Garcia-Reverter J., Bourne, M.C., and Mulet, A. (1994). *Journal of Food Science,* 59 (6), 1181.

Hammami, C. (1998). *Lyophilisation des produits végétaux: maîtrise de la qualité en relation avec les paramètres du procédé,* PhD dissertation, INRA-Grignon, France. In French.

Hendrickx, M., Engels, C., and Tobback, P. (1986). *Journal of Food Engineering,* 5, 269.

Hendrickx, M., Engels, C., and Tobback, P. (1987). *Journal of Food Engineering,* 6, 187.

Hsu, K.H. (1983a). *Journal of Food Science,* 48, 618.

Hsu, K.H. (1983b.) *Journal of Food Science,* 48, 1364.

Hsu, K.H., Kim, C.J., and Wilson, L.A. (1983). *Cereal Chemistry,* 60 (3), 208

Ilincanu, L.A., Oliveira, F.A.R., Drumond, M.C., Machado, M.F., and Gekas, V. (1997). In: *Proceedings of the First Workshop of the Copernicus project Process Optimisation and Minimal Processing of Foods,* Vol. 3 — Drying, Oliveira, J.C. and Oliveira, F.A.R. (Eds.), p. 64.

Ilincanu, L.A., Oliveira, F.A.R., Drumond, M.C., Mattea, M., and Gekas, V. (1998). In: *Proceedings of the Second Workshop of the Copernicus project Process Optimisation and Minimal Processing of Foods,* Vol. 3 — Drying, Oliveira, J.C. and Oliveira, F.A.R. (Eds.), p. 106.

Jayaraman, K.S., Das Gupta, D.K., and Babu Rao, N. (1990). *International Journal of Food Science and Technology,* 25, 47.

Kabbert, R., Herrmuth, K., and Kunzek, H. (1993). *Zeitschrift Lebensmittel Untersuchung und Forschung,* 196 (3), 219.

Khan, C.J. (1995). *Foods and Biotechnology,* 4(1), 47–54.

Le Loch-Bonazzi, C., Wolff, É., and Gibbert, H. (1992). *Lebensmittel Wissenchaft und Technologie,* 28, 334.

Lee, J.W., Lee, S.K., and Do, J.H. (1995). *Korean Journal of Food Science and Technology,* 27 (5), 724.

Lerici, C.R., Pepe, M., and Pinnavaia, G. (1977). *Industria Conserve,* 52 (2), 125.

Lerici, C.R., Mastrocola, D., and Nicoli, M.C. (1998). *Acta Alimentare Polonica,* 14, 35.

Li Shing Tat, B. and Jelen, P. (1987). *Food Microstructure,* 6 (2), 135.

Lim, L.T., Tang, J., and He, J. (1995). *Journal of Food Science,* 60 (4), 810.

Liu, K., Hung, Y.C., and Philips, R.D. (1993). *Food Structure,* 12 (1), 51.

Lomauro, C. J., Bakshi, A.S., and Labuza, T.P. (1985). *Journal of Food Science,* 50 (2), 397.

Lozano, J.E., Rotstein, E., and Urbicain, M.J. (1980). *Journal of Food Science,* 45 (5), 1403.

Lozano, J.E., Rotstein, E., and Urbicain, M.J. (1983). *Journal of Food Science,* 48 (6), 1497.

Lu, R., Siebenmorgen, T.J., and Archer, T.R. (1994). *Journal of Food Processing Engineering,* 17, 141.

Madamba, P.S., Driscoll, R.H., and Buckle, K.A. (1994). *Journal of Food Engineering,* 23, 309.

Mattea, M., Urbicain, M.J., and Rotstein, E. (1989). *Chemical Engineering Science,* 44 (12), 2853.

McLaughlin, C.P. (1996). *Dissertation Abstracts International,* 57 (4), 1355.

Medas, M. and Simatos, D. (1971). *Revue Générale du Froid,* 7, p. 695.

Misra, M.K. and Brooker, D.B. (1980). *Transactions of the ASAE,* 23 (5), 1254.

Misra, M.K., Young, J.H., and Hamman, O.D. (1981). *Transactions of the ASAE,* 24 (3), 751.

Mizza, G. and Le Maguer, M. (1980). *Journal of Food Technology,* 15 (2), 181.

Neubert, A.M., Wilson III, C.W., and Miller, W.H. (1968). *Food Technology,* 22 (10), 94.

Neumann, H.J. (1972). *Journal of Food Science*, 37 (2), 437.

Peleg, M. (1988). *Journal of Food Science*, 53 (4), 1216.

Peppas, N.A. and Brannon-Peppas, L. (1994). *Journal of Food Engineering*, 22, 189.

Plat, D., Shalom, N.B., and Levi, A. (1991). *Food Chemistry*, 39 (1), 1.

Preparata, F.P. and Shamos, M.I. (1985). *Computational Geometry*, Springer-Verlag, New York.

Roetenberg, K. (1995). *Cereal Foods World*, 40 (6), 427.

Roman, G.N., Rotstein, E., and Urbicain, M.J. (1979). *Journal of Food Science*, 44 (1), 193.

Rotstein, E. and Cornish, A.R.H. (1978). *Journal of Food Science*, 43 (3), 926.

Roudot, A.C., Duprat, F., and Pietri, E. (1990). *Food Structure*, 9, 215.

Saravacos, G.D. (1967). *Journal of Food Science*, 32 (1), 81.

Schwimmer, S. (1969). *Food Technology*, 23 (7), 115.

Sefa-Dedeh, S. and Stanley, D.W. (1979). *Cereal Chemistry*, 56, 379.

Shomer, I., Paster, N., Lindner, P., and Vasiliver, R. (1990). *Food Structure*, 9, 139.

Singh, B.P.N. and Kulshrestha, S.P. (1987). *Journal of Food Science*, 52 (6), 1538.

Sinnamon, H.I. (1968). Effects of prior freezing on dehydration and rehydration of apple half segments, *Food Technology*, 21, 1033–1035.

Sopade, P.A. and Obekpa, J.A. (1990). *Journal of Food Science*, 55 (4), 1084.

Sopade, P.A., Ajisegiri, E.S., and Badau, M.H. (1992). *Journal of Food Engineering*, 15, 269.

Staal, G., Noz, R.J., and van Laarhoven, G.J.M. (1990). *Voedingsmiddelentechnologie*, 23 (16/17), 15

Sterling, C. and Shimazu, F. (1961). *Journal of Food Science*, 26 (5), p. 479.

Summers, S. (1994). *Cereal Foods World*, 39 (10), 746.

Suzuki, K., Kubota, K., Omichi, M., and Hosaka, H. (1976). *Journal of Food Science*, 41, 1180.

Thakor, N.J., Sokhansanj, S., Patil, R.T., and Deshpande, S.D. (1995). *Journal of Food Engineering*, 18, 233.

Von Baumunck, E. and Hondelman, W. (1969). *Industrial Obst und Gemuse Verwertung*, 128.

Whittier, T.B. and Young, J.H. (1972). *Transactions of the ASAE*, 15 (1), 163.

Willis, C.A. and Teixeira, A.A. (1988). *Journal of Food Science*, 53 (1), 111.

Zhang, T-Y., Bakshi, A.S., Gustafson, R.J., and Lund, D.B. (1984). *Journal of Food Science*, 49, 246.

Zogzas, N.P., Maroulis, Z.B., and Marinos-Kouris, D. (1994). *Drying Technology*, 12 (7), 1653.

chapter thirteen

Recent advances in the drying of apples under variable process conditions

Dariusz Piotrowski and Andrzej Lenart

Contents

Summary 229
1 Introduction 230
2 Drying kinetics 232
   2.1 Drying under constant conditions 232
   2.2 Drying under variable conditions...234
3 Properties of the dried material 241
4 Conclusions..243
References..246

Summary

The introduction of sudden changes in the drying conditions during the drying process can assist the optimization of both economical and product quality factors. A still limited number of publications in this field justifies further research. This chapter studies the drying kinetics of apples under constant and variable drying conditions in a cabinet drier. Processes are compared on the basis of the temperature and water content evolution in the samples, drying rate curves, time required to reach specific water content levels and selected physical properties of the final product (density, shrinkage, resistance to compression, rehydration and leakage after rehydration). Experiments were performed in the temperature range of 50 to 90°C and velocity range 0.7 to 3 m/s. For both constant and variable drying conditions, the effect of changing the air temperature was greater than that of changing the air velocity. Thus, establishing variable temperature processes for improving drying efficiency and product quality seems to be more interesting than defining variable velocity conditions.

1 Introduction

Assessing the effect of changes on the drying parameters on the process, energy consumption, and final product properties can contribute to achieving economic efficiency, particularly for high-quality products (Strumillo and Adamiec, 1996). It is particularly attractive to analyze easy-to-implement changes in the parameters of the inlet air: temperature, velocity and humidity, and their application to minimize drying time or energy costs while striving for beneficial product characteristics. By introducing a sudden change of the air temperature set point at a specific process time, a progressive passage from low-temperature to high-temperature kinetics has been observed (Lebert et al., 1992). Drying of apples could be divided into steps by introducing sudden changes in the inlet air parameters. The kinetics of the drying process can be described by the recorded evolution of the average sample temperature or average water content as a function of time or of the drying rate as a function of water content (Pabis, 1982; Strumillo, 1983).

Among the basic parameters that influence the drying kinetics, the air flow velocity and temperature are the more frequently quoted (Lee and Pyun, 1993; Rocha et al., 1993; Soporonnarit et al., 1997).

A higher temperature of the air drying medium in a first stage has been recommended for fruit puree (Molys, 1986) and coconuts (Sankat et al., 1992). The air velocity has been attributed a significant influence in the first stage of drying, due to its role in increasing the intensity of heat and mass transfer (Lewicki, 1990). Experiments involving a step change of a chosen parameter of the drying medium have shown significant shifts of the drying curves, regardless of the water content of the materials at the time of introducing the step change (Laguerre et al., 1992). The time required to reach steady surface and center temperatures of the samples depends on the moment of introduction of the step change and on the temperature after the change (Piotrowski and Lenart, 1996a). When causing a rectangular pulse during the second hour of drying, the drying curves were visibly different from the standard process even in the latter part of the process (Piotrowski and Lenart, 1997a). Drying of plant tissue can be divided in stages by introducing interstage tempering or resting times. In this method, two or more stages are divided by pulses with lowered temperatures to ambient conditions (Farkas and Rendik, 1996; Hayashi, 1996; Pan et al., 1996).

The physical properties of the materials change during drying (Zogzas et al., 1994; Sjhölm and Gekas, 1995) as a function of process parameters (Rocha et al., 1993; Ratti, 1994; Wang and Brennan, 1995) or as a combined result of different pretreatments (Barbanti et al., 1991; Lenart and Iwaniuk, 1993; Rajchert and Lewicki, 1997). Several properties are relevant for quantifying product quality attributes (Bhardwaj and Lal Kaushal, 1990; Iciek and Krysiak, 1995; Szentmarjay et al., 1996).

The final porosity of dried celery, chosen to evaluate plant tissue quality, was higher for low processing temperatures (5°C) than for higher ones (60°C), although volume shrinkage was similar. While drying at low constant temperatures is too slow, a process using high-temperature pulses for short times should keep the quality improvement while maintaining reasonable drying times (Karathanos et al., 1993). An attempt to optimize carotene retention in dried carrots led to a three-stage process: 92°C for 28 min, followed by 56°C for 160 min, and 60°C for 100 min (Domagala et al., 1993).

Better color and flavor retention of apples have been reported for two-stage drying processes: 102°C for 2 h followed by 58°C for 3.5 h compared to drying at the constant temperatures of 94°C for 5.5 h (Bains et al., 1989). The color of apple slices was also improved using a linearly decreasing temperature profile from 100 to 70°C in 2 h (Özilgen et al., 1995). This was attributed to thermal inactivation of the enzymes involved in the first stage.

Comparison of the sensory properties of peas and French beans dried at a constant temperature of 110°C for 8 to 10 h with those obtained in a three-stage process — 110°C for 1.5 h followed by 65°C for 1.5 h, and then 45°C for 5h, also indicated a significant superiority of the option with variable conditions (Sinesio et al., 1995). Better results were obtained for color, off flavor, fibrousness, and texture homogeneity. Notwithstanding, freeze-dried samples were of a superior quality in all respects.

Extended analysis of the optimization of drying conditions are not very common in scientific and technical literature. The optimization approach requires the definition of search regions for the control variables, under several constraints, to achieve the objective defined. An example is given by Lee and Pyun (1993) for minimizing energy consumption in radish drying. The parameters that affect the drying rate, energy consumption, and product quality can be related by mathematical models to the relevant response factors, as shown by Soponronnarit et al., (1997), who considered inlet air temperature and flow rate and fraction of exit air recycled as independent variables. Madamba (1997) studied as response factors the final water content and rehydration ratio, applying a factorial design at three levels with four factors (drying temperature, relative humidity, air velocity, and sample thickness), thus determining the optimum conditions for drying garlic.

The limited number of publications reporting results of drying experiments under variable conditions justify research in this field (Laguerre et al., 1992; Jayaraman and Das Gupta, 1992; Kudra and Mujumdar, 1995). It is interesting in a first approach to analyze the temperature and water evolution in the samples and the quality characteristics of the final product when applying specific sudden changes (steps, pulses).

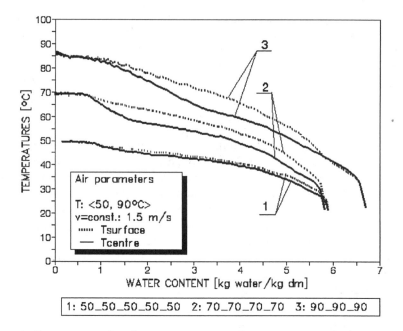

Figure 1 Temperature of apples vs. water content during convective drying at constant temperatures (Piotrowski, 1998). Printed with kind permission from Marcel Dekker Inc. Each number in the bottom legend indicates the air temperature for 1 h of process.

2 Drying Kinetics

2.1 Drying under constant conditions

A set of results concerning the drying of apple slices is discussed here.

Obviously, the selection of a constant temperature of inflowing air between 50 and 90°C will have a major impact on the sample temperature. Figure 1 shows that it is possible to divide the temperature evolution in three stages, as a function of water content: fast increase, medium increase, and slow increase (tending to zero). In the second stage, the temperature increase is approximately linear — the data presented were fitted to straight lines with correlation coefficients R ranging between 0.96 and 0.99 (Piotrowski and Lenart, 1998). The temperature histories, of which Figure 2 shows an example, indicated a slow and gradual increase of the temperature at the center, stabilizing slightly below the air temperature.

The advantageous effect of higher drying temperatures in decreasing the processing times required to reach specified levels of water content was observed. The water contents considered were typical of intermediate moisture foods (35%) and of shelf-stable dried products (17%). Results are shown in Table 1.

The higher temperature of 90°C reduced the times required to reach 35% and 17% water content by 62%, compared to the lowest temperature (50°C).

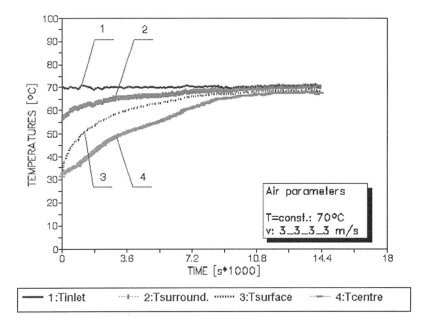

Figure 2 Kinetics of temperature evolution during convective drying of apples under constant temperature and velocity.

Table 1 Effect of Constant and Variable Temperatures on the Time Required to Obtain the Specified Levels of Water Content (Piotrowski, 1995)

No.	Experiments Air temperature [°C] * v = const. = 1.5 m/s	Time required to obtain water content 35% [h.min.]	17% [h.min.]
1	50_70_70_70_70_70	3 h. 08 min.	3 h. 42 min.
2	70_50_70_70_70_70	2 h. 58 min.	3 h. 34 min.
3	70_70 l 50_50_50_50	2 h. 58 min.	3 h. 52 min.
1	50_50_50_50_50	4 h. 56 min.	6 h. 13 min.
2	60_60_60_60_60	3 h. 53 min.	4 h. 45 min.
3	70_70_70_70	2 h. 42 min.	3 h. 17 min.
4	80_80_80_80	2 h. 27 min.	2 h. 56 min.
5	90_90_90	1 h. 59 min.	2 h. 22 min.
1	50_70_70_70_70_70	3 h. 08 min.	3 h. 42 min.
2	70_50_70_70_70_70	2 h. 58 min.	3 h. 34 min.
3	70_70 l 50_50_50_50	2 h. 58 min.	3 h. 52 min.

* Each number between dashes indicates the inlet air temperature during 1 h. Two numbers separated by a stroke (e.g., 70 l 50) mean that the air temperature changed from the first to the second in that hour.

Table 2   Effect of Constant and Variable Velocities on the Time Required to Obtain the Specified Levels of Water Content (Piotrowski, 1995)

	Experiments	Time required to obtain water content	
No.	Air velocity [m/s]* T = const. = 70°C	35% [h. min.]	17% [h. min.]
1	0.7_1.5_1.5_1.5_1.5	3 h. 08 min.	3 h. 45 min.
2	1.5_0.7_1.5_1.5_1.5	2 h. 54 min.	3 h. 31 min.
3	0.7_0.7ǀ1.5_1.5_1.5_1.5	3 h. 09 min.	3 h. 45 min.
4	1.5_1.5ǀ0.7_0.7_0.7_0.7	2 h .52 min.	3 h. 35 min.
1	0.7_0.7_0.7_0.7_0.7	3 h. 45 min.	4 h. 24 min.
2	1.5_1.5_1.5_1.5	2 h. 20 min.	2 h. 52 min.
3	3_3_3	2 h. 11 min.	2 h. 49 min.
1	3_1.5_1.5_1.5	2 h. 26 min.	3 h. 07 min.
2	1.5_3ǀ1.5_1.5_1.5	2 h. 24 min.	2 h. 59 min.
3	3_3ǀ1.5_1.5_1.5	2 h. 08 min.	2 h. 42 min.
4	1.5_1.5ǀ3_3_3	2 h. 30 min.	3 h. 00 min.

* Each number between dashes indicates the inlet air velocity during 1 h. Two numbers separated by a stroke mean that the velocity changed from the first to the second in that hour.

At 1.5 m/s of air velocity, decreasing the process temperature from 70 to 60°C increased the drying times required to reach the specific water content (by 45%) more than the reduction of drying times achieved by increasing the temperature from 70 to 80°C (10%) (Piotrowski and Lenart, 1996b).

Figure 3 shows that the maximum drying rates were attained usually after just a few minutes. The maximum drying rate was higher the higher the temperature.

The advantageous effect of higher air velocity on the reduction of drying times was also noticed. The maximum velocity of 3 m/s decreased the times required to reach water contents of 35% and 17% by 40% and 33%, respectively, compared to the minimum velocity (0.7 m/s). These results can be seen in Table 2. Reducing the velocity from 1.5 to 0.7 m/s had a greater effect on increasing the drying times (39–33%) than increasing the velocity from 1.5 to 3 m/s had on decreasing them (16–10%). The maximum drying rates occurred at an earlier time the lower the velocity, as shown in Figure 4.

2.2   Drying under variable conditions

Sudden steps or pulses of ±20°C were introduced in the inlet air temperature of 70°C, taken as reference. This caused an abrupt change of the sample temperature. The time required to reach steady surface and center temperatures depends on the temperature after the change and on the moment of its introduction. It was observed that negative temperature changes had a faster subsequent stabilization than positive ones (Piotrowski and Lenart, 1996a).

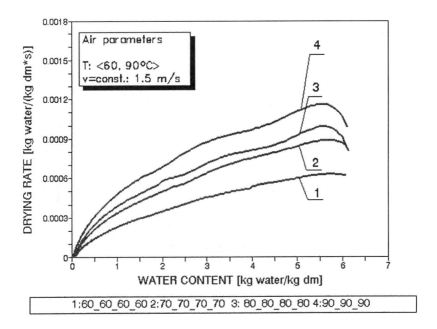

Figure 3 Rate of water loss in apples during convective drying at constant temperatures (Piotrowski and Lenart, 1997). Printed with kind permission from Sheffield Academic Press Ltd.

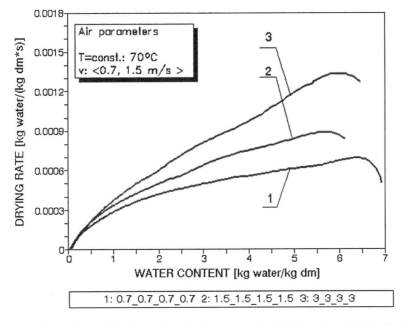

Figure 4 Rate of water loss in apples during convective drying at constant velocities.

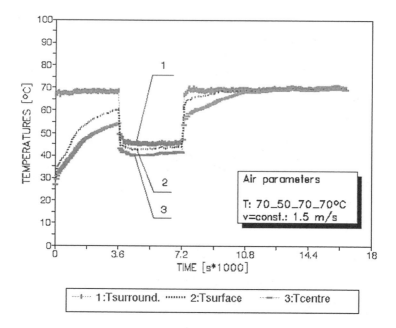

Figure 5 Kinetics of temperature evolution during convective drying of apples under variable velocities (air temperature rectangular pulse during the second hour) (Piotrowski, 1998). Printed with kind permission from Marcel Dekker Inc.

Figure 5 shows the effect of a rectangular pulse (–20°C followed by +20°C). The pulse was applied after 1 h of process, for a whole second hour. Temperature stabilization throughout the sample was fast in the second hour, due to the pulse. The times required to stabilize the temperatures after the pulse were overall similar to those of the experiment at constant temperature.

Figure 6 shows the effect of a sudden drop or increase of the temperature after one hour of process. Temperature evolution at the center of the samples when introducing the change was fast, and the remaining temperature histories were similar to the standard process. It can be seen that when starting at 50°C, after 1 hour the water content was 4 kg water/kg dry matter and when starting at 90°C after 1 hour the water content was 2.4 kg water/kg dry matter.

Similar sudden changes in the air velocity (at constant temperature) are shown in Figure 7. The standard reference is a constant air velocity of 1.5 m/s and the two starting options were 0.7 and 3 m/s (that is, around half and twice the standard). The difference in the sample center temperature at the time of introducing the step, that is, after the first process hour, amounted to 12°C. However, when looking at the evaluations of the sample center temperature as a function of water content, both curves were very similar (Figure 8).

A rectangular negative pulse for the second hour of the process was applied for the air velocity, by reducing from 1.5 to 0.7 m/s after one hour and increasing again to 1.5 m/s at the second hour (shown in Figure 9). It

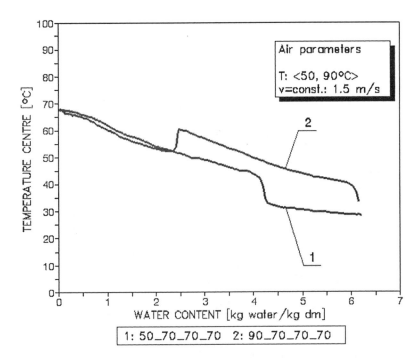

Figure 6 Temperature of apples vs. water content during convective drying under variable temperature (air temperature step change after the first hour) (Piotrowski, 1998). Printed with kind permission from Marcel Dekker Inc.

can be seen that the effect is much lower than that of a temperature pulse (Piotrowski and Lenart, 1997a).

Higher or lower air temperature pulses introduced during the second hour cause a visible difference in the drying curves (Figure 11). However, the effect of a sudden drop after 1 hour is more significant (Figure 10). Figure 12 shows an effect similar to that of Figure 10, obtained by causing a step change in the air velocity at the moment when the relative water content reached 0.6.

The drying curves obtained with a pulse in the air velocity during the second hour shown in Figure 13, indicated that although there were differences, they were lower than those caused by temperature pulses (Figure 11), as already observed in Figures 8 and 6, respectively.

It is interesting to analyze the effect of the variable conditions imposed on the times required to reach the specific values of water content (35% and 17%), shown in Tables 1 and 2. Using a rectangular pulse from 70 to 50°C during the second hour (rising then again to 70°C) decreased the drying times compared to starting at 50°C and rising to 70°C after 1 hour, though increasing in relation to constant processing at 70°C (by about 16 min). The greatest increase in drying time compared to the standard of 70°C constant

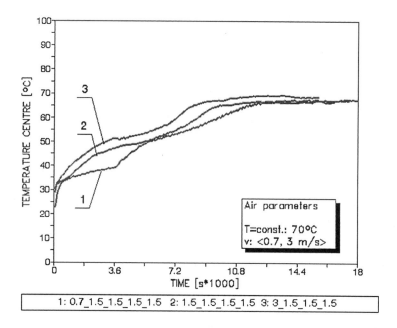

Figure 7 Temperature of apples vs. time during convective drying under variable velocities (air velocity step change after the first hour).

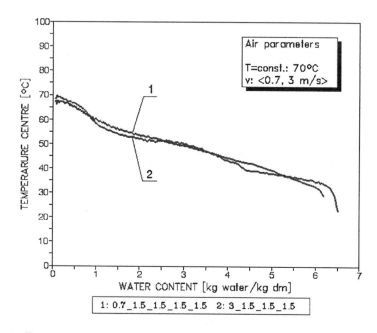

Figure 8 Temperature of apples vs. water content during convective drying under variable velocities (air velocity step change after the first hour, when reaching the water content of 3 kg water/kg dry matter) (Piotrowski, 1998). Printed with kind permission from Marcel Dekker Inc.

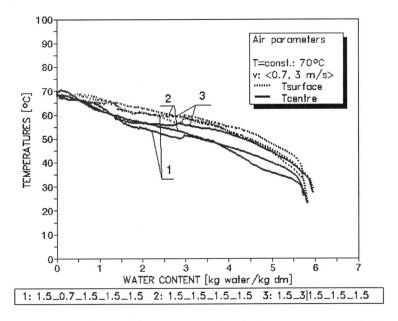

Figure 9 Surface and center temperatures of apples vs. water content during convective drying under variable velocities (rectangular pulses after the first hour, when reaching the water content of 3 kg water/kg dry matter) (Piotrowski and Lenart, 1997a). Printed with kind permission from Sheffield Academic Press Ltd.

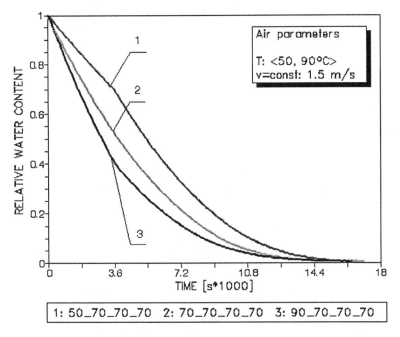

Figure 10 Influence of an air temperature step change after the first hour on water loss from apples.

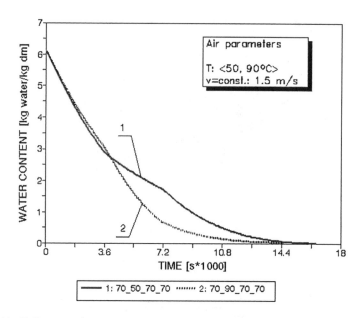

Figure 11 Influence of an air temperature rectangular pulse introduced during the second hour on water loss from apples (Piotrowski and Lenart, 1997a). Printed with kind permission from Sheffield Academic Press Ltd.

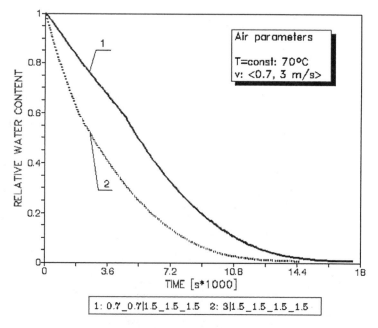

Figure 12 Influence of an air velocity step change introduced after a relative water decrease below 0.6 on water loss from apples (Piotrowski, 1998). Printed with kind permission from Marcel Dekker Inc.

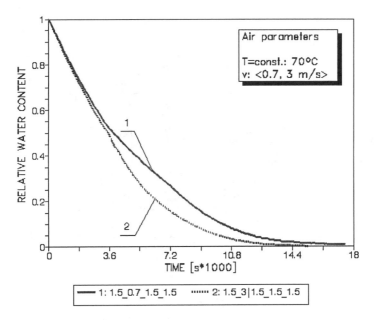

Figure 13   Influence of an air velocity rectangular pulse introduced after the first hour on water loss from apples convective drying at constant temperatures (Piotrowski and Lenart, 1997). Printed with kind permission from Sheffield Academic Press Ltd.

temperature was caused by the sudden step from 70 to 50°C after 1 hour, but only when taking 17% as target.

Similar observations can be made for the changes introduced in the air velocity, comparing to 1.5 m/s constant conditions as standard. Thus, drying times can be reduced by starting at 3 m/s for a sufficient time, before lowering to 1.5 m/s. Using a pulse of 3 m/s during the second hour actually increased the drying times, which is an opposite effect to what would be expected at first thought.

3   Properties of the dried material

Figure 14 shows that for constant process temperatures density decreased with increasing temperature while rehydration increased. The effect on leakage after rehydration was not so clear (see Figure 14). In relation to the effect of constant air velocities, a decrease of the rehydration and of the resistance to compression with increasing air velocity were the more clear effects (see Figures 15 and 16) (Piotrowski, 1995).

Figure 17 shows the effect of variable temperature (between 70 and 90°C), compared to constant temperatures, on density and rehydration. Thus, processing at 70°C with a pulse of 90°C during the second hour allowed to obtain the same rehydration as processing always at 90°C, while only the constant 90°C process gave a significantly lower density. Similar comparisons can be made in Figure 18 for variable air velocity conditions. In this

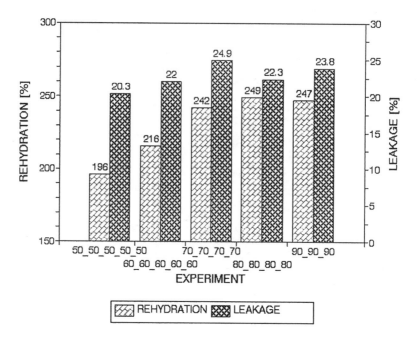

Figure 14 Effect of air constant temperatures on rehydration and leakage after rehydration of dried apples.

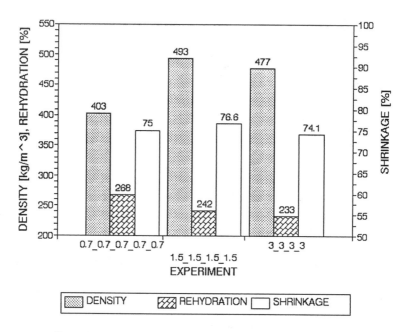

Figure 15 Effect of air constant velocities on density, rehydration, and shrinkage of dried apples.

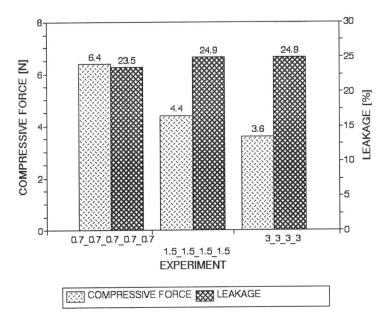

Figure 16 Effect of air constant velocities on compressive force and leakage of dried apples.

case, the constant air velocities had more leakage after rehydration than any of the variable conditions tested. The resistance to compression of a process initiated with 3 m/s and lowered to 1.5 after 1 hour was similar to that obtained at a constant 1.5 m/s velocity (Piotrowski and Lenart, 1997b).

Some sensory properties were selected for assessing the effects on organoleptic quality, as indicated in Tables 3 and 4. Clear differences were observed for color, appearance, and shape retention. The best overall quality was obtained at a constant temperature of 60°C: further temperature increase lowered the judgments for the appearance, shape retention, color, firmness, and flavor (Piotrowski and Lenart, 1997b). Compared to the standard of constant temperature of 70°C, positive and negative pulses caused a decrease of quality (appearance, shape retention, color, firmness, and taste).

In relation to the variable air velocity the best conditions were a constant air velocity of 1.5 m/s. Introducing sequences with lower velocity caused a decrease of quality, but less than those resulting from temperature changes.

4 Conclusions

The effect of constant and variable air temperatures and velocity in a cabinet dryer was detailed by analyzing drying curves, temperature evolution in the samples, and the drying rate curves. The effect of sudden changes, either a step caused after 1 hour of drying, or a pulse, lasting for a specific time, were considered. The analysis was completed by comparing the drying times

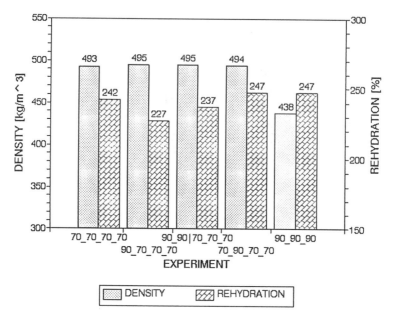

Figure 17 Influence of air temperature changes on density and rehydration of dried apples.

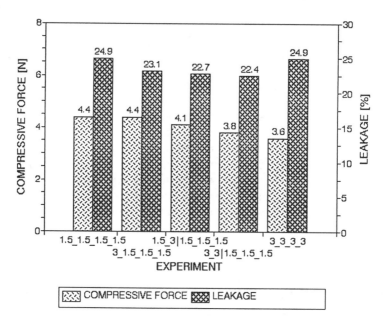

Figure 18 Influence of air velocity changes on compressive force and leakage after rehydration of dried apples.

Table 3 The Influence of Constant and Variable Temperatures on the Organoleptic Assessment of Selected Properties of Dried Apples (Piotrowski, 1995)

	Experiments	Organoleptic properties of dried apples					
No.	Air temperature [°C] v = const. = 1.5 m/s	Appearance	Color	Flavor	Firmness	Taste	Total judgment
1	50_70_70_70_70_70	4.4	4.7	3.6	4.3	4.3	21.3
2	70_50_70_70_70_70	3	2.7	3.1	3.3	3.6	15.7
1	50_50_50_50_50	3.4	3.9	3.7	3.3	3.9	18.2
2	60_60_60_60_60	4.7	4.9	4.7	4.3	4.6	23.2
3	70_70_70_70_70	4.3	4.7	3.6	4.1	4.6	21.3
4	80_80_80_80	4	4	3.7	4.1	3.9	19.7
5	90_90_90	3.9	3.7	3.1	4	4.4	19.1
1	90_70_70_70	4	4.4	3.7	4	3.9	20
2	70_90l70_70_70	3	2.3	4.6	3.3	4	17.2
3	70_90_70_70	3.7	3.6	3.3	3.9	3.9	18.4

Table 4 Influence of Constant and Variable Velocities on the Organoleptic Assessment of Selected Properties of Dried Apples (Piotrowski, 1995)

	Experiments	Organoleptic properties of dried apples					
No.	Air velocity [m/s] T = const. = 70°C	Appearance	Color	Flavor	Firmness	Taste	Total judgment
1	0.7_1.5_1.5_1.5_1.5	4.1	4.3	3.7	3.7	4	19.8
2	1.5_0.7_1.5_1.5_1.5	4	4.3	4.6	3.9	4.3	21.1
1	0.7_0.7_0.7_0.7_0.7	2.9	2.4	3.4	3.1	3.7	15.5
2	1.5_1.5_1.5_1.5_1.5	4.3	4.7	3.6	4.1	4.6	21.3
3	3_3_3_3	3.9	3.9	3.7	4	3.9	19.4
1	3_1.5_1.5_1.5	3.7	3.9	3.7	4.1	4.4	19.8
2	1.5_3l1.5_1.5_1.5	4.1	4.3	3.6	4.3	4.3	20.6

required to reach specific levels of water content and selected physical properties and sensory characteristics of the dried products.

Sudden changes of the air temperature ±20°C cause a rapid change in sample temperature, while sudden changes in the air velocity ±0.8 or 1.5 m/s cause more gentle evolutions. The introduction of negative pulses or steps had a greater effect on increasing drying times than the decrease caused by positive pulses or steps.

The analysis of the product properties showed that variable conditions can lead to comparable products to the optimum at constant conditions, with optimized drying times. In particular, changes in the air velocity did not cause significant changes in most of the attributes considered. In sensory terms, variable conditions did not visibly affect color, flavor, firmness, and taste, with evident influence only on appearance and shape retention.

References

Bains, M.S., Ramaswamy, H.S., and Lo, K.V. (1989). Tray drying of apple puree. *Journal of Food Engineering*, 9, (3), 195.

Barbanti, D., Mastrocola, D., Pinnavais, G., Severini, C., and Dalla Rosa, N. (1991). Air drying of fruit: effects of different pretreatments on drying rate and product quality. In: *Drying'91*, Mujumdar A.S. and Filkoma I. (Eds.), Elsevier Science Publishers, Amsterdam, Netherlands, p. 471.

Bhardwaj, J.C. and Lal Kaushal, B.B., (1990). A study on drying behaviour of rings from different apple cultivates of Himachal Pradesh, *Journal of Food Science and Technology*, 27 (3), 144.

Domagala, A., Gawrysiak-Witulska, M., and Janus, P. (1993). Wplyw niektórych czynników na jakosc suszu na przykladzie suszenia marchwi (The influence of selected parameters on quality of dried product for example the drying carrot). Presented at the XXIV Research Annual Meeting of Polish Food Technologies Association KTiChZ PAN. Abstracts of lectures and posters contributions, Wroclaw, 126. In Polish.

Farkas, I. and Rendik, Z. (1996). Intermittent thin layer corn drying. In: *Drying'96*, Strumillo, C. and Pakowski, Z. (Eds.), Drukarnia, Papaj, Lódz, Poland, p.B-905.

Hayashi, S. (1996). Persimmons drying with the aid of electrically controlled dryer — a case of persimmons drying at Toyama Prefecture in Japan. In: *Drying'96*, Strumillo, C. and Pakowski, Z. (Eds.), Drukarnia, Papaj, Lódz, Poland, p.B-921.

Iciek, J. and Krysiak, W. (1995). Convection drying conditions of potatoes and its effect on quality of dried material. In: *Proceedings of the Sixth Seminar on the Properties of Water in Foods*, Lewicki P.P. (Ed.), Warsaw Agricultural University Press, Warsaw, Poland, p. 38.

Jayaraman, K.S. and Das Gupta, D.K. (1992). Dehydration of fruits and vegetables — recent developments in principles and techniques, *Drying Technology*, 10 (2), 1

Karathanos, V., Anglea, S., and Karel, M. (1993). Collapse of structure during drying of celery. *Drying Technology*, 11 (5), 1005.

Kudra, T. and Mujumdar, A.S. (1995). Special drying techniques and novel dryers. In: *Handbook of Industrial Drying*, Vol. 2, Marcel Dekker, New York, p. 1087.

Laguerre, J.C., Techasena, O., Lebert, A., and Bimbenet, J.J. (1992). Compartmental modelling in pineapple drying. In: *Drying'92*, Part A, Mujumdar, A.S. (Ed.), Elsevier Science Publishers, Amsterdam, Netherlands, p. 805.

Lebert, A., Tharrault, P., Rocha, T., and Marty-Audouin, C. (1992). The drying kinetics of mint (Mentha spicata Hus.). *Journal of Food Engineering*, 17, 195.

Lee, D.S. and Pyun, Y.R. (1993). Optimisation of operating conditions in tunnel drying of food. *Drying Technology*, 11 (5), 1025.

Lenart, A. and Iwaniuk, B. (1993). Wlasciwosci rekonstytucyjne owoców i warzyw suszonych sposobem osmotycznokonwekcyjnym. (Reconstitution properties of fruits and vegetables dried in osmo-convective way). *Przemysl Spozywczy*, 97 (1), 11. In Polish.

Lewicki, P.P. (1990). *Inzynieria procesowa i aparatura przemyslu spozywczego*, Tom 2 — Procesy prezenoszenia ciepla i masy (Process Engineering and Food Industry Equipments, Vol. 2 — Processes of Heat and Mass Transfer), WNT, Warszaw, Poland, p. 275. In Polish.

Madamba, P.S. (1997). Optimisation of the drying process: an application to the drying of garlic. *Drying Technology*, 15 (1), 117.

Molys, A.L. (1986). A case study on modification and operation of two commercial fruit-leather dryers. *Canadian Agricultural Engineering,* 28 (1), 61.

Özilgen, M., Guvenc, G., Makaraci, M., and Tummer, I. (1995). Colour change and weight loss of apple slices during drying. *Zeitschrift fur Lebensmittel — Untersuchung und — Forschung,* 201, 40.

Pabis, S. (1982). *Teoria konwekcyjnego suszenia produktów rolnicych* (Convective Drying Theory of the Agricultural Crops), WRiL, Warszawa, p. 60. In Polish.

Pan, Y.K., Wu, H., Li, Z.Y., Mujumdar, A.S., and Kudra, T. (1996). Effect of a tempering period on drying of carrot in a vibro-fluidized bed. In: *Drying'96,* Strumillo, C. and Pakowski, Z. (Eds.), Drukarnia, Papaj, Lódz, Poland, p.B-1016.

Piotrowski, D. (1995). Studia nad kinetyka suszenia konwekcjnego przy stalych i zmiennych parametrach procesu (Studies on the convective drying kinetics of apples under constant and variable parameters), Ph. D. thesis, Warsaw Agricultural University. In Polish.

Piotrowski, D. and Lenart, A. (1996a). The influence of temperature's sudden on the changes drying kinetics of apples. In: *Drying'96,* Strumillo, C. and Pakowski, Z. (Eds.), Drukarnia, Papaj, Lódz, Poland, p.B-1037.

Piotrowski, D. and Lenart, A. (1996b). The influence of step changes in air temperature and velocity on the drying kinetics of apples. In: *Proceedings of the First Main Meeting,* Project Process optimization and minimal processing of foods, Vol. 3 — Drying, Oliveira, J. and Oliveira, F. (Eds.), Published by Escola Superior de Biotecnologia, Porto, Portugal, p. 18.

Piotrowski, D. and Lenart, A. (1997a). The effect of sudden changes in air conditions on convective drying of plant tissue. In: *Engineering and Food at ICEF 7,* Jowitt, R. (Ed.), Sheffield Academic Press, Sheffield, U.K., p.G-37.

Piotrowski, D. and Lenart, A. (1997b). Drying of apples under constant and variable conditions: evaluation of drying time. In: *Proceedings of the Second Main Meeting,* Project Process optimization and minimal processing of foods, Vol. 3 — Drying, Oliveira, J. and Oliveira, F. (Eds.), Published by Escola Superior de Biotecnologia, Porto, Portugal.

Piotrowski, D. and Lenart, A. (1998a). The influence of constant and variable conditions on the drying kinetics of apples. *Drying Technology,* 16 (3&5), p. 761.

Piotrowski, D. and Lenart, A. (1998b). Properties of dried apples obtained under constant and variable conditions. In: *Proceedings of the Third Main Meeting,* Project Process optimization and minimal processing of foods, Vol. 3 — Drying, Oliveira, J. and Oliveira, F. (Eds.), Published by Escola Superior de Biotecnologia, Porto, Portugal. *In press.*

Rajchert, D. and Lewicki, P.P. (1997). Reconstitution properties of osmo-convective dried plant tissue. In: *Engineering and Food at ICEF 7,* Jowitt, R. (Ed.), Sheffield Academic Press, Sheffield, U.K., p.G-45.

Ratti, C. (1994). Shrinkage during drying of foodstuffs. *Journal of Food Engineering,* 23 (1), 91.

Rocha, T., Lebert, A., and Marty-Audouin, C. (1993). Effect of pretreatments and drying conditions on drying rate and colour retention of basil (*Ocimum basilicum*). *Lebensmittel Wissenschaft und Technologie,* 26 (5), 459.

Sankat, C.K., McGaw, D.R., and Bailey, A.M.H. (1992). Factors influencing the drying behaviour of copra. In: *Drying'92,* Part B, Mujumdar, A.S. (Ed.), Elsevier Science Publishers, Amsterdam, Netherlands, p. 1439.

Sinesio, F., Moneta, E., Sparato, P., and Quaglia, G.B. (1995). Determination of sensory quality of dehydrated vegetables with profiling. *Italian Journal of Food Science, 7* (1), 1.

Sjhölm, I. and Gekas, V. (1995). Apple shrinkage upon drying. *Journal of Food Engineering, 25* (1), 123.

Soponronnarit, S., Nathakaranakule, A., Piyarat, N., and Tipaporn, Y. (1997). Strategies for papaya glacé drying in tunnel. *Drying Technology, 15* (1), 151.

Strumillo, Cz. (1983). Fundamentals fo Drying Technology (Podstawy teorii I techniki suszenia). WNT, Warsaw, pp. 71–106. In Polish.

Strumillo, Cz. and Adamiec, J. (1996). Energy and quality aspects of food drying. *Drying Technology, 14* (2), 423.

Szentmarjay, T., Pallai, E., and Szekrenyessy, K. (1996). Product quality and operational parameters of drying. In: *Drying'96*, Strumillo, C., and Pakowski, Z. (Eds.), Drukarnia, Papaj, Lódz, Poland, p.B-839.

Wang, N. and Brennan, J.G. (1995). Changes in structure, density and porosity of potato during dehydration. *Journal of Food Engineering, 24* (1), 61.

Zogzas, N.P., Maroulis, Z.B., and Marinos-Kouris, D. (1994). Densities, shrinkage and porosity of some vegetables during air drying. *Drying Technology, 12* (7), 1653.

chapter fourteen

Process assessment of high-pressure processing of foods: an overview

Dietrich Knorr

Contents

Summary 250
1   Introduction 250
2   Advantages of high hydrostatic pressure application...........................250
3   Opportunities of high hydrostatic pressure 252
    3.1   Preservation of food and related substances 252
    3.2   Opportunities for high-pressure processing of foods...................253
    3.3   Modifications of foods and related substances............................257
    3.4   Phase transition in food systems..257
    3.5   Membrane permeabilization 257
4   Challenges for high pressure R&D in food science
    and technology 259
    4.1   Mechanism of high-pressure effects on biological systems
          (microbial morphology)..261
    4.2   Mechanisms of high-pressure effects on biological systems
          (plant cell culture model systems)...262
    4.3   Interactions between food components and high pressure..........263
    4.4   Technical/engineering challenges...264
Acknowledgments...264
References..266

Summary

Key process advantages of high-pressure application to food systems are the independence of size and geometry of the samples during processing, possibilities for low-temperature treatment, and the availability of a waste-free, environmentally friendly technology. Opportunities for effective and relevant utilization of the potential of high hydrostatic pressure center around preservation processes, product modifications, and processes based on phase transitions or membrane permeabilization. Scientific challenges are the lack of kinetic data, little understanding of mechanisms involved in high-pressure effects on food systems, limited knowledge regarding the role of food constituents, and storage-related changes of pressure-treated products. Technical challenges of commercial application of high-pressure technology include material handling, process optimization, sanitation, cleaning and disinfection as well as package design. Engineering aspects to be dealt with are process inhomogeneities such as heat transfer issues and temperature distribution within pressure vessels.

1   Introduction

Nonthermal processes are currently receiving considerable attention from consumers as well as from producers and researchers. Processes that are under evaluation or development include high hydrostatic pressure treatment (Balny et al., 1992; Hayashi et al., 1994; Ledward et al., 1995; Barbosa-Canovas et al., 1998), the utilization of high electric field pulses (Knorr et al., 1994; Quin et al., 1995), high-intensity light pulses (Dunn et al., 1995), the application of supercritical carbon dioxide (Haas et al., 1988; Lin et al., 1992a, 1992b), or the use of magnetic fields (Pothakamury et al., 1993).

In addition treatments with biopolymers (Popper and Knorr, 1990, 1993) or with natural antimicrobials (Gould, 1995) are being applied or attempted. Also various combinations of the above-mentioned unit operations with thermal processes are being evaluated. Excellent reviews on these subjects have recently become available (Leistner and Gorris, 1994; Gould, 1995).

It is the aim of this paper to provide — within the given framework — an assessment of the advantages, opportunities, and challenges of high-pressure processing of food.

2   Advantages of high hydrostatic pressure application

High hydrostatic pressure treatment has been known to be applicable for food processing purposes for almost a century. Hite's early experiments (Hite, 1899; Hite et al., 1914) proved that various foods could be preserved for an extended period of time (Table 1).

Subsequent work by Chlopin and Tammann (1903) regarding the effect of high pressure on microorganisms supported the findings by Hite (1899). The principles of pressure shift freezing and pressure thawing have been

Table 1 Effects of Pressure on the Preservation of Foods
(after Hite, 1899; Hite et al., 1914)

Food product	Processing conditions	Comments
Milk	500–700 MPa, 66–71°C, 1–3 hrs	Stayed sweet for 20 days
Grape juice	680 MPa, RT, 10 min	Stopped fermentation
Apple juice	410–820 MPa, RT, 30 min	Did not ferment
Peaches, pears	410 MPa, RT, 30 min	In good condition for 5 years
Tomatoes	680 MPa, RT, 60 min	Most spoiled
Blackberries, raspberries	Low pressure	Usually fermented
Peas, beans, beets	Not specified	Samples spoiled

RT: room temperature

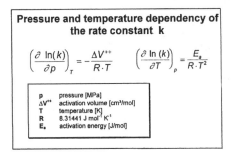

Pressure and temperature dependency of the rate constant k

$$\left(\frac{\partial \ln(k)}{\partial p}\right)_T = -\frac{\Delta V^{++}}{R \cdot T} \qquad \left(\frac{\partial \ln(k)}{\partial T}\right)_p = \frac{E_a}{R \cdot T^2}$$

p	pressure [MPa]
ΔV^{++}	activation volume [cm³/mol]
T	temperature [K]
R	8.31441 J mol⁻¹ K⁻¹
E_a	activation energy [J/mol]

Figure 1 Differential form of the pressure and temperature dependency of the rate constant of chemical reactions.

provided by Bridgman (1912). The thermodynamically derived equation concerning the effect of pressure on the rate of chemical reactions shows large similarity with the Arrhenius equation that relates the rate constant and temperature (Figure 1). The characteristic constant ΔV^{++} can take positive or negative values, producing a delayed or accelerated reaction rate with rising pressure, and is pressure independent in the ideal case.

In addition to advantages of the application of high pressure to foods and food constituents provided in the scientific literature (Cheftel, 1992; Hayashi et al., 1994; Tauscher, 1995), which include effects on reaction rates and reaction volumes, membrane permeabilization, and influences on phase transition, the instant transmittance of high pressure throughout food systems and the consequent independence of size and geometry of the samples represent a major advantage over conventional thermal processing where size and geometry can be process limiting factors. For example, size reduction required in conventional thermal processing to improve heat and mass transfer during processing is often accompanied by elevated losses of nutrients and subsequent environmental pollution (i.e., in hot water blanching processes). Such independence of size and geometry of the samples could not only reduce process severity and thus lead to higher product qualities, it could also increase process flexibilities and ultimately revolutionize food processing by making requirements for size reduction obsolete.

Another key advantage of high-pressure application is the possibility to perform processing at ambient or even lower temperatures. Indications exist that processing at subzero temperatures can be more effective with regard to inactivation on microorganisms or enzymes (Hayashi, 1988; Knorr, 1995). Low-temperature processing can help to retain nutritional quality and functionality of the raw materials treated and could allow maintenance of consistently low temperatures during postharvest treatment, processing, storage, transportation, and distribution periods of the life cycle of food systems.

Finally, the fact that high-pressure processing is environmentally friendly and a basically waste-free technology needs attention. For example, Eshtiaghi and Knorr (1993) obtained significantly less leaching of cell constituents after high-pressure blanching of potato cubes as compared to hot-water blanching. In addition the pressure processing related potential for future omission of size reductions of foods prior to processing could substantially reduce food-processing wastes (i.e., resulting from contents of ruptured plant or animal cells or tissues).

A summary of advantages and limitations of high-pressure treatment as related to food processing is provided in Table 2.

3   Opportunities of high hydrostatic pressure

3.1   Preservation of food and related substances

A vast amount of empirical information is available regarding the effects of high hydrostatic pressure on a wide range of vegetative microbial cells (Gould, 1995; Hoover, 1993; Knorr, 1995). Bottlenecks such as the baroresistance of microorganisms within environments of low water activity (Oxen and Knorr, 1993) could be overcome by combinations with mild heat or by pretreatment with ultrasound (Oxen-Bodenhausen, unpublished data). Work on pathogenic microorganisms is still scarce in the published literature (Patterson et al., 1995) and needs continued attention.

Increased pressure resistance of bacterial spores as compared to vegetative cells has been demonstrated repeatedly (Gould, 1995). Temperature or pressure-induced germination of spores and subsequent inactivation of germinating or germinated cells by treatment with high-pressure or combination processes is one route that is currently being considered (Knorr, 1995). Methodologies have been developed (Heinz and Knorr, 1995) that allow the study of germination processes via the release of dipicolinic acid or monitoring germination processes during pressure treatment via absorbance measurements of spores in a pressure cell with optical windows (Figures 2 and 3).

Successful preservation operations often depend on the effective reduction of enzyme activities during processing. Consequently, one of the requirements for high-pressure processing should include the effective reduction of undesirable enzyme activities (especially oxidases) to ensure high-quality, shelf-stable products. A vast amount of publications exist dealing with the effects of high pressure on food-related enzymes (Hara et al., 1990; Cheftel,

Table 2 Advantages and Limitations of High Hydrostatic Pressure Treatment for Food Processing Operations

Treatment	Advantages
Instant response	Immediate distribution throughout product (in the absence of gases)
Even distribution	Independence of sample size and geometry
Low/ambient temperature	Reducing thermally generated quality reduction/losses
Application affects (directly) mainly non-covalent bonds	Quality retention (i.e., flavor, color, nutrients)
Increased reaction rates	Increased bioconversion rates; increased metabolite production; improved separation processes
Affects phase transition	Process and product development (i.e., gelling, melting, crystallization)
Degassing	Improved heat transfer, reduced oxidation
Membrane permeabilization	Aids separation processes
Waste-free technology	Environmentally friendly process
Volume compression	Compacting, forming, coating
Affects enzyme activity	Food preservation
Affects microbial activity	Food preservation
Differs from thermal effects	Selective process/product development (i.e., pressure induced gelling)
Adiabatic heating	Additional temperature effect
pH reduction	Additional pH effect

Treatment	Limitations
Membrane permeabilization	Stress reaction (plants, microorganisms),texture effects
Residual enzyme activity	Quality effects
Incomplete microbial inactivation	Safety and quality effects
Reaction enhancement	Quality effects (i.e, enzymatic browning)
Temperature effects	Adiabatic heating, heat of fusion
Volume effects	Compression of water

1992; Seyderhelm et al., 1996) indicating that certain food enzymes can be reduced by high pressure to tolerable levels, but also containing a wide range of sometimes conflicting information (e.g., Anese et al., 1995).

It seems clear that food constituents are affecting baroresistance of enzymes (Ogawa et al., 1990; Seyderhelm et al., 1990; Asaka and Hayashi, 1991) and it also seems evident that when evaluating pressure effects on given enzyme systems under given conditions, a case by case approach is necessary.

3.2 Opportunities for high-pressure processing of foods

Processing opportunities integrating high-pressure treatment as a processing step can involve preservation as well as modification or separation processes

Figure 2 Prototype high-pressure cell with optical window.

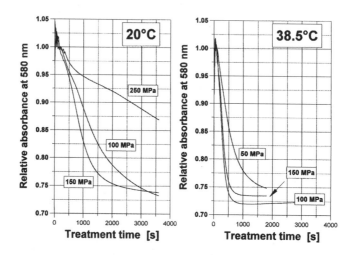

Figure 3 Relative absorbance of *Bacillus subtilis* spores (ATCC 9327) during pressure treatment at 20°C or 38.5°C measured at 580 nm (after Heinz and Knorr, 1995).

and can lead to product development opportunities. Some examples of such opportunities are given below.

The development of high-pressure blanching is based on the concept of designing a process with the advantages of conventional hot water or steam blanching but without the related leaching of nutrients, quality losses, and environmental effects (Eshtiaghi and Knorr, 1993). A summary of the results

Table 3 Comparison of the Effectiveness of Hot-Water and High-Pressure Blanching (after Eshtiaghi and Knorr 1993)

Potato cubes 2 × 2 × 2 cm	Hot water 150 sec at 100°C	Pressure 400 MPa, 15 min, 20°C
Microbial counts	3 logs	4 logs
Polyphenol oxidase activity (%)	0	50
		0#
Hardness (%)	65	70
Potassium leaching (%)	60	60 (in water)
		15 (vacuum packed)
Retention of ascorbic acid (%)	—	85

citric acid dipping to complex the cofactor (copper)

of comparative studies between high-pressure and hot water blanching of potato cubes is provided in Table 3.

High-pressure pretreatment of green beans, carrot dice, or potato cubes followed by freezing and fluidized bed drying gave good rehydration properties. High-pressure pretreatment followed by drying resulted in incomplete rehydration but combined with freezing, water uptake was between 2.1 and 4.8 ml·g^{-1}. Retention of cell wall structures of frozen samples during drying was presumed responsible for more efficient mass transfer. Pressure-treated samples had textures nearest to that of the raw material. No major differences in color were observed. Weight losses during fluidized bed drying varied with pretreatments and were highest for water blanched/frozen and/or pressure treated/frozen samples (Eshtiaghi et al., 1994).

High-pressure blanching of potatoes followed by freezing and frying resulted in fat contents of the french fries that were approximately 40% that of water-blanched, frozen samples. It seems of interest that the pressure-blanched samples had lower moisture contents than the water blanched-ones (Eshtiaghi and Knorr, unpublished data). Studies are currently under way to identify the mechanisms affecting mass transfer of water and oil.

Attempts to affect solid–liquid extractions with the aid of high pressure using coffee powder and water as models revealed pressure/temperature effects on caffeine concentration (Figure 4) as well as on pH values (Opitz and Knorr, unpublished data) as compared to extraction at 100°C and atmospheric pressure. These data suggest that temperature/pressure combinations could become a viable alternative to high-temperature extraction processes at atmospheric pressures.

Substantial amounts of data exist regarding the preservation effects of high pressure (Knorr, 1995, Cheftel, 1995). Commercially pressurized food products currently available in Japan include fruit-based products (strawberry, kiwi, apple jams, jellies, purees, fruit pieces such as sweet orange, grape, black cherry, sugar impregnated tropical fruits), grapefruit and mandarin juice, limonoid glucoside drink, raw beef or pork ham, raw squids, sea urchin paste, raw frozen fish, fish sausages and puddings, and "raw sake" (Cheftel, 1995).

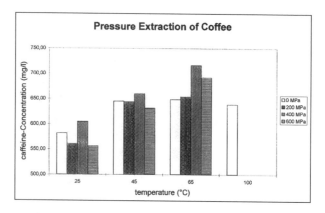

Figure 4 Effect of pressure/temperature treatment on the caffeine concentration of coffee powder-water mixtures (after Opitz and Knorr, unpublished data).

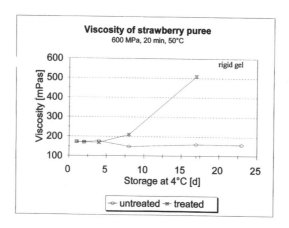

Figure 5 Gel formation of pressurized strawberry purees during cold storage (after Seyderhelm et al., unpublished data).

The extensive work that has been carried out in the area of pressure induced gelling of proteins and polysaccharides has been summarized by Cheftel (1995). A review by Ohshima et al. (1993) summarized high-pressure processing of fish and fish products with special emphasis on fish proteins. In addition pressure induced polysaccharide gels could be created during cold storage of pressurized kiwi or strawberry purees (Rovere, 1995; Seyderhelm et al., unpublished data, Figure 5).

High-pressure processing has been shown effective in supporting the reduction of allergens in milk (Okamoto et al., 1991). Recently a rice cake product became available in Japan with supposedly substantially lower allergenic potential than nonpressurized rice cakes (Cheftel, personal communication).

3.3 Modifications of foods and related substances

An extensive set of data exist on the gelling behavior of proteins, polysaccharides, and to some extent also on protein/polysaccharide combinations under high hydrostatic pressure conditions (Balny and Masson, 1993; Ohshima et al., 1993; Dumay et al., 1994). Because of the differences in functionality experienced between pressure and temperature induced gels (Ohshima et al., 1993), a wide field for product modifications via pressure or pressure/temperature treatments becomes available. Changes in composition and functionality of plant tissues have also been identified. For example, hardening of vegetable tissues (Eshtiaghi, unpublished data; Kasai et al., 1995) and the formation of solid gels during cold storage of kiwi or strawberry puree (Seyderhelm and Knorr, unpublished data; Rovere, 1995) has been observed. The most likely explanation seems a pressure-induced change of pectins, which could also be caused by residual activities of pressure-tolerant enzymes such as pectin esterase and by the release of calcium ions.

Within this context it appears also highly interesting to indicate the effects of high pressure on plant cell cultures as model systems for plant foods (Knorr, 1994). Current investigations in our laboratory on the stress response of cultured plant cells to high-pressure treatment indicate that treatment at 90 MPa and higher results in instant cell death without subsequent stress reactions of the cell. Lower pressures lead to a time-delayed stress response, suggesting pectin degradation and an elicitor effect of such degradation products (Dörnenburg and Knorr, 1995).

3.4 Phase transition in food systems

Processing-related effects of high pressure of importance on water are the compressibility of water (Table 4) and the impact on the phase transitions of water, such as freezing and melting (Figure 6).

It seems noteworthy that the phase transition of water to ice I is the only one resulting in an increase in volume (Table 5).

Pressure-induced phase transitions such as crystallization of lipids (Buchheim and Abu El-Nour, 1992) or thawing or freezing of high-moisture systems (Kalichevsky et al., 1995) offer numerous opportunities for process or product development. However, some engineering challenges such as the rapid removal of the heat of fusion (Figure 7) because of instant ice crystal formation during pressure shift freezing and the requirement for studies on the kinetics of ice nucleation, or crystal size, distribution and growth as well as on recrystallization still exist.

3.5 Membrane permeabilization

The permeabilization of membranes of vegetative microbial cells as well as of plant membranes has been demonstrated (Osumi et al., 1992; Dörnenburg

Table 4   Volume of Water (cm³/g) as a Function of Pressure
and Temperature (after Bridgman, 1912)

Pressure MPa approximated	Temperature				
	0°	20°	40°	60°	80°
0	1.0000	1.0016	1.0076	1.0168	1.0287
50	.9771	.9808	.9873	.9965	1.0075
100	.9578	.9630	.9700	.9791	.9896
150	.9410	.9471	.9544	.9632	.9732
200	.9260	.9327	.9403	.9489	.9585
250	.9133	.9203	.9279	.9363	.9457
300	.9015	.9087	.9164	.9247	.9337
350	.8907	.8979	.9056	.9138	.9226
400	.8807	.8880	.8956	.9037	.9123
450	.8717	.8788	.8864	.8945	.9028
500	.8632	.8702	.8778	.8858	.8940
550	.8554	.8621	.8698	.8777	.8858
600	.8480	.8545	.8623	.8702	.8781

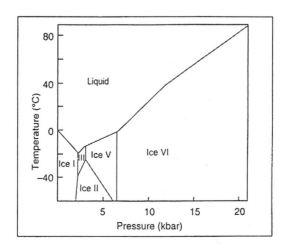

Figure 6   Phase diagram of water under various pressure/temperature conditions (after Kalichevsky et al., 1995).

and Knorr, 1994; Knorr, 1994). This has led to the inactivation of microbial cells and has opened new opportunities for process development. For example, mass transfer during dehydration of plant tissues (Eshtiaghi et al., 1994), during processing of french fries, during pasteurization of strawberries (Figure 8), or during high-pressure blanching (Eshtiaghi et al., 1994) could be affected. Work is under way in our laboratory to attempt to understand the mechanisms involved in the phenomena observed.

Table 5   Thermodynamic Properties of the Phase Transition of
Water (after Kalichevsky et al. 1995)

Phase transition	T(°C)	P (MPa)	ΔV(cm³.g⁻¹)	ΔH(kJ.kg⁻¹)
Liquid ◆ ice I	−20	193.3	+0,1313	−241
	−15	156.0	+0.1218	−262
	−10	110.9	+0.1122	−285
	−5	59.6	+0.1016	−308
	0	0.1	+0.0900	−334
Liquid ◆ ice III	−22	207.5	−0.0466	−213
	−20	246.2	−0.0371	−226
	−17	346.3	−0.0241	−257
Liquid ◆ ice V	−20	308.0	−0.0828	−253
	−15	372.8	−0.0754	−265
	−10	442.4	−0.0679	−276
	−5	533.7	−0.0603	−285
	0	623.9	−0.0527	−293
Liquid ◆ ice VI	−10	518.0	−0.0960	−264
	0	623.9	−0.0916	−295
	10	749.5	−0.0844	−311
	20	882.9	−0.0751	−320
	30	1038.9	−0.0663	−330

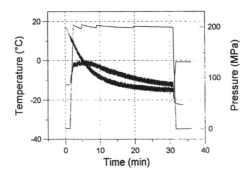

Figure 7   Pressure and temperature conditions during pressure shift freezing of potato cubes (Koch et al., 1996).

4   Challenges for high pressure R&D in food science and technology

Key areas where additional information are required include the need for kinetic data on the inactivation of microorganisms and enzymes as well as on the changes of food quality and functionality; a better understanding of the mechanisms involved during high-pressure treatments; experiments clarifying the interactions between food constituents and high-pressure

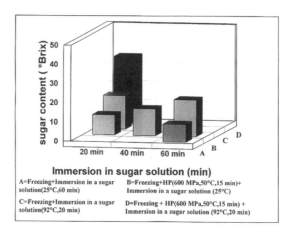

Figure 8 Total solids contents of pressure thawed, pasteurized vs. atmospheric pressure thawed, pasteurized strawberries (Eshtiaghi and Knorr, unpublished data).

effects on food systems; the necessity to gain more knowledge regarding interactions between high pressure and nutrients, toxins, or allergens; and finally compilation of data during postpressure-treatment storage periods of food materials.

Kinetic studies on the lethality of high-pressure treatment yielded different results and subsequent interpretations dependent on the microbial organisms under investigation. At this point no basic principle of pressure inactivation is available (Cheftel, 1995). Thermodynamically derived equations, especially the pressure dependency of the rate constant k must fail when the order of the reaction is not uniform. First-order inactivation may occur, and in that case logarithmic-linear survivor curves can be expressed in D-values, but very often deviations from this behavior are obtained, leading to logarithmic S-shaped curves if the calculation of D-values is not possible. Due to the more or less pronounced tailing and shoulder-formation, variation in reaction order was observed even within the same bacterial strain of *Bacillus subtilis* (Heinz and Knorr, 1996).

In Figure 9 the inactivation kinetics of *Bacillus subtilis* show a typical tailing and an initial lag-phase indicating a certain time-dependent resistance mechanism that can be extensively prolonged in the case of the 250 MPa treatment.

A description of these curves is possible by a first-order inactivation of a metastable intermediate state that is reached after a certain time of resistance and is assumed to be distributed due to the diversity in the bacterial population. This homeostatic mechanism is thought to be an active transport of protons through the membrane of the bacterial cell as a consequence of the pressure-induced displacement of internal dissociation equilibria. The membrane-bound ATPase (F_0F_1 ATPase) as the energy supplying system of this proton extrusion can be affected by pressure leading to an internal drop of pH (Smelt, 1996). This intermediate state may be the place of combined

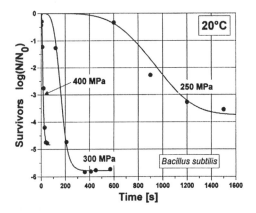

Figure 9 Inactivation kinetics of *Bacillus subtilis* ATCC 9372 suspended in Ringer's solution. Initial bacterial count: 10^6 CFU/ml. (after Heinz and Knorr, 1996). Printed with permission of the Editor.

pressure-temperature-pH action at different, not yet known, targets inside the bacterial cell.

The lethality of the treatment concerning the vegetative forms can be increased extensively by raising the pressure whereas the spores remain unaffected (Figure 9). The final values of the kinetic curves indicate a certain, cultivation dependent portion of spores typical for spore-forming bacteria. This part of the population can be affected at much lower pressures by inducing the germination (Sale et al., 1997), producing a loss of the heat and pressure resistance. The optimum of this reaction seems to be located within the range of 100 to 150 MPa. Data presented in Figure 3 show a steep decrease in optical density in the case of 150 MPa/20°C and 100 MPa/38°C, which can be interpreted as a loss of refractility in consequence of changes in spore structure during germination. Raising the temperature up to about 50°C during pressure treatments (150 MPa) yields more complete germination of the spore population. At higher temperatures, the inactivating seems to be dominant (Figure 10).

4.1 Mechanisms of high-pressure effects on biological systems (microbial morphology)

Hoover et al. (1989) stated that high hydrostatic pressure induces a number of changes to the morphology, biochemical reactions, genetic mechanisms, and cell membrane and wall of microorganisms. Some of these aspects have been covered by Smelt (1996). Pressure effects on the cellular morphology are presented in Figure 11, indicating severe pressure effects. It is of interest that — most likely due to structural differences of the cell wall and cell membrane — pressure-induced separation of the two takes place. In addition, pressure effects seem to be compartmentalized. This can be due to differences in compressibility within different compartments of the microbial

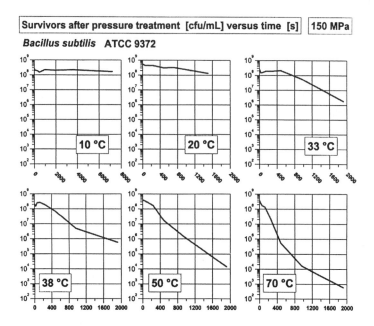

Figure 10 Pressure-induced germination of *Bacillus subtilis* spores (ATCC 9372) suspended in Ringer's solution recorded by absorbance measurement during treatment in a pressure cell with optical windows (after Heinz and Knorr, unpublished data).

cell. The volume of water at the pressure conditions selected (600 MPa, 25°C) is 0.8564 cm³/g as compared to 1.0028 cm³/g at atmospheric pressure (Bridgman, 1912).

4.2   Mechanisms of high-pressure effects on biological systems (plant cell culture model systems)

The effects of high hydrostatic pressure on plant membranes of plant tissue cultures have been demonstrated by Dörnenburg and Knorr (1994), suggesting permeabilization of the tonoplast of *Chenopodium rubrum* cultures at approximately 100 MPa and subsequent loss of cell viability (due to release of toxic and acidic vacuole contents into the cytoplasma). Permeabilization of the outer cell membrane and release of secondary metabolites in the immersion medium did not occur until pressures of 250 to 350 MPa. Such findings are essential in optimizing minimal processing of plant tissues (i.e., maintaining tissue texture). The use of plant and cell tissue cultures as model systems for monitoring the effects of unit operations on plant systems has been suggested (Knorr, 1995) in order to avoid systematic errors when studying plant tissues due to the stress and wound responses of the plants (i.e., release of phytoalexins and antimicrobial enzymes, enzymatic browning) to the size reduction (i.e., cutting, slicing) operations commonly performed. For example, the physiologic responses of tomato or potato cell cultures to high-pressure

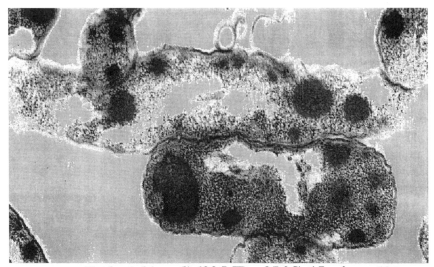

Escherichia coli,600 MPa, 25 °C, 15 min. *120x*

Figure 11 Effect of high hydrostatic pressure on the morphology of microorganisms (after Kanchanakanti and Knorr, unpublished data).

treatment suggest a pressure-induced, time-delayed stress response such as H_2O_2 production (Apostal et al., 1989) at low pressures due to degradation of pectins (Figure 12) and subsequent stress response to the degradation products of pectin (Dörnenburg and Knorr, 1994). This has significance because it indicates that higher pressures (i.e., 90 MPa) result in loss of cell viability prior to stress responses. Considering the increased interest in minimal processing of plant food materials the physiologic responses to processing (i.e., production of undesirable metabolites such as phytoalexins) and methods to monitor such responses, require increased attention. Also increased enzymatic browning was observed in potato cell cultures after pressure treatment. Here various mechanisms, including increased reaction rates, increased interactions between enzymes and substrates, breaking of glucosidic bonds, "solubilization" of polyphenol oxidases, or increased activities of polyphenol oxidase and/or glucosidases are being considered (Wille and Knorr, unpublished data).

4.3   Interactions between food components and high pressure

The combined effects of food constituents affecting water activity and high pressure on microorganisms (Oxen and Knorr, 1993) and on enzymes (Seyderhelm et al., 1996) are briefly exemplified.

The relationship between water activity of the immersion medium during pressure treatment and cell viability of *Rhodotorula rubra* is demonstrated in Figure 13. At 25°C and 400 MPa complete baroresistance at water activity of 0.92 and lower could be observed. Temperatures higher than 30°C in

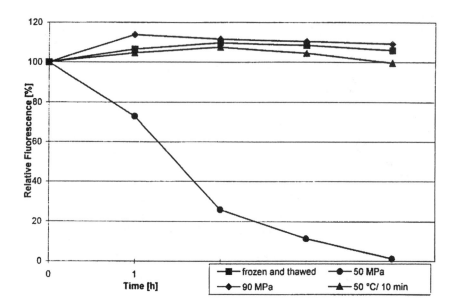

Figure 12 Stress response (reduction in fluorescence) of tomato cell cultures after treatment at 50°C, freeze-thaw cycle or pressures of 50 MPa and 90 MPa (after Schreck et al., 1996). Printed with permission of the Editor.

combination with pressure (400 MPa, 15 min) were necessary to overcome this baroresistance (Oxen and Knorr, 1993). Also a pretreatment of the yeast cells with ultrasonic waves in media with low water activity led to subsequent pressure inactivation of the cells (Figure 14).

4.4 *Technical/engineering challenges*

Technical challenges of commercial application of high-pressure technology are, according to Mertens (1995), material handling, package design, sanitation, cleaning, and disinfection of high-pressure equipment; bulk or in-container processing, and "high pressure short time" processing or "low temperature long time" processing. In addition, heat transfer within pressure transferring media, temperature distribution within pressure vessels and pressure distribution within food materials — due to differences in compressibilies because of the complex composition of foods (and other biological systems such as microorganisms) — are engineering issues that require attention.

Acknowledgments

Parts of this work have been funded by grants from the European community (EC-AIR CT92-0296, EC-AIR CT-96-1175), the German Research Foundation (DFG Kn 260/3-1, 3-2, 3-3) and the FEI (Forschungskreis der Ernährungsindustrie e.V., Bonn), the AIF and the Ministry of Economics

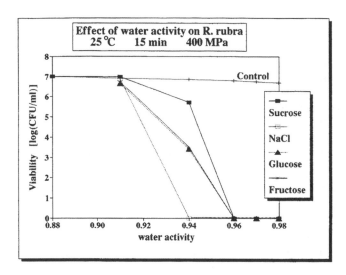

Figure 13 Effects of water activity of the immersion medium adjusted with various food constituents on the pressure (400 MPa,25°C,15 min) tolerance of *Rhodotorula rubra* (after Oxen-Bodenhausen and Knorr, unpublished data).

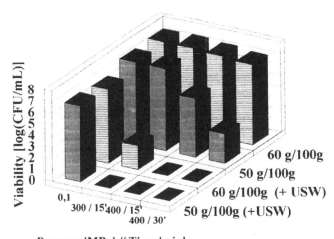

Figure 14 Effects of pretreatment of *Rhodotorula rubra* immersed sucrose solutions at various water activities (50 g/100 g, 60 g/100 g) with ultrasonic waves (usw) on their pressure resistance (after Oxen-Bodenhausen and Knorr, unpublished data).

(Project No.: AIF-FV 8774, AIF-FV 9918). Parts of this paper have been published in Hayashi, R. and Balny, C., 1996. High Pressure Bioscience & Biotechnology, Elsevier Science, Amsterdam, as well as in Reid, D., 1998, The Role of Water in Foods, Elsevier Science, Amsterdam.

References

Anese, M., Nicoli, M.C., Dall'Aglio, D., and Lerici, C.R. (1995). *Journal of Food Biochemistry*, 18, 285.

Apostal, I., Heinstein, P.F., and Low, P. (1989). *Plant Physiology*, 90, 109.

Asaka, M. and Hayashi, R. (1991). *Agricultural and Biological Chemistry*, 55, 2440.

Balny C. and Masson P. (1993). *Food Review International*, 9, 611.

Balny, C., Hayashi, R., Heremans, K., and Masson, P. (1992). *High Pressure and Biotechnology*, John Libbey Eurotext, Montrouge.

Barbosa-Canovas, G.V., Pothakamury, P.E., and Swanson, B.G. (1998). *Nonthermal Preservation of Foods*, Marcel Dekker, New York.

Buchheim, W. and Abou El-Nour, A.M. (1992). *Fat Science and Technology*, 94, 369.

Bridgman, P.W. (1912). *Proceedings of the American Academy of Arts and Sciences*, 47 (13), 439.

Cheftel, J.C. (1992). In: *High Pressure and Biotechnology*, Balny, C., Hayashi, R., Heremans, K. and Masson, P. (Eds), John Libbey Eurotext, Montrouge.

Cheftel, J.C. (1995). *Food Science and Technology International*, 1, 75

Chlopin, G.W. and Tammann, G. (1903). *Z. Hygiene Infektionskrankh.*, 45, 171. In German.

Dörnenburg, H. and Knorr, D. (1994). *Food Biotechnology*, 8, 57.

Dörnenburg, H. and Knorr, D. (1995). *Enzyme and Microbial Technology*, 17, 674.

Dumay, E., Kalichevsky, M.T., and Cheftel, J.C. (1994). *Journal of Agricultural Chemistry*, 42, 1861.

Dunn, J., Ott, T., and Clark, W. (1995). *Food Technology*, 49 (9), 95.

Eshtiaghi, M.N. and Knorr, D. (1993). *Journal of Food Science*, 58, 1371.

Eshtiaghi, M.N., Stute, R., and Knorr, D. (1994). *Journal of Food Science*, 59, 1168.

Gould, G.W. (1995). *New Methods of Food Preservation*, Blackie Academic & Professional, London.

Haas, G.J., Prescott, H.E., Duddley, E., Dik, R., Hintlian, C., and Keane, L. (1988). *Journal of Food Safety*, 9, 253.

Hara, A., Nagahama, G., Ohbayashi, A., and Hayashi, R. (1990). *Nippon Nogeikagaku Kaishi*, 64, 1025.

Hayashi, R. (1988). In: *The Role of Water in Foods*, Reid, S. (Ed.), Elsevier Science, Amsterdam, Netherlands.

Hayashi, R., Kunugi S., Shimads, S., and Suzuki, A. (1994), *High Pressure Bioscience*, San-Ei Suppan Co., Kyoto.

Heinz, V. and Knorr, D. (1995). Annual Report, EC Project *High Pressure Processing of Foods* (AIR-CT92-0296), European Commission, Brussels, Belgium.

Heinz, V. and Knorr, D. (1996). *Food Biotechnology*, 10 (2), 149.

Hite, B.H. (1899). *West Virginia Agricultural Experimental Station Bulletin*, 58, 15.

Hite, B.H., Giddings, N.J., and Weakley, C.E. (1914). *West Virginia Agricultural Experimental Station Bulletin*, 146, 3.

Hoover, D.G. (1993). *Food Technology*, 47 (6), 150.

Hoover, D.G., Metrick, C., Papineau, A.M., Farkas, D.F., and Knorr, D. (1989). *Food Technology*, 43 (3), 99.

Kalichevsky, M.T., Knorr, D., and Lillford, P.J. (1995). *Trends in Food Science and Technology*, 6, 153.

Kasai, M., Hatae, K., Shimada, A., and Iibuchi, S. (1995). *Nippon Shokuhin Kagaku Kogaku Kaishi*, 42, 594.

Knorr, D. (1994). *Trends in Food Science and Technology*, 5, 328.

Knorr, D. (1995). In: *New Methods of Food Preservation*, Gould, G.W. (Ed.), Blackie Academic & Professional, London, U.K, p. 159.

Knorr, D., Geulen, M., Grahl, T., and Sitzmann, W. (1994). *Trends in Food Science and Technology*, 5, 71.

Koch, H., Seyderhelm, I., Wille, P., Kalichevski, M.T., and Knorr, D. (1996). *Nahrung-Food*, 40, p. 125

Ledward, D.A., Johnston, D.E., Earnshaw, R.G., and Hasting, Q.P.M. (1995). *High Pressure Processing of Foods*, Nottingham University Press, Nottingham, U.K., p. 7.

Leistner, L. and Gorris, L.G.M. (1994). Final Report, *FLAIR Concerted Action no. 7*, Subgroup B, European Commission, DG XII, Brussels, Belgium.

Lin, H., Yang, Z., and Chen L. (1992a). *Biotechnology Progress*, 6, 458.

Lin, H., Yang, Z.Y., and Chen L. (1992b). *Biotechnology Progress*, 8, 165.

Mertens, B. (1995). In: *New Methods of Food Preservation*, Gould, G.W. (Ed.), Blackie Academic & Professional, London.

Ogawa, H., Fukuhisa, K., Kubo, J., and Fukumoto, H. (1990). *Agricultural and Biological Chemistry*, 54, 1219.

Ohshima, T., Ushio, H., and Koizumi, C. (1993). *Trends in Food Science and Technology*, 4, 370.

Okamoto, M., Hayashi, R., Kaminogawa, S., and Yamanchi, K. (1991). *Agric. Biol. & Chem.*, 55: 1253–1257.

Osumi, M., Yamada, N., Sato, M., Kobori, H., Shimada, S., and Hayashi, R. (1992). In: *High Pressure and Biotechnology*, Balny C., Hayashi R., Heremans K., and Masson P. (Eds), John Libbey Eurotext, Montrouge, p. 9.

Oxen, P. and Knorr, D. (1993). *Lebensmittel Wissenchaft und Technologie*, 26, 220.

Patterson, M.F., Quinn, M., Simpson, R., and Gilmour, A. (1995). In: *High Pressure Processing of Foods*, Ledward, D.A., Johnston, D.E., Earnshaw, R.G., and Hasting, Q.P.M. (Eds.), Nottingham University Press, Nottingham, U.K., p. 47.

Popper, L. and Knorr, D. (1990). *Food Technology*, 7, 84.

Popper, L. and Knorr, D. (1993). *Bioengineering*, 9, 27.

Pothakamury, P.E., Barbosa-Canovas G.V., and Swanson B.G. (1993). *Food Technology*, 12, 85.

Quin, B, Pothakamury, P.E, Vega, H. Martin, O., Barbosa-Canovas G.V., and Swanson, B.G. (1995). *Food Technology*, 49, 55.

Rovere, P. (1995). *Tecnologie Alimentaire*, 4, 1. In French.

Sale, A.J.H., Gould, G.W., and Hamilton, W.A. (1997). *Journal of General Microbiology*, 60, 323.

Seyderhelm, I., Boguslawski S., Michaelis G., and Knorr, D. (1996). *Journal of Food Science*, 61, 308.

Schreck, S., Dornenburg, H., and Knorr, D. (1996). *Food Biotechnology*, 10, p. 163

Smelt, J. (1996). Presented at the ISOPOW 6 meeting, St. Rosa, March 2–8.

Tauscher, B. (1995). *Z. Lebensm.Unters.Forsch.*, 200, 3.

chapter fifteen

A comparison between pressure and temperature effects on food constituents

K. Heremans, F. Meersman, P. Rubens, L. Smeller, J. Snauwaert, and G. Vermeulen

Contents

Summary 270
1 Introduction 270
2 Pressure compared with temperature effects...............................271
3 Stability diagrams of proteins..271
   3.1 Thermodynamics of denaturation..271
   3.2 Kinetics of denaturation 272
   3.3 Role of solvent: glass transitions..272
4 *In situ* observation of protein denaturation, aggregation,
   and gel formation 273
5 Case studies 274
   5.1 Stability diagrams of amylases..274
   5.2 Pressure-assisted cold denaturation of metmyoglobin 275
   5.3 Pressure effects on emulsions and inverted micelles 276
   5.4 Pressure-induced gelation of starch...277
6 Conclusions..277
Acknowledgment...279
References..279

Summary

Enzyme inactivation and protein quality are essential aspects of the high-pressure processing of foods and therefore an in-depth knowledge of the effect of high pressure on protein molecules is needed. This chapter overviews the current knowledge of the mechanisms of heat- and pressure-induced changes on protein conformations and functionality. The application of Fourier transform infrared spectroscopy in the diamond anvil cell has a fundamental role in elucidating these issues, with the ability to analyze *in situ* transformations at pressures up to 1 GPa or more, with or without additional heating. Some case studies are provided, namely, the stability diagram of amylases, cold denaturation of metmyoglobin, pressure effects on emulsions and inverted micelles, and gelatinization of starch.

1 Introduction

Proteins can be denatured by high hydrostatic pressure as well as by temperature. This opens up possibilities for creating new textures in food materials. At the beginning of this century Bridgman observed that egg white is coagulated after a pressure treatment at 700 MPa for 30 min. The texture of the pressure-induced coagulum was quite different from the coagulum induced by temperature. Hayashi et al. (1989) observed that the yolk becomes solid at a lower pressure than the white. It is well known from the preparation of a hard-boiled egg that the yolk becomes solid at a slightly higher temperature than the white.

It is now clear that these observations are the consequence of the unique behavior of proteins. The stability diagram for the conditions under which the native and the denatured conformation occur has an elliptic shape in the temperature–pressure plane. This has been observed in many instances for the denaturation of proteins, the inactivation of enzymes, as well as the killing of bacteria (Suzuki, 1960; Heinisch et al., 1995; Ludwig et al., 1996). Most observations are *ex situ*, i.e., analysis of the sample is made after heat treatment or compression.

In situ observations of the pressure-induced conformational transitions and denaturation of proteins can conveniently be studied with Fourier transform infrared (FTIR) spectroscopy in the diamond anvil cell. In most cases the difference between the temperature and the pressure denaturation of proteins is very pronounced. This is not only true for the conformation of the polypeptide chain but also for the intermolecular interactions of the proteins in the denatured state. Temperature-denatured proteins show an extensive intermolecular hydrogen bond network that is absent in most pressure-denatured proteins.

In this chapter we discuss some of the topics that are under investigation in our laboratories. For a more complete discussion and a correlation between various experimental approaches we refer the reader to a recent review (Heremans et al., 1997). Rich sources of information on the pressure behavior of biosystems of interest for the food industry can also be found in the proceedings of recent conferences (Balny et al., 1992; Ledward et al., 1995; Hayashi and Balny, 1996; Heremans, 1997).

2 Pressure compared with temperature effects

The unique properties of biomolecules that constitute the foodstuffs are determined by the delicate balance between internal interactions of the biopolymers that compete with interactions with the solvent. Because of the imperfect packing, the specific folding of the polymers gives rise to a free volume, which may be expected to decrease with increasing pressure. As temperature effects act via an increased kinetic energy as well as via free volume, it follows that pressure effects should be easier to interpret than temperature effects. The energy stored in the compression of a given amount of liquid water to 1 GPa is much less than that of heating to 100°C. Whereas the compressibility decreases with increasing pressure, the heat capacity is nearly temperature independent.

3 Stability diagrams of proteins

The biologically active structure of a protein is only stable within restricted conditions of temperature, pressure, and solvent composition. Outside this range, unfolding or denaturation takes place. At sufficiently high concentration, aggregation and/or gel formation may take place.

3.1 Thermodynamics of denaturation

Following the initial experiments of Bridgman on egg white, detailed investigations by Suzuki (1960) have shown that these observations may be put together in a phase diagram for the denaturation of proteins. As shown in Figure 1, at high temperature, pressure stabilizes the protein against temperature denaturation. At low temperature, increasing temperature stabilizes the protein against pressure denaturation. The fact that one can "cook" an egg with pressure results in the unique phase diagram of proteins. There is also evidence that this phenomenon occurs in starch. For proteins, an equilibrium between the native and denatured form is assumed:

Native (Folded, Active) ↔ Denatured (Unfolded, Inactive)

The change in the free energy difference for this process depends on the temperature and pressure:

$$d(\Delta G) = -\Delta S dT + \Delta V dP$$

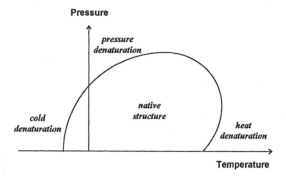

Figure 1 Schematic stability diagram showing the relation between cold-, pressure-, and heat-denaturation of proteins. The specific features of these processes are explained in the text.

Integration of this equation, assuming that the entropy (ΔS) and volume change (ΔV) are temperature- and pressure-dependent, gives an elliptic equation that contains information on the change in thermal expansion, compressibility, and heat capacity upon denaturation (Smeller and Heremans, 1997). Thus, it is possible to model the experimental data starting from a physical model. This has advantages over a pure mathematical modeling.

3.2 *Kinetics of denaturation*

In many cases kinetic data as a function of temperature and pressure are obtained for the inactivation of enzymes or the denaturation of proteins. The activation parameters calculated from the rate constants are pressure- as well as temperature-dependent. This was first observed by Suzuki (1960) for the denaturation of ovalbumin and carbonylhemoglobin. It was found that the activation volume and energy become negative at low temperatures. This was interpreted by a two-step mechanism for the denaturation:

$$N \leftrightarrow D \rightarrow I$$

The first step ($N \leftrightarrow D$), which predominates below room temperature, is a reversible unfolding. This step gives rise to negative volume and energy parameters. At higher temperature the second step ($D \rightarrow I$), an irreversible aggregation, takes place, which gives rise to the usual positive activation energy and volume. As will be shown below, temperature and pressure induce different conformational changes in proteins.

3.3 *Role of solvent: glass transitions*

The structure of a protein is the result of a delicate balance between the intramolecular interactions in the polypeptide chain that compete with the solvent interactions. Proteins in the dry state are found to be extremely resistant to temperature- and pressure-denaturation. The pressure effect on dry proteins may easily be studied in the diamond anvil cell. The consequence for the application of the high-pressure technique in the food industry is evident. Slade and Levine (1995) have emphasized the role of water as a plasticizer in food ingredients via its effect on the glass-rubber transition (T_g). The fundamental and generic similarities between synthetic polymers and food molecules has greatly stimulated research in this field because of the importance of T_g on all aspects of food processibility, properties, and safety. In view of the nonequilibrium nature of the glass-transition temperature, this transition is governed by activation rather than equilibrium parameters. Limited studies on the effect of pressure on the glass transitions in polymers suggest a change of *ca.* 22 K/100 MPa for nonhydrogen-bonded systems. The effect of pressure on the hydrogen-bonded system sorbitol shows a much weaker dependence of 4 K/100 MPa (Atake and Angell, 1979). This suggests that similar orders of magnitude may be expected for glass transitions in proteins.

4   *In situ observation of protein denaturation, aggregation, and gel formation*

The diamond anvil cell (DAC) high-pressure technique allows the observation of biomacromolecules, *in situ*, during the compression and decompression phase. With this technique pressures of 1 GPa and more can easily be obtained. The technique can easily be used with Fourier transform infrared spectroscopy and has distinct advantages for the observation of pressure-induced changes in biomolecules. Protein solutions are mounted in a stainless steel gasket of the diamond cell. Two gem-cut diamonds, with polished off culets, are compressed against each other with a thin metal gasket, leaving a small hole between them. The sample volume is of the order of 20 μl and the minimum amount of protein needed per run is about 1 mg. The DAC allows the observation of infrared spectral changes induced by pressure and/or temperature. Visual inspection of the cell content under the microscope makes it possible to follow, *in situ*, the pressure-induced gel formation in proteins and polysaccharides such as starch.

Infrared spectroscopy probes the vibration of chemical bonds within the molecules. The CO vibration of the peptide group of proteins is dependent on the secondary structure. A detailed analysis of the shape of the amide I' band allows a closer look to the pressure or temperature induced changes in the secondary structure. This is done with a combination of self-deconvolution and band-fitting procedure. Figure 2 gives the infrared spectrum in the amide I' region for α-amylase from *B. subtilis*. The figure shows the spectrum after resolution enhancement by Fourier self-deconvolution. Fitting of this spectrum to a sum of Gaussian bands indicates that the spectrum is composed of two main bands, which can be assigned to β-structures and

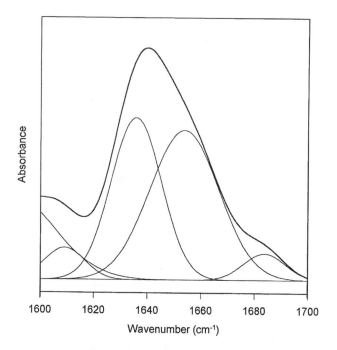

Figure 2 Deconvoluted spectrum in the amide I' region of *B. subtilis* amylase at ambient conditions. The two main fitted components can be assigned to β-structures and α-helix/unordered structures.

α-helix/unordered structures. In the next section, the effect of temperature and pressure on the spectrum of proteins will be discussed and the important differences will be highlighted.

5   Case studies

5.1   Stability diagrams of amylases

The pressure and temperature effect on the activity of amylases from *Bacillus* species was determined in order to probe the differences in stability (Weemaes et al., 1996). Detailed analysis of the frequency maximum of the amide I' band also allows the characterization of the pressure and temperature of denaturation. The data show an elliptical outline when plotted in the temperature-pressure plane as shown in Figure 3. The enzyme from *B. licheniformis* shows a higher stability than the enzymes from *B. subtilis* and *B. amyloliquefaciens*. The pressure stability seems to show less pronounced differences near room temperature.

The pressure-induced spectral changes at room temperature shows a broad band with an absorption maximum at 1640 cm⁻¹, which can be assigned to the unordered structure. At high temperature and ambient pressure the amide I' band shows the formation of two new bands at 1615 and

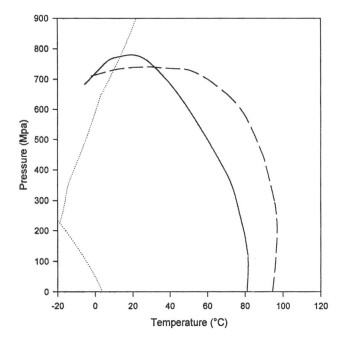

Figure 3 Stability diagram for amylases from *Bacillus* species: *B. licheniformis*: dashed line; *B. subtilis* and *B. amyloliquefaciens*: solid line; limits for the existence of fluid D$_2$O: dotted line.

1680 cm^{-1}. These bands are assigned to the formation of intermolecular β-sheet, which indicates aggregation (Mozhaev et al., 1996). This is in contrast with the pressure-induced changes where only a broadening of the amide I′ band is observed. The spectrum at high pressure and temperature reveals a clear band at 1615 cm^{-1}. Due to the absence of the band at 1680 cm^{-1}, it cannot be assigned to intermolecular β-sheet aggregation. The origin of this band needs further investigation. These observations may have important consequences for combined pressure-temperature treatments.

5.2 Pressure-assisted cold denaturation of metmyoglobin

Figure 4 shows the changes in the amide I′ region (1600–1700 cm^{-1}) of the deconvoluted spectra of metmyoglobin upon cooling down from 17°C to –28°C at 200 MPa. Due to ice formation, the denaturation is, however, not complete. The ice formation gives rise to spectral changes that no longer allow us to assign the observed changes solely to cold denaturation. Techniques are available that should allow us to cool down to –40°C without solidification of the solvent (Franks, 1995).

The spectral changes reflect the conformational changes that the protein undergoes upon cooling. It can be seen that the intense peak near 1650 cm^{-1}, which is characteristic for α-helix, becomes less intense at lower temperatures,

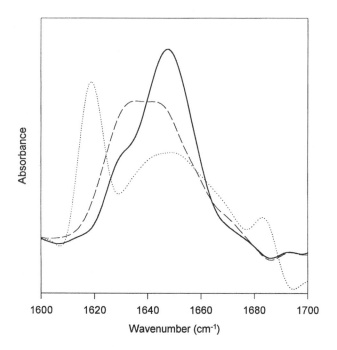

Figure 4 Infrared spectra of metmyoglobin at pH 7.6 and 30°C: solid line; heat denatured at 90°C: dotted line; cold denatured (partial transition) at –28°C: dashed line. Notice the absence of bands that are typical for heat-denatured proteins in the cold-denatured state.

while the initial shoulder at 1636 cm^{-1}, characteristic for intramolecular β-sheet, shows an increase in intensity. The figure also shows the absence of two distinct absorption bands near 1615 and 1683 cm^{-1}, which are characteristic for intermolecular hydrogen bonding. As shown in the figure, the bands do appear in case of the heat denaturation. This suggests that the mechanisms of cold and heat denaturation are different.

5.3 Pressure effects on emulsions and inverted micelles

The versatility to observe pressure-induced phenomena in proteins can also be illustrated on emulsions and inverted micelles. Figure 5 shows that β-lactoglobulin denatures at about 200 MPa in solution whereas in an oil/water emulsion, a cooperative change in the spectrum occurs at about 600 MPa. No such differences are observed upon heating. Even more interesting is the behavior of α-chymotrypsin in inverted micelles. The enzyme denatures at about 800 MPa, but in contrast to the solution behavior bands that are typical for the intermolecular hydrogen bonding that usually appear upon heating show up *after* the pressure is released. It remains to be investigated whether this peculiar behavior is dependent on the water content of the inverted micelles.

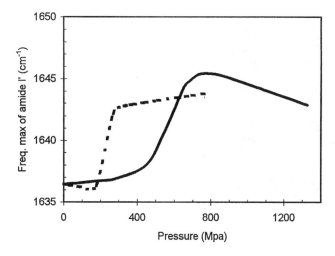

Figure 5 Effect of pressure on the maximum of the amide I′ band of *β-lactoglobulin*. Dotted line: in solution; solid line: in an emulsion of 20% w/w of soybean oil/D_2O.

5.4 Pressure-induced gelation of starch

The effect of pressure on the gelation temperature of starch suggests phase diagrams similar to those of proteins (Thevelein et al., 1981). Pressure-induced gelation of a potato starch granule suspension shows changes in the infrared spectrum in regions that are sensitive to the polymer conformation. The intensity ratio of bands at 1157 and 1127 cm^{-1} shows a transition above 500 MPa that corresponds to the gelation process observed under the optical microscope. From Figure 6 it can be seen that no swelling of undamaged granules can be observed below 500 MPa. Kinetic investigation pointed out that no swelling could be detected in undamaged granules after 91 hours at 400 MPa. Our results are in agreement with the *ex situ* observations of Ezaki and Hayashi (1992).

6 Conclusions

The use of pressure as an alternative to temperature treatment has brought about the need for fundamental studies on the pressure–temperature behavior of macromolecular food constituents. The examples discussed above prove the versatility of FTIR spectroscopy as a tool for studying the effects of pressure, temperature, or a combination of the two on the conformation of proteins and polysaccharides. The appearance of infrared bands characteristic of temperature-induced gel formation in proteins points at differences in the properties of pressure- and high-temperature-induced gels. For the cold denaturation no such intermolecular bands are observed. We therefore tend to conclude that there is also a difference between the mechanisms of heat and cold denaturation. The observed correlation with visible microscopy and FTIR

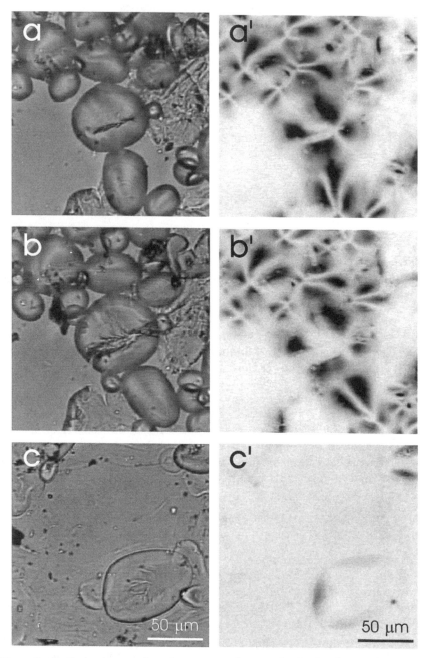

Figure 6 Optical micrographs of potato starch granules in the diamond anvil cell (a, a' = 70 MPa; b, b' = 490 MPa; c, c' = 640 MPa). *Left* (a, b, c): normal image; *right* (a', b', c'): negative of the image taken between crossed polarisers to show the Maltese cross.

spectroscopy for starch gelation point to a very fruitful combined approach. The application to the formation of protein gels and mixed protein polysaccharide gels is in progress in our laboratory.

Acknowledgment

This research was supported by grants of the European Community, AIR1-CT92-0296, FAIR-CT96-1175 and CIPA CT94-0195. Author Smeller also thanks the Hungarian Academy of Sciences and F.W.O. Flanders (Belgium) for financial support.

References

Atake, T. and Angell, C. A. (1979). Pressure dependence of the glass transition temperature in molecular liquids and plastic crystals, *Journal of Physical Chemistry*, 83, 321.

Balny, C., Hayashi, R., Heremans, K., and Masson, P. (1992). *High Pressure and Biotechnology*, John Libbey Eurotext Ltd, Montrouge.

Ezaki, S. and Hayashi, R. (1992). High pressure effect on starch: structural change and retrogradation. In *High Pressure and Biotechnology*, Balny, C., Hayashi, R., Heremans, K., and Masson, P., (Eds.), John Libbey Eurotext Ltd, Montrouge, p. 163.

Franks, F. (1995). Protein destabilization at low temperatures, *Advances in Protein Chemistry*, 46, 1055.

Hayashi, R. and Balny, C. (1996). *High Pressure Bioscience and Biotechnology*, Elsevier, Amsterdam, Netherlands.

Hayashi, R., Kawamura, Y., Nakasa, T., and Okinaka, O. (1989). Application of high pressure to food processing: pressurization of egg white and yolk, and properties of gels formed, *Agricultural Biological Chemistry*, 53, 2935.

Heinisch, O., Kowalski, E., Goossens, K., Frank, J., Heremans, K., Ludwig, H., and Tauscher, B. (1995). Pressure effects on the stability of Lipoxygenase: FTIR and enzyme activity studies, *Z. Lebensm. Unters. Forsch.*, 201, 562.

Heremans, K. (1997). *High Pressure Research in Bioscience and Biotechnology*, Leuven University Press, Leuven, Belgium.

Heremans, K., Van Camp, J., and Huyghebaert, A. (1997). High-Pressure effects on proteins. In: *Food Proteins and their Applications*, Damodaran, S. and Paraf, A., (Eds.), Marcel Dekker, New York, p. 473.

Ledward, D.A., Johnston, D.E., Earnshaw, R.G., and Hasting, A. (1995). *High Pressure Processing of Foods*, Nottingham University Press, Nottingham, UK.

Ludwig, H., Scigalla, W., and Sojka, B. (1996). Pressure and temperature induced inactivation of microorganisms. In: *High Pressure Effects in Molecular Biophysics and Enzymology*, Markley, J.L., Royer, C., and Northrup, D., (Eds.), Oxford University Press, Oxford, UK, p. 346.

Mozhaev, V.V., Heremans, K., Frank, J., Masson, P., and Balny, C. (1996). High pressure effects on protein structure and function, *Proteins: Structure, Function, and Genetics*, 24, 81.

Slade, L. and Levine, H. (1995). Glass transitions and water-food interactions, *Advances in Food and Nutritional Research*, 38, 103.

Smeller, L. and Heremans, K. (1997). In: *High Pressure Research in Bioscience and Biotechnology*, Leuven University Press, Leuven, Belgium, p. 55.

Suzuki, K. (1960). Studies on the kinetics of protein denaturation under high pressure, *Review of Physical Chemistry of Japan*, 29, 91.

Thevelein, J., Van Assche, J. A., Heremans, K., and Gerlsma, S. Y. (1981). Gelatinisation temperature of starch as influenced by high pressure, *Carbohydrate Research*, 93, 304.

Weemaes, C., De Cordt, S., Goossens, K., Ludikhuyze, L., Hendrickx, M., Heremans, K., and Tobback, P. (1996). High pressure, thermal, and combined pressure-temperature stabilities of α-amylases from *Bacillus* species, *Biotechnology and Bioengineering*, 50, 49.

chapter sixteen

High-pressure treatment of fruit, meat, and cheese products — equipment, methods and results

Monika Fonberg-Broczek, Jacek Arabas, Ewa Kostrzewa, Arnold Reps, Jacek Szczawiński, Janussz Szczepek, Bozena Windyga, and Sylwester Porowski

Contents

Summary 282
Introduction 282
1 Effect of ultrahigh pressure on vegetative microorganisms
   and spores of chosen bacteria and molds.................................... 283
   1.1 Experimental details 283
   1.2 Results.. 284
   1.3 Conclusions... 285
2 Effect of high pressure on selected strains of lactic acid bacteria,
   raw cow's milk, and ripening cheeses 285
   2.1 Experimental details 285
   2.2 Results.. 286
   2.3 Conclusions... 287
3 Comparison of the influence of high-pressure treatment on
   survival of *Listeria monocytogenes* in minced meat, sliced, cured ham,
   and ripened, sliced cheeses... 288
   3.1 Experimental details 289
   3.2 Results.. 289
   3.3 Conclusions... 290
4 Quality studies of high-pressure processed fruit products 290
   4.1 Experimental details 291
   4.2 Results.. 292
   4.3 Conclusions... 294

5   High-pressure experimental techniques 294
    5.1   Principle of operation 296
    5.2   Pressure vessel 297
    5.3   Pressure medium 297
    5.4   Pressure and temperature measurements 297
    5.5   Pressure packaging 298
References 298

Summary

The increasing importance of high-pressure processing in food preservation requires an extended body of research. This chapter overviews a series of studies performed at the High Pressure Research Center of Poland and describes the equipment developed for these applications. Namely, Section 1 describes the effects of different high-pressure treatments on laboratory strains of microorganisms Gram (+) and Gram (−), identified in literature as components of pathogenic and saprophytic microflora of food, as well as on spores of *Bacillus subtilis*, selected molds and yeasts. In Section 2, the effects of high pressure on selected strains of lactic acid bacteria in raw milk and ripening cheeses are presented. In Section 3, the D-values and z values are calculated for pathogenic *Listeria monocytogenes* after inactivation of this microorganism in minced meat, cured ham, and ripening cheeses. Section 4 reports the studies on physicochemical changes in high-pressure processed strawberry desserts sweetened with aspartame. In Section 5, the high-pressure facilities developed at UNIPRESS for the reported studies are described. New equipment engineered by UNIPRESS for high-pressure investigation in biological and food science is also outlined.

Introduction

Recently, high pressure has increasingly found new applications in biological systems and has been proposed as a promising method for food preservation, alternative to conventional thermal processing. The first high-pressure-processed food products appeared in Japan in the early 1990s as a result of the interdisciplinary research program devoted to introducing high pressure in food processing.

In Europe, high-pressure-processing of foods is still in the stage of research or pilot production. Fundamental and applied research in this area is carried out in many institutions.

In Poland, research on high pressure for food preservation was initiated by the High Pressure Research Center, Polish Academy of Sciences, in 1992. At the beginning, an interdisciplinary research group was created, including members from 14 universities and research institutes.

The first research program devoted to high-pressure-processing of fruit products and the development of high-pressure food processors started by

UNIPRESS in 1993 in close cooperation with the National Institute of Hygiene. This project was then continued and extended in cooperation with the Warsaw Agricultural University, Institute of Agricultural and Food Biotechnology, and Olsztyn University of Agriculture and Technology.

This chapter describes the results of studies and the achievements in the development of the equipment for high-pressure treatment in the framework of the collaborative programs coordinated by the UNIPRESS.

1 Effect of ultrahigh pressure on vegetative microorganisms and spores of chosen bacteria and molds

Food products are excellent environments for pathogen growth, which may cause foodborne diseases. Quality and shelf life of food products depend greatly on the properties of microorganisms contaminating the food. Many methods of food preservation are used for providing microbiologic safety, among which ultrahigh pressure (UHP) seems very promising.

The mechanism of killing microorganisms with UHP is thought to be based on the destruction of the microbial cell membrane caused by changes of the cell volume. Literature data on the parameters of UHP used for inactivation of microorganisms differ considerably. The aim of this study was to investigate the UHP effect on several microorganisms exposed to similar pressure and temperature conditions.

1.1 Experimental details

The studies were carried out for the following microorganisms:

Gram (–) bacteria: *Salmonella enteritidis, Escherichia coli* NCTC 8196 and *Proteus mirabilis,*
Gram (+) bacteria: *Staphylococcus aureus* NCTC 4163 and *Bacillus cereus* ATCC 10876,
yeasts: *Candida albicans* and *Saccharomyces cerevisiae,*
molds: *Aspergillus flavus.*

Suspensions with a specified density of microorganisms in phosphate-buffered saline were prepared. Initial counts were 10^6 cells/ml for *Staphylococcus aureus* and *Proteus mirabilis,* 10^7 cells/ml for other bacteria, yeasts and conidia of *Aspergillus flavus,* 10^5 cells/ml for spores of *Bacillus cereus.* Microbiologic investigations were carried out according to the ISO method used for total number of bacteria, molds, and yeasts.

Culture of bacteria were treated at pressures from 100 to 1000 MPa for 5, 10, and 15 minutes at temperatures of 10 and 20°C. Spores of *Bacillus cereus* were treated in the pressure range from 200 to 1000 MPa for 10 minutes at 10, 20, and 50°C. The experiments were carried out in the Stands for Microbiological Cultures, described in Section 5. Distilled water was used as pressure medium. The samples were closed in PTFE or polypropylene ampoules.

Table 1   Pressure Inactivation of Microorganisms in Phosphate Buffered
Saline (10^6–10^7 cfu/ml.) (Fonberg-Broczek et al., 1996)

Microorganisms	Pressure, MPa	Temperature, °C	Time, min.
Gram (–)			
Salmonella	500	20	10
Escherichia coli	500	20	10
Proteus mirabilis	500	20	10
Gram (+)			
Staphylococcus aureus	800	20	10
Staphylococcus aureus	600	10	10
Spores			
Bacillus cereus	800	50	10
Yeasts			
Candida albicans	400	20	10
Saccharomyces cerevisiae	500	20	10
Molds			
Aspergillus flavus	400	20	10

1.2   Results

The results of the inactivation achieved after 10 minutes of treatment were used to compare the response of the various microorganisms to pressure (Table 1). The inactivation of microorganisms was found to be directly correlated to the level of high pressure, but significant differences between Gram (+) and Gram (–) bacteria were observed. Pressure of 500 MPa applied for 10 minutes reduced the population of Gram (–) bacteria: *Salmonella, E. coli, Proteus mirabilis*, by 6–7 log. Population of 10^6 cells/ml of Gram (+) bacteria — *Staphylococcus aureus* — was inactivated at 800 MPa, 20°C or at 600 MPa, 10°C (Table 1).

Even more resistant were spores of *Bacillus cereus*. The treatment of 1000 MPa, 20°C, 10 minutes, reduced the count of spores by about 1 log. The same effect was observed for 800 MPa applied for 10 minutes. Pressure of 800 MPa, 10 minutes, was sufficient to destroy all spores when the temperature was increased to 50°C. These results are in agreement with other studies of Knorr (1993) and Gould and Tytul (1995), indicating that it is necessary to combine high pressure and temperature treatments for inactivation of all bacteria spores. Conidia of *Aspergillus flavus* were destroyed at 400 MPa, 20°C, 10 minutes.

Inactivation of bacterial cells was irreversible. The ability of bacteria to grow was investigated immediately after the UHP treatment, 24 hours and 48 hours after the treatment. No growth was observed in these conditions. A combined treatment: 800 MPa, 50°C, 10 minutes, killed all *Bacillus cereus* spores — 10^6 cells/ml. For inactivation of conidia of *Aspergillus flavus* the application of a pressure of 400 MPa, 20°C in 10 minutes was effective. Yeasts were inactivated at a pressure of 400 or 500 MPa applied in 10 minutes at 20°C.

1.3   Conclusions

Our studies confirmed a significant variation in pressure sensitivity between Gram (+) and Gram (–) bacteria. A combined treatment with UHP and temperature was found to be necessary for inactivation of *Bacillus cereus* spores. Conidia of *Aspergillus flavus* were inactivated in conditions similar to those for Gram (–) bacteria.

2   Effect of high pressure on selected strains of lactic acid bacteria, raw cow's milk, and ripening cheeses

The survival of microorganisms depends on pressure, time and temperature of processing, composition of medium, and growth phase of microorganisms (Oxen and Knorr, 1993, Earnshaw, 1995; Patterson et al., 1995). Evaluation of the difference between lethality caused by high pressure to microorganisms in buffers and real food systems needs special attention. Some authors observed that milk has a protective effect to microorganisms during pressure treatment.

The objective of this work was to study the effect of high pressure on: (1) selected strains of lactic acid bacteria, (2) changes in raw milk microflora, (3) activity of enzymes and total bacterial count in Camembert, Gouda, and Kurpiowski cheeses at different stages of ripening.

2.1   Experimental details

The effect of high pressure on growth and biochemical activity was studied for the following bacteria: *Lbc. acidophilus T6/1, Lbc. paracasei T140/1, Lbc. rhamnosus T13, Lbc. plantarum T81/1, Lc. lactis ssp. lactis T1/1, Lc. lactis ssp. cremoris T155/1, Lc. lactis ssp. diacetilactis T56/1, Bifidobacterium breve 558.*

After 24 h of cultivation, the strains (MRS, skim milk, or broth with glucose) were treated with pressure ranging from 200 to 1000 MPa at 200 MPa intervals, for 15 minutes at ambient temperature. Control and experimental samples were assayed for population number and fermentation activity in medium containing 1% cellobiose, esculin, galactose, glucose, lactose, maltose, mannose, salicin, or trehalose. The fermentation activity of the strain was examined after 6 to 216 hours of culturing.

Raw milk from a local dairy was used in the experiments. The milk was treated with pressure from 200 to 1000 MPa at 200 MPa intervals for 15 minutes at ambient temperature. The changes in the population of total number of microorganisms, acidic bacteria, proteolytic microorganisms, psychrotrophs, coliforms, heat-resistant bacteria as well as molds and yeasts were determined for evaluating the effect of pressure on milk microflora. Simultaneously, microbiologic analyses were performed in raw milk and in milk pasteurized at 72°C for 15 sec.

Pressurization was conducted using the stand for treatment of microbial cultures and the stand for food samples, described in Section 5.

Cyclic pressure in the range of 200 to 1000 MPa with 200 MPa intervals, lasting 3.5 minutes at ambient temperature, was applied to cheeses of different ripening stage: 2- and 6-week-old Gouda, 2- and 5-week-old Kurpiowski (Emmentaler type), and 5- and 10-day-old Camembert.

The following microorganisms were determined in the cheeses: proteolytic microorganisms, acidic bacteria, *colibacilli*, *enterococci*, spores, and total microbial count. Microbiological analyses were performed immediately after decompression and following a 12-week cheese storage at 5°C.

The proteolytic (substrate casein), aminopeptidase and endopeptidase activities were studied in the protein extracts obtained by centrifugation of citric buffer-homogenized pressurized cheeses and in pressurized extracts of cheese proteins.

2.2   Results

The results obtained show the possibility of an almost total reduction of lactic acid bacteria, cultured in skimmed milk and MRS medium. Although the reduction rate of microorganisms content in skimmed milk at 400 MPa was very high, from the practical point of view, the number of survived bacteria may unfavorably affect the quality of stored fermented products, e.g., yogurt. It seems that only pressures of 600 to 800 MPa could possibly reduce the number of the strains analyzed in the media studied by almost 100%.

The effect of pressure on the inactivation of lactic acid bacteria depends on their initial number in skimmed milk. It was observed that when the survival rate of *Lbc. paracasei* and *Lc. lactis* ssp. *diacetilactis* was 0.0006% and 0.0007%, respectively, the skimmed milk contained 8000 and 9500 cfu/ml, respectively. For *Lbc. rhamnosus*, the milk contained 2000 cfu/ml at the survival rate of 0.0007%. Some strains lost their fermentation activity at 800 to 1000 MPa. In the case of other strains, the inhibition of fermentation activity was dependent on the culture medium. For example, a pressure of 1000 MPa did not inhibit the fermentation activity of *Lbc. paracasei* being cultured in skimmed milk. However, the strain cultured in MRS medium lost their fermentation activity at 800 to 1000 MPa.

Bifidobacterium breve and *Lc. lactis* ssp. *diacetilactis* lost their fermentation activity at 800 to 1000 MPa, irrespective of the culture medium. *Lbc. acidophilus* and *Lc. lactis* ssp. *cremoris* maintained their fermentation activity in the pressure range studied.

It was stated that the level of inactivation of the microorganisms analyzed in raw milk increased with the pressure employed. The significant decrease in the total microbial count appeared at pressures above 200 MPa. At 400 MPa, the total microbial count in the milk decreased about 1.5 log cycles. In milk, subjected to the pressure of 800 MPa, the total microbial count decreased by more than 90%, irrespective of the initial number of microorganisms.

In the milk analyzed, the sensitivity of microorganisms to the applied pressure was dependent on their type. It was found that the groups of microorganisms that are resistant to temperature were also less sensitive to pressure.

It was observed that the number of yeasts and molds, psychrotrophs, and coliforms decreased considerably more rapidly with the rise of pressure, as compared to the acidic and termoduric bacteria and proteolytic microorganisms. For example, at pressure as low as 200 MPa, the reduction varied from 91.5 to 98.95% for psychrotrophs, 97.25 to 98.75% for coliforms and only 40.7 to 47.23% for acidic bacteria. These results indicate that in milk samples Gram (–) bacteria are more sensitive to pressure than Gram (+) microorganisms. A similar effect was observed for model conditions.

Among the microbial groups analyzed, yeasts and molds were the most pressure-sensitive: they were totally killed at 400 MPa.

Psychrotrophs and coliforms were absent in milk subjected to the pressure treatment at 600 MPa, irrespective of their initial level. In milk with a relatively low number of coliforms and psychrotrophs, their almost complete inactivation was obtained at pressure as low as 400 MPa.

At pressures above 600 MPa, the microbiologic contamination of milk was comparable to that observed in the process of pasteurization at 72°C in 15 sec.

It was stated that the degree of microorganisms inactivation in cheeses increased with the rise of pressure, except for the spores. A considerable decrease of total microbial count was observed at pressures above 400 MPa. The application of cyclic pressure increased the reduction of microorganisms. The 100% inactivation of acidic bacteria and proteolytic microorganisms was not obtained, however, although the degree of their reduction at 600 MPa exceeded 99.5%.

Coliforms and *enterococi*, as well as yeasts and molds, were absent in the cheese samples subjected to the pressure of 600 MPa. It was noted that in cheeses exposed to pressure, in which the microbial groups analyzed were totally inactivated, these groups were also absent after cheese storage.

Protease inactivation in the cheeses examined and in their extracts increased with pressure. At pressures above 800 MPa the proteases in cheeses lost their capability to hydrolyze the substrates analyzed. Any significant effect of high pressure on the degradation of Gouda cheese proteins was not stated. In Camembert and Kurpiowski cheeses the degree of proteolysis was dependent on the pressure applied and on the degree of maturity.

The favorable effect of high pressure on the organoleptic properties of cheeses and, especially, on their consistency, was verified.

2.3   Conclusions

The inactivation level of selected acid bacteria by pressure up to 1000 MPa depended on their genera and species and even strain. The fermentation activity was dependent on the pressure level and bacterial strain. Some strains maintained their fermentation activity even at 1000 MPa. High-pressure treatment improves the microbiologic quality of raw milk. Milk pressurized above 600 MPa revealed microbiologic quality similar to that of milk pasteurized at 72°C in 15 sec. Yeasts and molds were inactivated at 400 MPa and psychrotrophic bacteria and coliforms at 600 MPa.

The level of microorganism inactivation depended more on the initial microorganism count than on the cheese type and ripening stage. *Colibacilli, enterococci,* yeasts, and molds were absent in cheese subjected to 600 MPa. Acidic bacteria and proteolytic microorganisms were not inactivated 100% within the range of pressure applied, although their reduction exceeded 99.5% at 600 MPa. The applied pressure caused irreversible inactivation of microorganisms. Proteases present in cheeses lost the hydrolyzing activity at pressures above 800 MPa.

3 Comparison of the influence of high-pressure treatment on survival of Listeria monocytogenes in minced meat, sliced, cured ham, and ripened, sliced cheeses

During industrial processing of food, particularly during slicing and packaging, secondary contamination is practically unavoidable (Berends et al., 1995). Microbial contamination consists of various microorganisms including pathogenic bacteria: *Staphylococcus aureus, Salmonella spp., Listeria spp.* (Kwiatek et al., 1992, Berends et al., 1995). Contamination with *L. monocytogenes* seems to be particularly dangerous. Contrary to the majority of pathogens, *L. monocytogenes* (classified as a psychrotrophic microorganism) has the ability to multiply in vacuum-packaged food products, stored in a refrigerator (Grau and Vanderline, 1990). *Listeria monocytogenes* is responsible for severe food poisonings and may cause death among sensitive persons. Foodborne listeriosis outbreaks have been linked to the consumption of dairy products including raw and pasteurized milk, ice cream, and a variety of cheeses, meat, and meat products as well as chicken and fish (McLauchlin, 1992).

Various control measures have been applied to reduce the incidence of Listeria in food, including an improvement of GMP by introduction of the HACCP concept; however, a complete elimination of the pathogen by those means seems to be impossible (Tompkin, 1990). Properly conducted heat treatment during preparation of food at home, as well as an industrial pasteurization or sterilization, should cause complete inactivation of *Listeria monocytogenes* (Stańczak and Szczawiński, 1996); however, vacuum-packaged food products cannot mostly be subjected to heat treatment.

The aims of this study were: (1) to compare the effect of high pressure on the survival of *Listeria monocytogenes* in minced beef, cured ham, and in three different ripened, hard cheeses (Edamski, Gouda, Podlaski — the most common cheeses in Poland), sliced and vacuum-packaged in polyamide-polyethylene bags; (2) to compare the effect of high pressure on the sensory properties of treated samples.

3.1 Experimental details

Lean beef meat was minced, divided into 10-gram portions, and placed in polyamide-polyethylene bags (Multiseven 78 TOP, Wipak, Finland). In order to restrict the growth of saprophytic microflora (which might affect the determination of *Listeria monocytogenes*), the samples were decontaminated by gamma irradiation at a dose of 10 kGy. Cured, sliced, pasteurized pork ham and hard, sliced cheeses (Edamski, Gouda, Podlaski), were divided into 10 g portions. Every portion, made of two adjoining slices, was placed in commercially used polyamide-polyethylene bags. These samples were not exposed to ionizing radiation. All samples were inoculated with *L. monocytogenes* (a mixture of strains isolated from milk, obtained from the Polish National Veterinary Institute, incubated on BHI at 37°C for 24 h) and vacuum-packed (Henkovac 1000).

The samples were pressurized using a high pressure food processor, described in Section 5.

The samples of minced beef and cured ham were subjected to pressures of 100, 150, 200, 300, and 400 MPa, and the samples of cheeses at 200, 350, and 500 MPa. All samples were treated at 20°C, for 5, 10, and 15 minutes. After treatment the samples were stored at 4°C for 18 hours. The number of surviving *Listeria* in each sample per gram of product was determined by the tenfold dilution method and plating onto Oxford Agar (Oxoid). The plates were incubated for 72 hours at 30°C.

The bacterial counts were logarithmized and statistically analyzed.

Additionally, some control (uninoculated) samples for sensory studies were treated with high pressure. The appearance, color, smell, and taste of samples were evaluated by a trained panel (6 persons) using 9-point quality scores.

3.2 Results

In our previous studies (Szczawiński et al., 1996; and Szczawiński et al., 1997) it was found that the mathematical parameters applied in thermobacteriology, i.e., D-value (time required for decimal reduction of *L. monocytogenes* at given pressure) and z-value (coefficient of resistance of *L. monocytogenes* to high-pressure), can be also useful for predicting the effects of high-pressure treatment on microflora present in food. Because it is generally accepted that the reduction of *L. monocytogenes* by 6 log units in slightly contaminated food is large enough for consumers' safety, 6D values (time required for reduction of bacteria by 6 log cycles at given pressure) were calculated on the basis of the results previously obtained.

The results presented in Table 2 demonstrate that the resistance of *L. monocytogenes* to high pressure is different in various foods. In hard cheeses bacteria are much more resistant to pressure than in minced meat and cured ham. This could be explained by lower water activity in the cheeses than in beef and ham (Johnston, 1995). Much more difficult is to find a reason for large differences

Table 2   6D-Values
(time required for 6 decimal reductions of *L. monocytogenes* at given pressure)

Pressure	Podlaski cheese (min)	Gouda cheese (min)	Edamski cheese (min)	Raw beef (min)	Cured ham (min)
100 MPa	2460.09	1641.92	263.91	745.80	190.80
200 MPa	873.60	607.68	128.88	153.60	169.80
300 MPa	180.95	143.73	70.01	58.80	34.80
400 MPa	49.07	42.52	36.06	15.60	14.40
500 MPa	17.40	15.72	17.64	8.57	6.06

in the 6D values found for three kinds of cheeses, because all of them had similar basic chemical composition (fat, water, and protein content), pH, and physical structure. Those differences are particularly visible at the lowest (100 MPa) pressure (6D value ranged from 264 to 2460 minutes). The 6D values for various products tend to level off with the increase of pressure. Therefore, it seems that the effect of high-pressure treatment on microflora can be predicted better at higher levels of pressure. At 500 MPa the 6D values for particular food products ranged from only 6 to 17 minutes.

Raw meat samples that were subjected to high-pressure treatment revealed undesirable organoleptic changes, especially visible after pressurization at 300 and 400 MPa. Even before opening the experimental bags, changes in color and consistency of the samples and also leakage could be observed. The organoleptic changes of raw meat subjected to high pressure were reported by other researchers (Johnston, 1995).

Samples of cured, sliced, and pasteurized pork ham subjected to the pressure of 400 MPa for 15 minutes did not differ statistically from the untreated samples in general appearance, smell, and taste. High pressure (400 MPa for 15 minutes) had a slight impact only on the color of cured ham.

High-pressure treatment (up to 500 MPa) did not have any influence on appearance, smell, and taste of the ripened, hard cheeses. Only once, in the case of Gouda cheese, was a positive effect of high pressure on the color of samples found (statistically significant difference at $P < 0.05$).

3.3   Conclusions

These results point out that the application of high-pressure technology to the processing of vacuum-packed cured, pasteurized pork ham and hard cheeses can cause reduction in *Listeria monocytogenes* contamination by 6 logarithmic cycles, with practically unchanged organoleptic qualities.

4   Quality studies of high-pressure processed fruit products

Many experimental works (Hoover et al., 1989; Knorr, 1993; Fonberg-Broczek et al., 1996) report that most vegetative forms of microorganisms are pressure-sensitive, especially at the low pH characteristic of fruit products. Under

Table 3 Parameters of Thermal Preprocessing and Pressurization

Sample No.	Preprocessing Temperature	Pressure processing		
		Pressure	Temperature	Time
1	20°C	400 MPa	20°C	10 min
2	75°C	400 MPa	20°C	10 min
3	20°C	400 MPa	50°C	10 min
4	75°C	400 MPa	50°C	10 min

these conditions, spore forms are unable to grow. However, the research reports available do not cover many other important aspects of pressure inactivation of microorganisms in foods.

One of these is the investigation of pressure-induced changes of significant factors determining the acceptability of food products: color, flavor, consistency, and taste, caused by pressure treatment and storage period.

This section describes microbiologic and sensorial investigation performed for selected fruit desserts preserved using UHP methods. More detailed studies, including physicochemical parameters, were carried out for strawberry desserts. Strawberries were chosen due to their high sensitivity to process parameters.

4.1 Experimental details

Pressure processing was carried out using the high pressure food processor described in Section 5. Two series of experiments were performed (series A and B).

In series A, fresh strawberries, raspberry, blueberry, cherry, apricot, pears, and apples of excellent quality were used as raw material. The fruits were crumbled, mixed with pectin, citric acid, and sugar (5 to 15%), warmed up to 60°C for 5 minutes, and packed in 100 ml or 250 ml polystyrene packages, closed with thin metal foil. Then, the packages were placed in the high-pressure vessel and subjected to a pressure of 400 MPa for 15 minutes at ambient temperature. After pressure processing, the fruit desserts were stored for 3 months at temperatures +3° to +5°C.

For series B, desserts were prepared from strawberry *Wega* of excellent quality. The fruits were crumbled by cutting and mixed with a solution of pectin and Aspartame (0.1 g/100 g), preprocessed at 20°C or 75°C and packed in commercial polystyrene packages of 100 and 250 ml closed with metal foil. The desserts were then processed at a pressure of 400 MPa, at temperatures of 20 or 50°C (Table 3).

Quality analysis of the products were performed before pressure-processing (control samples), directly after pressure-processing and after 3 months' storage at +3 to +5°C.

The following measurements of physicochemical parameters were performed: pH, soluble solids (°Brix at 20°C), total titratable acidity (g citric acid/100 g f.w.), total anthocyanins (ACN as pelargonidin-3-glucoside —

PGN), degradation index (Di) using the Fuleki method, instrumental color values: L*, a*, b*, and dominating wavelength λ — using Minolta Chroma Meter CR200, and Aspartame content.

In addition, the evaluation of appearance, color, smell, taste, and consistency was performed by a trained panel (six persons) using 5-point quality scores.

4.2 Results

For series A, microbiologic investigations and sensorial tests of fruit desserts were carried out directly after processing and after 3 months' storage. Microbiologic investigations for content of molds, yeast, and total count of microorganisms (cfu) proved a very good microbiologic quality. Also, sensorial tests revealed high quality of the products, gelatinous consistency, almost natural color, and fresh fruit aroma.

Physicochemical parameters for series B are summarized in Table 4. For all samples tested, pH, soluble solid, and total acidity were not influenced by high-pressure treatment and remained also almost constant during 3 months' storage. No significant changes of total anthocyanins were found directly after high-pressure treatment, but after 3-months storage this parameter decreased by about 50% in all samples tested. The degradation index (Di) remained unchanged directly after pressurization but increased significantly after the storage period. This observation is in agreement with our previous reports (Fonberg-Broczek et al., 1997a; Fonberg-Broczek et al., 1997b) and other authors (Cano et al., 1997). Aspartame contents decreased slightly after pressure treatment (3 to 8%), and after the storage period the decrease was more significant (17 to 22%).

The instrumental color values L*, a*, b*, and dominating wavelength λ were diversely affected by pressure and storage period. The L* value revealed no change both directly after processing and after 3 months' storage. Also no significant changes of the dominating wavelength λ were observed after pressurization and storage period (610 to 613 nm). On the other hand, visible decreases of the parameters a* and b* were observed. For a* it means a shift on the red-green axis, and for b* this decrease is related with a shift on the yellow-blue axis. In spite of the changes of a* and b*, the color of the samples remained satisfactory.

The sensorial evaluation (color, smell, taste, and consistency) in the strawberry desserts was performed after pressurization and after 3 months' storage using 5-point quality scores. The results of postprocessing evaluation were very good, i.e., 5 points for all parameters for samples 1, 2, and 3. The results for sample 4 were slightly lower: taste — 4.5, smell — 4.6, color — 4.7, consistency — 5. This result is probably related to higher temperatures in preprocessing and processing more than to the pressure level. Sensorial evaluation after 3 months' storage revealed only an insignificant decrease of the results. The mean values dropped down to 4.3 for color, 4.6 for smell, 4.3 for taste, and 4.8 for consistency.

Table 4 Physicochemical Parameters of Strawberry Desserts (arithmetic means from 5 estimates)

Sample	When tested	pH	Soluble solid (°Brix at 20°C)	Total acidity (g citric acid/100g)	ACN (mg/100g)	Di	Aspartame (g/100g)
1	Before pressurization	3.5	6.3	0.69	31.7	1.40	0.100
1	After pressurization	3.5	6.3	0.69	31.4	1.40	0.097
1	After 3 months storage	3.4	6.3	0.78	13.6	2.57	0.080
2	Before pressurization	3.5	6.3	0.69	28.8	1.46	0.100
2	After pressurization	3.5	6.3	0.69	28.4	1.48	0.095
2	After 3 months storage	3.5	6.3	0.70	12.8	2.24	0.074
3	Before pressurization	3.4	6.4	0.80	36.2	1.44	0.100
3	After pressurization	3.4	6.4	0.80	36.2	1.47	0.094
3	After 3 months storage	3.4	6.4	0.80	19.0	1.86	0.076
4	Before pressurization	3.4	6.7	0.76	32.9	1.47	0.100
4	After pressurization	3.4	6.7	0.76	31.7	1.53	0.092
4	After 3 months storage	3.4	6.7	0.77	15.2	2.12	0.072

4.3   Conclusions

In our studies, high-pressure treatment did not influence pH, soluble solids, and total acidity, which remained constant or changed only insignificantly, also after 3 months' storage at temperatures +3 to +5°C. Anthocyanins did not suffer from high pressure but decreased significantly during the storage period. Aspartame contents, almost unchanged after pressurization, decreased after 3 months' storage by 20%. Instrumental color values a*, b* appeared to decrease after storage, while L* and dominating wavelength λ did not change. Sensorial evaluation did not reveal any significant changes in the taste, smell, color, and consistency of the strawberry desserts tested.

5   High-pressure experimental techniques

In the early 1990s, the High Pressure Research Center UNIPRESS involved itself in high-pressure research in bioscience and food science. Taking advantage of the high-pressure experimental technique developed earlier for studies in solid state physics and material technology, UNIPRESS created several facilities for new research programs in biologic and food science. Experience from the application of these facilities in own research programs as well as cooperation with West European and Japanese laboratories benefited the development of new custom-oriented equipment for research in food- and bioscience.

For high-pressure treatment of biologic materials and foods the following facilities were used: Stands for Treatment of Microbial Cultures (Arabas and Fonberg-Broczek, 1996), Stands for Treatment of Food Samples (Szczepek and Arabas, 1997), High Pressure Food Processor (Szczepek et al., 1996a) and Gas Pressure Cells (Szczepek et al., 1996b). The main parameters of the facilities are summarized in Table 5.

The experience collected while working with these facilities allowed us to develop and commercialize custom-oriented high-pressure equipment for food- and bioscience:

1. Set-up for pressure research in chemistry and biology U101 (Szczepek and Arabas, 1997),
2. Apparatus for kinetics investigation of pressure-temperature inactivation of enzymes U111.

Set-up U101, Figure 1, is dedicated to high-pressure experiments at pressures up to 1600 MPa and temperatures from 5 to 120°C. From one to eight sample ampoules of volumes between 2 to 25 cc can be placed in the high-pressure vessel. The high pressure in the vessel is generated by a manual hydraulic press of 300 kN. The temperature in the vessel is maintained by a thermostatic jacket or elastic electric tape. Pressure and temperature are measured directly in the vessel using a Manganin pressure gauge and thermocouples Cu-Constantan, displayed digitally. Time of pressurization can

Table 5 Parameters of the High Pressure Facilities

	Stands for treatment of microbial cultures	Stand for treatment of food samples	High pressure food processor	Gas pressure cells
Pressure	2000 MPa	500 MPa	700 MPa	1300 MPa
Temperature	+5 ...120°C	+5..120°C	−20...+80°C	−272...+100°C
Volume	30...100 cc	400...600 cc	1500 cc	2...10 cc
Inner diameter	13. 16. 20 mm	50 mm	110 mm	7...16 mm
Pressure medium	Mixture hexane +pentane. white spirit (water)	Water	Water	Inert gases
Configuration	Piston type	Piston type	Piston type (plunger)	Pump type
Source of pressure	Hydraulic presses	Hydraulic press 2500 kN	Hydraulic press 10 MN	Gas compressor 1500 MPa
Pressure measurement	High-pressure gauge or low-pressure manometer	Low-pressure manometer	High-pressure gauge	High-pressure gauge
Temperature measurement	High-pressure thermocouple or outer thermocouple	Outer thermocouple	High pressure thermocouple	High-pressure thermocouple
Heating/cooling	Thermostatic jacket	Thermostatic jacket	Inner spiral heat exchanger	Cryostat. bath
Other possibilities				Optical accesses. Magnetic fields

Figure 1   Set-up for pressure research in chemistry and biology.

be from several minutes (in biologic tests) up to several days in long-lasting experiments (e.g., in chemical reactions).

The application of the U111 Apparatus for kinetic investigation is to pressurize five samples simultaneously up to 700 MPa, closed in flexible tubes or envelopes, at constant temperatures from 233K to 373K and in variable times. The main feature of the apparatus is the measurement of pressure and temperature using a Manganin gauge and a thermocouple, respectively, in each hydrostatic cell close to the sample or even in the middle of the sample ampoule. The pressure and temperature sensors are linked to the computer data acquisition system. Due to this advantage, the accurate pressure-temperature-time dependence in the sample during treatment are available.

The essential features are subsequently detailed.

5.1   Principle of operation

All high-pressure apparatuses described above are based on the piston–cylinder technique. In this technique, high pressure is generated by the application of a one-dimensional thrust into a pressure-transmitting medium confined in a stiff container. Pressure can be generated directly in the sample vessel (piston type) or in an external generator and piped to the sample vessel (pump type). In the former case, the working volume depends on piston position, while in the latter the sample can reach almost the entire volume of the vessel. The piston type apparatus allows faster compression than pump type systems.

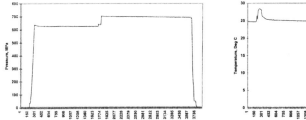

Figure 2 Example of (a) pressure-time and (b) temperature–time dependence measured in the high-pressure vessel.

5.2 Pressure vessel

In most cases, the heart of all high-pressure systems is a multilayer vessel in the form of a compound cylinder made of steel. From one side the vessel is closed with a steady plug, usually provided with high-pressure gauge and thermocouples. From another side, the vessel is closed by a piston or plug incorporating a moveable plunger (in the high-pressure food processor). For pressures up to 700 MPa, plugs and pistons are sealed with multiuse seals. For higher pressures single-use seals are used.

5.3 Pressure medium

Distilled water was customarily used as the pressure-transmitting medium. The main advantage of this medium is that it does not contaminate the sample packaging during and after pressurization. The sample packagings do not need troublesome washing.

Silicone grease was applied to piston and seals for lubrication and improvement of tightness. Other pressure-transmitting media, e.g., mixture of hexane and pentane, extraction naphtha (white spirit), were used for the highest pressures and at low temperatures to guarantee perfect hydrostatic conditions.

5.4 Pressure and temperature measurements

Due to the high density of the transmitting medium under high pressure and to the high thermal inertia of pressure devices, accurate measurements of pressure and temperature require that the sensors are located inside the pressure vessel as close to the sample as possible. Figure 2 presents typical results obtained with such an arrangement. The variation of temperature in relation to pressure can be essential for the analysis of the mechanism of inactivation and as a consequence for designing the equipment.

For high-pressure measurements three kinds of gauges were used: Manganine Pressure Gauge, strain gauge pressure transducer, and Bourdon type manometer. The Manganin Pressure Gauge type MPG10 is in the form of a Manganin wire coil of about 6 mm diameter and resistance 70 to 80 Ohm.

The resistance increases linearly by 23,4×10^{-3} Ohm for 1000 MPa with accuracy 1%. The Manganin gauge works properly in nonpolar liquids and neutral gases. Many strain gauge pressure transducers up to 700 MPa are commercially available. The high pressure food processor was equipped with a transducer with accuracy of 1% mounted on the lower plug of the vessel. The Bourdon type manometer was used for direct measurement of high pressure up to 1000 MPa in the high pressure food processor vessel. In the Stand for Treatment of Food Samples, high pressure could not be measured in the vessel and it was evaluated from the low pressure in the hydraulic presses measured with a Bourdon type manometer.

The temperature in the vessels was measured in three ways. In the Stands for Microbial Cultures and Gas Pressure Cells, the vessel plug was equipped with multi-leads feed-through containing up to seven Cu-Constantan thermocouples. In the high pressure food processor, a lower plug was provided with a high-pressure sheathed thermocouple of 1 mm diameter, tightened using standard cone-to-cone seal. In the Stand for Treatment of Food Samples the temperature was measured using a thermocouple inserted into the thermostatic jacket.

5.5   Pressure packaging

For pressurizing, the samples were placed in containers. These containers were made of flexible material, with tight closure and the possibility of air evacuation. For microbial cultures, commercially available polypropylene-polyethylene tubes 1 and 2 cc or special PTFE ampoules 2, 4, 6, 8, 10, 25 cc, manufactured by UNIPRESS, were used.

Multilayer plastic foil of food grade, heat-sealed, was mainly used for fruit and meat product samples. For cheese samples, multilayer plastic alimentary foil (used in production of sausages) was tied with a cord which guaranteed good tightness. Fruit product samples, pressurized in the high pressure food processor, were packed in commercial polystyrene packages of 100 and 250 ml of 95 mm diameter.

References

Arabas, J. and Fonberg-Broczek, M. (1996). The High Pressure Research Center in Warsaw, Poland. In: *Proceedings of the Second Main Meeting*, Project Process Optimization and Minimal Processing of Foods. Vol. 4 — High Pressure. Oliveira, J.C., Oliveira, F.A.R., and Knorr, D. (Eds.), published by Escola Superior de Biotecnologia, Porto, Portugal, p. 9.

Berends, B., Burt S., and Snijders J. (1995). Critical points in relation to breaking *Salmonella* and *Listeria* cycles in pork production. In: *New challenges in meat hygiene: specific problems in cleaning and disinfection*. Burt, S. A., Bauer, F. (Eds.), ECCEAMST Foundation, Utrecht, Netherlands, p. 9.

Cano, M. P., Hernandez, A., and De Ancos, B. (1997). High pressure and temperature effects on enzyme inactivation in strawberry and orange products, *Journal of Food Science*, 62, 85.

Earnshaw, R. (1995). High pressure microbial inactivation kinetics. In: *High Pressure Processing of Foods*, Ledward, D.A., Johnson, D.E., Earnshaw, R.G., and Hasting, A.P. (Eds.), Nothingham University Press, Loughborough, U.K., p. 37.

Fonberg-Broczek, M., Arabas, J., Górecka, K., Grochowska, A., Karlowski, K., Kostrzewa, E., Szczepek, J., Sciezyñska, H., Windyga, B., Zdziennicka, D., Zurkowska-Beta, J., and Porowski, S. (1997a). High-pressure processed apple and strawberry desserts. Presented at the XXXVth Meeting of the European High Pressure Research Group, Reading, U.K., September 7 — 11.

Fonberg-Broczek, M., Arabas, J., Karlowski, K., Górecka, K., Grochowska, A., Karlowski K., Kostrzewa, E., Szczepek, J., Sciezynska, H., Windyga, B., Witczak, Z., Zdziennicka, D., Zurkowska-Beta, J., and Porowski, S. (1997b). Quality studies of high pressure processed fruit products — preliminary results. In: *Proceedings of the Third Main Meeting*, Project Process Optimization and Minimal Processing of Foods. Vol. 4 — High Pressure. Oliveira, J.C., Oliveira, F.A.R., and Knorr, D. (Eds.), published by Escola Superior de Biotecnologia, Porto, Portugal.

Fonberg-Broczek, M., Windyga, B. Sciezyñska, H. Górecka, K., Grochowska, A., Napiórkowska, B., Karlowski, K., Arabas, J., Jurczak, J., Podlasin, S., Porowski, S., Salañski, P., and Szczepek, J. (1996). The effect of high pressure in vegetative bacteria and spores of *Aspergillus flavus* and *Bacillus cereus*, In: *Proceedings of the Joint XV AIRAPT & XXXII EHPRG International Conference*, Warsaw, Poland, Trzeciakowski, W.A. (Ed.), World Scientific Publishing Co. Ltd., Singapore, p. 892.

Gould, G.W. and Tytul A. (1995). In: *High Pressure Processing of Foods*, Ledward, D.A., Johnson, D.E., Earnshaw, R.G., and Hasting, A.P. (Eds.), Nothingham University Press, Loughborough, U.K., p. 27.

Grau, F. H. and Vanderlinde, P.B. (1990). Growth of *Listeria monocytogenes* on vacuum-packaged beef, *Journal of Food Protection*, 53, 739.

Hoover, D., Metrick, C., Pepineau, A.M., Farkede, F., and Knorr, D. (1989). Biological effects of high hydrostatic pressure on food microorganisms, *Food Technology*, 43, 99.

Johnston, D.E. (1995). High pressure effects on milk and meat. In: *High Pressure Processing of Foods*, Ledward, D.A., Johnson, D.E., Earnshaw, R.G., and Hasting, A.P. (Eds.), Nothingham University Press, Loughborough, U.K., p. 99.

Knorr, D. (1993). Effect of high hydrostatic pressure processes on food safety and quality, *Food Technology*, 47, 156.

Kwiatek, K., Wojtoñ, B., Rola, J., and Rózañska, H. (1992). The incidence of *Listeria monocytogenes* and other *Listeria* spp. in meat, poultry and raw milk., *Bulletin Veterinary Institute in Pulawy*, 35, 7.

McLauchlin, J. (1992). Listeriosis. Declining but may return. *British Medical Journal*, 304, 1583.

Oxen, P. and Knorr, D. (1993). Baroprotective effects of high solute concentrations against inactivation of *Rhodotorula rubra*, *Lebensmittel Wissenchaft und Technologie*, 26, 220.

Patterson, M. F., Quinn, M., Simpson, R., and Gilmour, A. (1995). Effect of high pressure on vegetative pathogens. In: *High Pressure Processing of Foods*, Ledward, D.A., Johnson, D.E., Earnshaw, R.G., and Hasting, A.P. (Eds.), Nothingham University Press, Loughborough, U.K., p. 47.

Stañczak, B.J. and Szczawiñski, J. (1996). Cieploopornoœæ *Listeria monocytogenes* w mleku i smietance o róznej zawartosci tluszczu, *Medycyna Weterynaryjna*, 52 (6), 389. In Polish.

Szczawiński, J., Pconek, J., Szczawińska, M., Porowski, S., Fonberg-Broczek, M., and Arabas, J. (1996). High pressure inactivation of *Listeria monocytogenes* in meat and meat products. In: *Proceedings of the Second Main Meeting*, Project Process Optimization and Minimal Processing of Foods. Vol. 4 — High Pressure. Oliveira, J.C., Oliveira, F.A.R., and Knorr, D. (Eds.), published by Escola Superior de Biotecnologia, Porto, Portugal, p. 51.

Szczawiński, J., Szczawińska, M., Stańczak, B., Fonberg-Broczek, M., Arabas, J., and Szczepek, J. (1997). Effect of high pressure on survival of *Listeria monocytogenes* in ripened, sliced cheeses at ambient temperature. In: *High Pressure Research in Bisoscience and Biotechnology, Proceedings of the XXXIVth Meeting of the European High Pressure Research Group*, Heremans, K. (Ed.) Leuven University Press, Leuven, Belgium, p. 295.

Szczepek, J. and Arabas, J. (1997). Experimental technique for high pressure investigation in chemistry, biology and food science. In: *High Pressure Research in Bisoscience and Biotechnology, Proceedings of the XXXIVth Meeting of the European High Pressure Research Group*, Heremans, K. (Ed.) Leuven University Press, Leuven, Belgium, p. 471.

Szczepek, J., Arabas, J., Fonberg-Broczek, M., Zurkowska-Beta, J., and Porowski, S. (1996a). Equipment for laboratory high pressure processing of fruit products in the High Pressure Research Centre. In: *Proceedings of the Second Main Meeting*, Project Process Optimization and Minimal Processing of Foods. Vol. 4 — High Pressure. Oliveira, J.C., Oliveira, F.A.R., and Knorr, D. (Eds.), published by Escola Superior de Biotecnologia, Porto, Portugal, p. 46.

Szczepek, J. Arabas, J., and Porowski, S. (1996b). High Pressure laboratory gaseous systems. In: *Advances in Instrumentation*, Proceedings of the International Conference on Instrumentation, Bangalore, India, Rampsrad, B.S., Asokan, S., Rajana, K., and Shivaprash, N.C. (Eds.), New Age International Limited Publishers, Special Issue J. Inst. Soc. India, 26, (3), 20.

Tompkin, R. B. (1990). The use of HACCP in the production of meat and poultry products. *Journal of Food Protection*, 53, 795.

chapter seventeen

Combined high-pressure/temperature treatments for quality improvement of fruit-derived products

Pilar M. Cano, Almudena Hernández, and Begoña De Ancos

Contents

Summary 301
1 Introduction 302
2 Materials and methods 303
    2.1 High-pressure equipment and treatments.................................303
    2.2 Experimental design..304
3 Effects of combined high pressure/temperature treatments
    on enzyme activities...305
4 Effects of combined high-pressure/temperature treatments
    on microbiologic quality..308
5 Quality and stability of pressurized fruit-derived products................309
6 Conclusions..310
Acknowledgments..311
References..311

Summary

This chapter describes a study performed on the effect of high-pressure treatments, alone or combined with temperature treatments, on enzyme inactivation, microbial lethality, and selected sensory factors in fruits and fruit-derived products, namely orange, strawberry, kiwi fruit, tomato and orange juice, and strawberry, kiwi fruit and tomato purees. The inactivation of peroxidase, polyphenoloxidase, and pectin methylesterase was studied. At conditions able to cause enzyme inactivation, microbial activity was not found in orange juice, but in other products microbial cultures were detected in some cases. The selection of the adequate pressure, temperature, and processing time factors to ensure a high-quality product must consider the subsequent storage and distribution conditions.

1 Introduction

Fruit-derived products are traditionally produced by thermal processing, which inhibits microbial spoilage and reduces enzymatic activity. Although ensuring safety and shelf life, thermal processing dramatically affects quality factors, particularly aroma and vitamin and volatile content. High-pressure processing has been used as an alternative with the ability to preserve much better quality of the fresh products (Donsi et al., 1996).

High hydrostatic pressure treatment reduces microbial counts and enzyme actitvity and affects product functionality (Hoover et al., 1989; Cheftel, 1990; Farr, 1990; Traff and Bergmann, 1992). This provides a good potential basis for development of new processes for food preservation or product modifications (Mertens and Knorr, 1992). The first commercial products made using high-pressure treatments have been almost exclusively plant or products containing plants (Knorr, 1995). In 1990, Meidi-ya Food Co. (Osaka, Japan) introduced into the market apple, strawberry, and kiwi jams that had been sterilized using pressure alone (Mozhaev et al., 1994). Ogawa et al. (1989a, 1990) reported that the high-pressure treatment could sterilize juices from various citrus fruits without major changes in their nutritive components, natural flavor, and taste. Also, these authors described that complete inactivation of pectinesterase was not attained after pressurization at 3000 or 4000 bar for 10 min, but that this partly inactive enzyme did not recover under ordinary conditions of storage and transportation.

High-pressure effects on microorganisms are one of the most important tools in food processing. Early studies on the effects of high hydrostatic pressure on foods and food microorganisms showed that subjecting milk to a pressure of 680 MPa for 10 min reduced the bacterial count from 10^7 cells/ml to 10^1 to 10^2 cell/ml (Hite, 1899). Fruits, such as peaches and pears, were shown to remain preserved for five years after pressure treatments (Hite et al., 1914); the stability of the products was maintained because spoilage microorganisms, yeast and lactic acid bacteria, are barosensitive and the low pH of fruit did not support spore outgrowth (Hoover et al., 1989). Citrus juices sterilization (Ogawa et al., 1989b) or fresh cut pineapple pasteurization (Aleman et al., 1994) are two examples of high-pressure application to the preservation of fruit-derived products. Pressures exceeding 300 MPa have been reported ineffective in destroying bacterial spores (Sale et al., 1970). Treatment of *Bacillus* spp. spores at pressures of 100 to 300 MPa were more lethal than higher pressure (up to 11,800 MPa). Lower pressures (100 to 300 MPa) may induce spore germination and the outgrowth stage may be more sensitive to pressure (Hoover et al., 1989).

High-pressure effects on enzymes are another of the most important tools in the development of new processed foods (Hendrickx et al., 1998). Effects of high pressure on enzymes may be divided in two classes: (1) low pressures ($\cong$ 100 MPa) have been shown to activate some enzymes (Curl and Jansen, 1950; Asaka et al., 1993; Jolibert et al., 1994), explained by the stimulation effect of pressure on monomeric enzymes; (2) pressures higher than 100 MPa, which induce enzyme inactivation (Curl and Jansen, 1950). Miyagawa et al. (1964) explained that food enzyme inactivation may be distinguished depending on the kind of enzyme, and based on their loss and recovery of its activity. The enzyme activation can arise from pressure-induced decompartmentalization, which can be destroyed upon the application of low pressure (Jolibert et al., 1994; Butz et al., 1994). Hendrickx et al. (1998) reported that there is a minimum pressure below which no or little enzyme inactivation occurs. When pressure exceeds this value, enzyme inactivation (within a specified time interval) increases until completed at a certain pressure. This pressure inactivation range is strongly dependent on the type of enzyme, pH, medium composition, temperature, etc. (Cheftel, 1990; Balny and Masson, 1993).

Application of combined high-pressure and temperature treatments on processing of fruit-derived products without any additives is an important objective of the food industry. This chapter reviews some of the results obtained from Spanish research projects on the application of high-pressure technology to food. The studies conducted on fruit-derived products were related both to basic and technologic aspects of this physical food processing.

Plant materials employed in these studies were orange (*Citrus aurantium*, cv. Salustiana), strawberry (*Fragaria ananassa*, cv. Chandler), kiwi fruit (*Actinidia chinensis*, cv. Hayward) and tomato (*Lycopersicum esculentum*, cv. Pera). Fruit-derived products were juice from oranges and purees from the other three fruits (tomato, strawberry, and kiwi fruit). In all cases, any previous treatments or addition of preservatives were made before pressurization.

2 Materials and methods

2.1 High-pressure equipment and treatments

High-pressure treatments were performed in a hydrostatic pressure unit with 2350 mL capacity, a maximum pressure of 500 MPa, and a potential maximum temperature of 95°C (Gec Alsthom ACB 900 HP, Type ACIP n°. 665, Nantes, France). Pressure was increased and released at 2.5 MPa/s. The pressure treatment was divided into two steps: the first lasting for 15 min and the second following subsequently for 1 to 15 extra min. The temperature of the hydrostatic medium (initial sample at atmospheric pressure: 20°C) ranged between 20 and 60°C. The general scheme of the experiments conducted is shown in Figure 1.

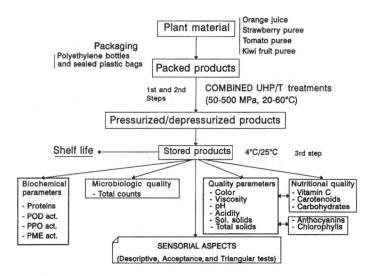

Figure 1 General scheme of the HP/T treatments of fruit-derived products performed.

Table 1 Levels of Variables for HP/T
Fruit-Derived Products According
to the Experimental Design

Pressure (MPa)	Temperature (°C)
50	20.0
101.2	25.8
225.0	40.0
348.7	54.1
400.0	60.0

2.2 Experimental design

High-pressure treatments were carried out in triplicate. Values from chemical
and biochemical analyses are averages of three independent determinations.
The STATGRAPHICS (Statistical Graphics system, vers. 7.0, USA) software
was employed for statistical data analysis and graphical presentation.

Surface response methodology (SRM) was used to study the simultaneous effect of two processing variables. The experiments were designed
according to a central rotatable design (Cochran and Cox, 1957). The variables (factors) studied were pressure and temperature (in the 1st step). Five
levels of each variable were chosen in accordance with the principles of the
central composite design (Table 1), for enzymatic studies. Thirteen combinations of two variables were performed following the designs of Cochran and
Cox (1957). The error was determined from replication, as suggested in the
design.

For each factor assessed a second-order polynomial equation was fitted:

$$y = b_o + \sum_{i=1}^{K} b_i x_i + \sum_{i=1}^{K} b_{ii} x^2 + \sum_{i=1}^{K} \sum_{i=1}^{K} b_{ij} x_i x_j \tag{1}$$

where y is the estimated response, b_0, b_i, b_{ii} are the equation parameter estimates (constant, b_0; parameter estimates for linear terms, b_i, for quadratic terms, b_{ii}; for interaction terms, b_{ij}), x_i, x_j are levels of factors, and k the number of factors. For each factor the variance was partitioned into linear, quadratic, and interaction components in order to assess the adequacy of the second-order polynomial function and the relative importance of the components. The significance of the equation parameters for each response variable was assessed by a t-test.

3 Effects of combined high pressure/temperature treatments on enzyme activities

High-pressure/temperature treatments of fruit-derived products differed in inactivation of enzyme activities, depending on the plant characteristics, the kind of enzyme, and the processing conditions. Peroxidase (POD) in strawberry puree was increasingly inactivated up to 300 MPa for treatments carried out at room temperature (20°C) and 15 minutes of pressurization. HP/T treatments above 300 MPa were similar to those of previously reported studies (Balny and Masson, 1993; Anese et al., 1995). Temperature affected POD activity in a similar way, and POD was inactivated by HP/T and temperature treatments up to 45°C and pressures up to 280 MPa. Above that temperature a decrease in inactivation was observed for all pressures (Cano et al., 1997). The effects of HP/T combined treatments on POD activity in orange juice were slightly different. POD activity underwent a continuous decrease up to 400 MPa at room temperature (20°C), process time 15 minutes. A 25% reduction of initial POD activity was observed in these samples. However, HP/T treatments conducted at 32 to 60°C adversely affected the enzyme activity, producing a strong activation under these conditions (Cano et al., 1997). This behavior was also observed in fresh-made tomato puree. A reduction of POD activity (25%) was obtained in this product treated at 350 MPa/20°C, but a combination of higher pressures and mild temperatures (30 to 60°C) produced an enhancement of this activity (Hernández and Cano, 1998). In the fourth assayed fruit-derived product, kiwi fruit puree, the combination of HP/T treatments affects the POD activity in a very different way. This enzyme was continuously inactivated at room temperature up to 500 MPa (see Figure 2). Combinations of pressure with higher temperatures (30 to 60°C) did not render better results.

Polyphenol oxidase activity (PPO) was also differently affected by HP/T treatments depending on the fruit-derived product. In tomato pureee, this

control: 1.30 OD/min/mg protein

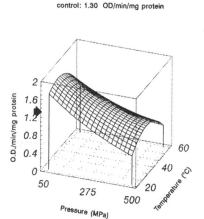

Figure 2   Effect of combined HP/T treatments on peroxidase activity (POD) in kiwi fruit puree.

control: 2,09 OD/min/g f.w.

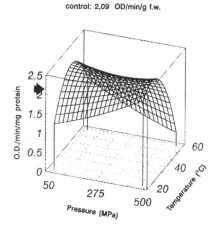

Figure 3   Effect of combined HP/T treatments on polyphenol oxidase activity (PPO) in kiwi fruit puree.

activity did not show any significant change due to the treatments, only a combination of 200 MPa/20°C seemed to produce a significant loss (10%) in PPO activity. However, in strawberry puree the PPO activity strongly diminished up to 285 MPa, at room temperature and 15 minutes of treatment (Anese et al., 1995). Strawberry samples treated at different pressures combined with temperatures ranging from 30 to 60°C showed a significant increase in PPO activity. Kiwi fruit PPO activity was affected in a different way. The activity increased up to 500 MPa working at room temperature (Figure 3), but treatments combining temperatures near 60°C produced a significant PPO inactivation (80%).

control: 2.58 OD/min/g f.w

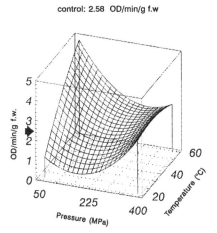

Figure 4   Effect of HP/T treatments on pectin methylesterase activity (PME) in straw-berry puree.

Pectin methylesterase activity (PME) is also influenced by HP/T treatments in fruit-derived products. Tomato puree PME was reduced to 35% of initial activity using a combined treatment of 150 MPa/30°C for 15 min (Hernández and Cano, 1998). This enzyme is the most affected by high-pressure treatments in tomato products. However, PME of orange juice was activated when pressure was applied at room temperature and only a reduction of 25% was observed in juice treated by a combination of 200 MPa/30°C. Increasing the processing temperature strongly affected the observed activation. Only combinations of low pressures and mild temperatures significantly inactivated PME in orange juice (Cano et al., 1997). A very different behavior was observed in strawberry PME (Figure 4), that was continuously activated by combined treatments of high pressure and temperature. The enzyme activation observed at moderate high pressures could be attributed to reversible configuration and/or conformation changes of the enzyme and/or substrate molecules (Ogawa et al., 1990; Balny and Masson, 1993; Anese et al., 1995), while the enzyme activation at high pressures could be most probably related to the liberation of isoenzyme forms linked to cellular membranes damaged by the pressure treatments.

Differences in the effectiveness of combined HP/T treatmemts in POD and PPO activities were related to the fruit pH and soluble solids content. Orange juice had lower pH (3.7) and higher soluble solids (11.3°Brix) than strawberry puree (pH 3.9; soluble solids 8.7° Brix). Since strawberry pH was higher, enzyme inactivation may have been more favored than in orange juice (lower pH) (Anese et al., 1995). However, tomato puree having higher pH (4.08) and lower soluble solids (5.6°C) than orange juice showed a greater PME inactivation using combinations of HP/T. From these results we can conclude that other factors in addition to pH and soluble solids must contribute to the effectiveness of combined HP/temperature treatments in enzyme inactivation of fruit-derived products.

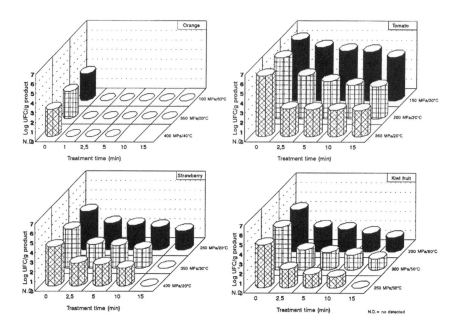

Figure 5 Effects of combined HP/T treatment on microbiologic quality of some fruit-derived products.

4 Effects of combined high-pressure/temperature treatments on microbiologic quality

Microbiologic quality is a very important aspect in the preservation of fruit-derived products. In the present study on the high-pressure effects on orange, strawberry, tomato and kiwi fruit products, the total plate counts of pressurized products was evaluated. In this assay, a previous selection of the combined HP/T treatments selected in terms of enzyme inactivation effectiveness was carried out, in order to study the individual effect of pressurization time on the microbiologic quality (see Figure 5). In all four fruit products there was a significant decrease in the total plate counts when the treatment time increased up 15 min. In fresh orange juice the microbial population after pressurization at any selected combined treatment was not detectable. However, in tomato puree any combined treatment selected attending to its enzyme inactivation capacity did not produce total inactivation of microorganisms. Only the application of a combination of 350 MPa/20°C for 15 min produced a tomato puree with near 50% reduction of its microbial population. Strawberry and kiwi fruit purees showed an intermediate behavior between orange juice and tomato puree. It is possible to obtain nondetectable counts using combinations of 350 MPa/30°C for 15 min or 400 MPa/20°C for 15 min in strawberry puree and a combination of 350 MPa/50°C for 15 min in kiwi fruit puree to obtain a pressurized/depressurized product without any detectable microbial population (see Figure 5).

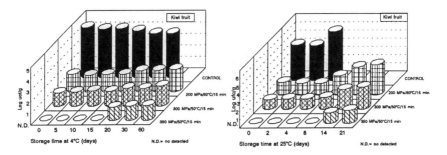

Figure 6   Evolution of microbiologic quality of pressurized kiwi fruit puree during storage.

Other "hurdles" must be combined together with pressure and temperature in order to increase the effectiveness of preservation in tomato-derived products, taking into account its especially high microbial population and their nature (including yeasts, molds, and bacterial spores, very resistant species to HP/T).

5   Quality and stability of pressurized fruit-derived products

Quality and stability of pressurized/depressurized fruit-derived products were evaluated in terms of sensorial, microbiologic, and nutritional aspects, as shown in Figure 1. Storage of pressurized/depressurized products was carried out at 4 to 5°C and 25°C. Physical (color, viscosity,...), physicochemical (pH, acidity, soluble solids, moisture content,...), and chemical (vitamins, sugars, acids,...) parameters were analyzed together with the enzyme activities and microbial plate counts employed for HP/T treatment selection. Also, descriptive and overall acceptance sensorial tests were conducted to study the consumer response to the pressurized products throughout storage. Three combined HP/T treatments were selected for each fruit product, as indicated in Section 3.

Figure 6 shows the evolution of microbial plate counts in HP/T treated kiwi fruit puree when stored at 4 and 25°C for 60 and 21 days, respectively. Combined treatment (350 MPa/50°C/15 min) produced a very stable kiwi fruit puree for up to 60 days of storage. This treated product only showed a slight increase in the microbial population at 20 days of storage, but this increase did not produce a significant microbiologic problem for the safety and stability of the product. Higher storage temperatures (25°C) severely limit the shelf life of the product; only pressurized kiwi fruit puree can be successfully stored for 8 days at this temperature.

Figure 7 shows the changes in vitamin C content in pressurized treated tomato products during cold storage (4°C). In this product, vitamin C suffered a continuous decrease during product storage, but this vitamin loss

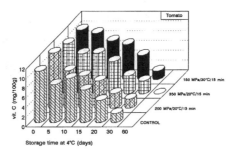

Figure 7 Changes in vitamin C content in pressurized tomato puree during cold storage at 4°C.

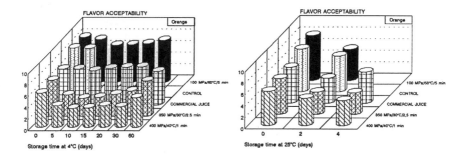

Figure 8 Changes in flavor acceptability of pressurized orange juice during storage.

was signficantly greater in the control (nonpressurized samples) than in HP/T treated ones, especially in tomato puree treated by a combination of 350 MPa/20°C/15 min. This treated sample showed a twofold greater vitamin C content than the control at 30 days of storage.

Regarding the sensorial acceptance of the pressurized/depressurized fruit-derived products in terms of flavor acceptability, the treated orange juice can be cold stored for a period of 30 days at 4°C (see Figure 8). However, the storage time for this product was severely limited when storing at 25°C. Fresh squeezed orange juice pressurized at the right conditions showed higher flavor acceptability than other novel commercial orange juices made from reconstituted juices and preserved by HTST technology.

6   Conclusions

Results from the first study conducted at the Plant Foods Science & Technology Department of Instituto del Frío (CSIC) through a national Spanish-financed project on the Application of High Pressure Technology to Plant Foods, led to some conclusions:

- Combined high-pressure/temperature treatments affect soluble proteins depending on the vegetable matrix. Peroxidase (POD), polyphenol oxidase (PPO), and pectin methylesterase (PME) activities are also modified by HP/T treatments and their changes are related to their intrisic protein characteristics and to the plant matrix (pH, sugars, organic acids,...). The major results obtained were: PME inactivation (strawberry>orange>kiwi fruit>tomato); PPO inactivation (kiwifruit>strawberry>tomato); and POD inactivation (kiwi fruit>strawberry>orange>tomato).

- Stabilization of fruit-derived products can be successfully obtained by HP/T treatments, but selection of the right conditions for each product must be previously made. Combination of HP/T treatment must be selected attending to the subsequent conditions of storage and distribution of the pressurized/depressurized product (shelf life, packaging, etc...).

- Flavor and color of pressurized fruit products are not always unchanged. Relatively high pressures can produce off-flavors and combinations of high pressure/mild temperatures can also cause fruit pigment modifications.

- Further research is needed to improve the basic knowledge of the relationship between plant constituents and the effects of high pressure in plant foods.

Acknowledgments

This study was conducted through Spanish financed projects no. ALI94-0786 and no. ALI97-0759 of Plan Nacional de Tecnología de Alimentos, and project no. 06G/053/96 of Comunidad de Madrid, Spain.

References

Aleman, G., Farkas, D.F., Torres, J.A., Wilhensen, S., and McIntyre, S. (1994). Ultra-High pressure pasteurization of fresh cut pineapple. *Journal of Food Proct.*, 57, 931.

Anese, M., Nicoli, M.C., Dall Aglio, G., and Lerici, C.R. (1995). Effect of high-pressure treatments on peroxidase and polyphenoloxidase activities. *Journal of Food Biochemistry*, 18, 285.

Asaka, M., Aoyama, Y., Ritsuko, N., and Hayashi, R. (1993). Purification of a latent form of polyphenol oxidase from La France pear fruit and its pressure-activation. *Bioscience, Biotechnology and Biochemistry* 58, 1486.

Balny, C. and Masson, P. (1993). Effects of high pressure on proteins. *Food Rev. International*, 9, 611.

Butz, P., Koller, W.D., Tauscher, B., and Wolf S. (1994). Ultrahigh pressure processing of onions: chemical and sensory changes. *Lebensmittel Wissenchaft und Technologie*, 27, 463.

Cano, M.P., Hernandez, A., and De Anco, B. (1997). High pressure and temperature effects on enzyme inactivation in strawberry and orange products. *Journal of Food Science*, 62, 85.

Cheftel, J.C. (1990). Applications des hautes pressions en technologie alimentaire. *Act. Ind. Aliment. Agro-Aliment.*, 108, 141. In French.

Cochran, W.G. and Cox, G.M. (1957). *Experimental design*, 2nd Ed. Wiley International, New York.

Curl, L. and Jansen, E.F. (1950). Effect of high pressures on trypsin and chymotrypsin, *Journal of Biology and Chemistry*, 184, 45.

Donsi, G., Ferrari, G., and Di Matteo, M. (1996). High pressure stabilization of orange juice: evaluation of the effects of process conditions. *Italian Journal of Food Science*, 2, 99.

Farr, D. (1990). High pressure technology in the food industry. *Trends in Food Science and Technology*, 1, 14.

Hendrickx, M., Ludikhuyze, L., Van den Broek, I., and Weemans C. (1998). Effects of high pressure on food quality related enzymes. *Trends in Food Science and Technology*, In press.

Hernández, A. and Cano, M.P. (1998). High-pressure and temperature effects on enzyme inactivation on tomato puree. *Journal of Agriculture and Food Chemistry*, 46, 266.

Hite, B.H. (1899). The effects of pressure in the preservation of milk. *West Virginia Agricultural Experiment Station Bulletin*, 58, 15.

Hite, B.H., Giddins, N.J., and Weakly, C.E. (1914). The effects of pressure on certain microorganisms encountered in the preservation of fruits and vegetables. *West Virginia Agricultural Experiment Station Bulletin*, 146, 1.

Hoover, D.G., Metrick, C., Papineau, A.M., Farkas, D.F., and Knorr, D. (1989). Biological effects of high hydrostatic pressure on food microorganisms, *Food Technology*, 43, 99.

Jolibert, F., Tonello, C., Sageh, P., and Raymon, J. (1994). Les effects des hautes pressions sur la Polyphénol oxydase des fruits, *Bioscience et Boissons*, 251, 27. In French.

Knorr, D. (1995). High pressure effects on plant derived foods. In: *High Pressure Processing of Foods*, Earnshaw, D.E. and Hasting, A.P.H. (Eds.), Nottingham University Press, Nottingham, U.K., p. 123.

Mertens, B. and Knorr, D. (1992). Development of nonthermal processes for food preservation. *Food Technology*, 46, 124.

Miyagawa, K., Sannoe, K., and Suzuki, K. (1964). Studies on Taka-Amylase A under high-pressure treatment; part II: recovery of enzymic activity of pressure inactivated Taka-Amylase A and its enhancement by retreatment at moderate pressure. *Archives of Biochemistry and Biophysics*, 106, 467.

Mozhaev, V.V., Heremans, K., Frank, J., Masson, P., and Balny, C. (1994). Exploiting the effects of high hydrostatic pressure in biotechnological applications. *Tibtech.*, 12, 493.

Ogawa, H., Fukuhisa, K., and Fukumoto, H. (1989a). In: *Use of High Pressure in Food*, Hayashi, R. (Ed.), San-ei Shuppan Co., Kyoto, Japan, p. 57.

Ogawa, H., Fukuhisa, K., Fukumoto, H., Hori, K., and Hayashi, R. (1989b). Effects of hydrostatic pressure on sterilization and preservation of freshly squeezed non-pasteurized citrus juices. *Nippon Nogeikaagaku Kaihi*, 63, 1109.

Ogawa, H., Fukuhisa, K., Kubo, Y., and Fukumoto, H. (1990). Pressure inactivation of yeasts, molds, and pectin methylesterase in Satsuma mandarin juice: effects of juice concentration, pH, and organic acids, and comparison with heat inactivation. *Agriculture Biology and Chemistry*, 54, 1219.

Sale, A.J.H., Gould, G.W., and Hamilton, W. A. (1970). Inactivation of bacterial spores by hydrostatic pressure, *Journal of General Microbiology*, 60, 323.

Traff, A. and Bergman, C. (1992). High pressure equipment for processing. In: *High Pressure and Biotechnology*, Balny, C., Hayashi, R., Heremans, K., and Masson, P. (Eds.), J. Libbey Eurotext Ltd, London. p. 509.

chapter eighteen

Influence of culturing conditions on the pressure sensitivity of Escherichia coli

Christian Schreck, Günter van Almsick, and Horst Ludwig

Contents

Summary 313
1 Introduction 314
2 Materials and methods 314
   2.1 Materials 314
      2.1.1 Microorganisms 314
      2.1.2 Substances 314
   2.2 Methods 314
3 Results and discussion 315
   3.1 *E. coli* strain ATCC 11303 315
   3.2 *E. coli* strain ATCC 39403 319
4 Conclusions 323
Acknowledgments 323
References 324

Summary

The kinetics of pressure inactivation was investigated for two strains of *Escherichia coli* at 25°C and 2.5 or 3 kbar. According to how the bacteria were grown, with or without oxygen, their sensitivity to pressure and the inactivation kinetics differed markedly. The characteristics of the kinetic curves also changed with pH. Bacteria from the exponential growth phase were more sensitive to pressure than those from the stationary phase. This fact and also the biphasic behavior of inactivation curves were shown to be caused very probably by various stages of the bacterial development in the cell cycle and by different amounts of protective proteins.

1   Introduction

Escherichia coli are facultatively aerobic bacteria. They grow very well in aerated media but can also develop and reproduce if oxygen is missing. Under anaerobic conditions the metabolism changes leading to alterations in the composition of the bacteria (Neidhard, 1987).

Therefore it might be expected that growth conditions and oxygen availability influence the sensitivity to pressure and the kinetics of inactivation.

2   Materials and Methods

2.1   Materials

2.1.1   Microorganisms

Escherichia coli ATCC 11303 were aerobically or anaerobically grown in Standard I nutrient broth. *Escherichia coli* ATCC 39403 were grown under aerobic conditions in M9 minimal medium. The bacteria were purchased from the Deutsche Sammlung von Mikroorganismen und Zellkulturen, Braunschweig, Germany.

2.1.2   Substances

Chloramphenicol (DAB 10, BP 93) was obtained from Eu-Rho-Pharma GmbH, Kamen-Heeren, Germany, and Trimethoprim (DAB 10) from Synopharm, Barsbüttel/Hamburg, Germany. Nutrients were from Difco, Detroit, U.S.A. or from Merck, Darmstadt, Germany. All the other substances used were of p.a. or DAB 10 quality.

2.2   Methods

The bacteria were cultured just before each experimental run. Usually they were harvested in the early stationary phase and thus, suspended in nutrient broth, used in high-pressure experiments. For these, they were enclosed in polyethylene tubes and then filled into the ten small pressure vessels of the high-pressure device (Figure 1). All vessels were thermostatted at 25°C and pressurized to 2.5 or 3 kbar. The single vessels were opened at different times to measure the kinetics of inactivation.

The surviving organisms were then counted as colony-forming units on agar plates (cfu = viable bacteria per ml).

The culturing flask was either almost filled with the liquid medium to obtain nearly anaerobic conditions with restricted oxygen supply or it contained a small amount of liquid and much of air for aerobic growth. In some

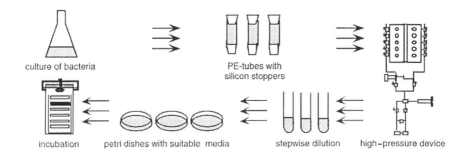

culture of bacteria

PE-tubes with
silicon stoppers

incubation   petri dishes with suitable media   stepwise dilution   high-pressure device

Figure 1   Scheme of experimental procedure.

cases of culturing *E. coli* ATCC 11303 additional oxygen was blown through the medium.

3   Results and discussion

3.1   E. coli strain ATCC 11303

Figure 2 shows the inactivation of *E. coli*, strain ATCC 11303, at 2.5 kbar and two different temperatures. The bacteria had been grown aerobically, the culture being gased with additional oxygen. In Figure 2 the logarithm of colony-forming units is plotted as a function of time. The controls had been at the same temperatures for the time indicated, but under normal pressure. The dashed line indicates the detection limit.

The inactivation curves are very different, depending on whether the temperature is low or high. At high temperature there is a simple first-order reaction, a straight line in the semilogarithmic plot, but at the lower temperature of 30°C or less we have a biphasic inactivation. Deviations from simple first-order reactions are also known for other inactivation methods (Rahn, 1929; Moats, 1971; Wallhäußer, 1995).

All the following results are at 25°C and thus show this biphasic behavior. Such curves can be described by three parameters: the D-values of the two linear parts, describing the slopes, and the ratio of sensitive to less sensitive specimens which is obtained by extrapolating the long-time linear line to the ordinate. The D-value is the time in minutes needed to kill 90% of the bacteria, i.e., to reduce them by one decade. The interpretation of biphasic behavior is that the population contains a small fraction of less sensitive specimens in spite of the fact that it has been grown from a single clone.

The initial pH value of the culture medium is 7.5 and it usually changes to 7 or somewhat smaller values in aerobic growth. However, in a growing culture of *E. coli* gased with additional oxygen, the pH reaches a steady value of 9.3 in the stationary phase (Figure 3).

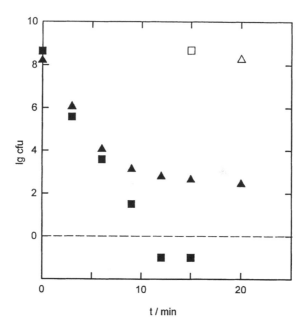

Figure 2   Inactivation of *E. coli* at 2.5 kbar, ▲30°C; and ■ 50°C; open symbols are controls.

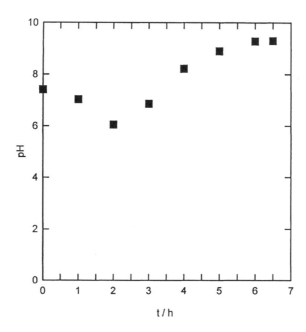

Figure 3   Development of pH during aerobic growth (gased with oxygen) of *E. coli* at 37°C in Standard I broth.

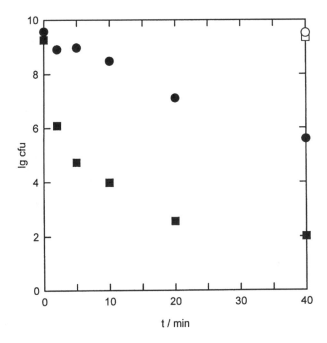

Figure 4 Pressure inactivation of aerobically grown *E. coli* (gased with oxygen) at 2.5 kbar and 25°C; ■ in the culture medium, pH 9.3; ● centrifuged and suspended in a new culture medium, pH 7.5; open symbols are controls.

The reason for this strange development of pH in the growing culture is the fact that the oxygen gas expels the carbon dioxide from the solution and thus changes the buffer system.

Figure 4 gives the inactivation of the bacteria at pH 9.3. If the bacteria are transferred to the original medium with pH 7.5 they are much more resistant. This is shown in Figure 4, too, where the fast inactivation of the sensitive part of the bacteria is 18 times slower at pH 7.5 compared with that at pH 9.3.

The inactivation curve changes dramatically if the bacteria have been grown anaerobically, deficient in oxygen. Now a distinct lag time appears before the inactivation of the insensitive fraction starts (Figure 5).

The description of such a curve needs four parameters: two slopes, less sensitive fraction, and duration of the lag time. In such an anaerobic culture the pH has become 6 because a lot of fatty acids have been built in the fermentation process. If the pH is carefully adjusted to a value of 8, the lag time disappears and the insensitive fraction of the bacteria becomes smaller (Figure 6). If the bacteria are resuspended in fresh medium adjusted to pH 6, the lag time comes back.

Bacteria that had been grown with oxygen do *not* show a lag time, even if they are suspended in a medium used for anaerobic growth with pH 6 (Figure 7).

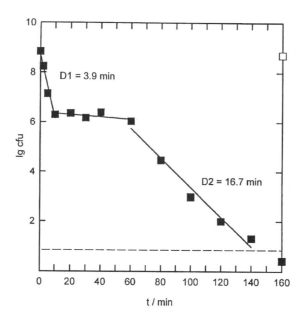

Figure 5 Pressure inactivation of *E. coli* at 2.5 kbar and 25°C, grown for 6 h with restricted oxygen supply, pH 6; open symbol is control.

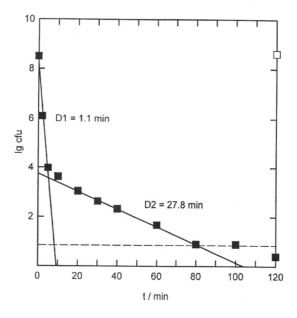

Figure 6 Pressure inactivation of *E. coli* at 2.5 kbar and 25°C, grown for 6 h with restricted oxygen supply, centrifuged, washed in 0.9% NaCI solution, centrifuged and resuspended in the same broth in which the bacteria had been grown but sterilized before and adjusted to pH 8; open symbol is control.

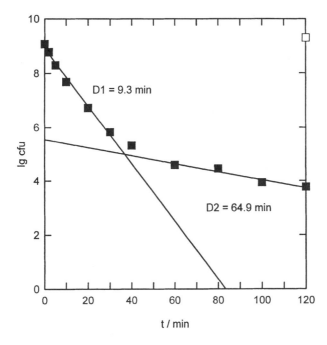

Figure 7 Pressure inactivation of *E. coli* at 2.5 kbar and 25°C, aerobically grown for 6 h (gased with oxygen), centrifuged and suspended in sterile broth that had been used before to grow *E. coli* for 6 h with restricted oxygen supply; pH 6; open symbol is control.

3.2 E. coli strain ATCC 39403

The following results are for *E. coli* strain ATCC 39403. It was aerobically cultured but without an additional supply of pure oxygen. It is able to grow in a well-defined minimal medium consisting of salts, glucose, proline, thiamine, and phosphate buffer. In this case the pH remains 7 during the exponential phase and reaches an end value of 6 in the stationary phase of the growth curve. These bacteria grow much slower than the strain given above. As can be seen from Figures 9 or 11 it needs at least 10 hours to reach the stationary phase.

Bacteria from the exponential phase have a much larger fraction of sensitive specimens than those of the stationary phase (Figure 8). This fact, which is also well known in temperature inactivation (Hansen and Rieman, 1963; Dabbah et al., 1969), gives the clue to reveal the underlying reasons for biphasic behavior.

For this purpose we added different amounts between 2 and 8 μg/ml of Chloramphenicol in the exponential phase (Figure 9), thus inhibiting competitively the synthesis of proteins in the bacteria. They were then harvested after 6 hours and pressurized.

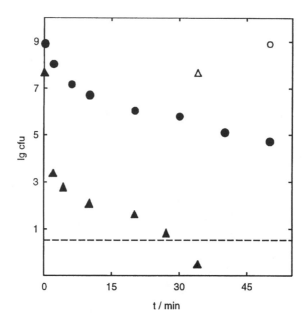

Figure 8 Inactivation of *E. coli* at 3 kbar and 25°C; ● culture in the stationary phase; ▲culture in the exponential phase; open symbols are controls.

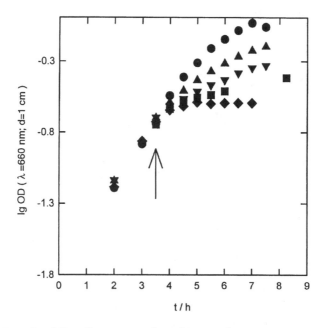

Figure 9 Growth of *E. coli* in minimal medium without oxygen aeration, arrow marks addition of Chloramphenicol after 3.5 h in concentrations of ▲2; ▼ 4; ■ 6; ◆ 8 µg/ml; ● without chloramphenicol.

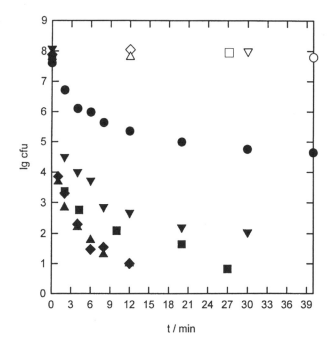

Figure 10 Pressure inactivation of *E. coli* at 3 kbar and 25°C, aerobically grown for different times (without oxygen aeration) of ■ 3.5 h; ▼ 6 h; ● 10 h; ▲6 h but 2 µg/ml chloramphenicol after 3.5 h; ♦ 6 h but 8 µg/ml chloramphenicol after 3.5 h; open symbols are controls.

Figure 10 shows again the inactivation of stationary phase cells and exponential phase cells after 3.5 and 6 hours. With chloramphenicol the cells are even more sensitive than the exponential phase cells, but there is no difference between 2 and 8 µg in spite of the fact that with 2 µg a lot of new cells grow up (see Figure 9). It seems that the biosynthesis of proteins is inhibited mainly for those proteins that are made during the stationary phase for the protection of the bacteria. Reeve et al. (1984) have shown that carbon-starved *E. coli K12* synthesize proteins that are needed to survive and that the synthesis of these proteins is inhibited by chloramphenicol.

Another experiment gives additional insight. We stopped the growing culture in the exponential phase by adding the substance Trimethoprim in different amounts (Figure 11).

The bacteria were then harvested after 7 hours. Trimethoprim is a competitive inhibitor of the important enzyme dihydrofolate reductase (Mutschler, 1991). If this enzyme is inhibited, tetrahydrofolic acid cannot be built, but the cell needs this molecule to synthesize purine and pyrimidine bases via the transfer of activated carbon atoms. In consequence, the synthesis of nucleic acids is slowed down. The results can be seen in the inactivation curves of Figure 12.

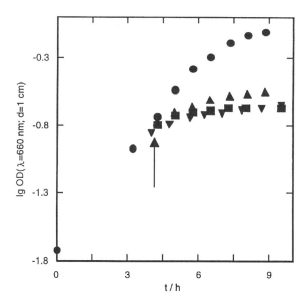

Figure 11 The growth of *E. coli* was stopped within the exponential phase by adding trimethoprim in different amounts of ▲1; ▼ 2; ■ 4 µg/ml; ● is control; arrow marks addition of trimethoprim; culturing condition: M9-medium; 37°C. Reprinted from Rostain, 1995, with kind permission from the editor.

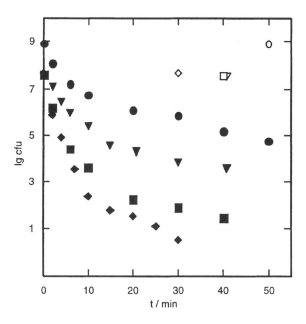

Figure 12 Inactivation of *E. coli* at 3 kbar and 25°C depending on the concentration of added Trimethoprim; ● 0; ▼ 2; ■ 4; ♦ 8 µg/ml; ● harvested in the stationary phase after 15 h.

The minimal inhibitory concentration (MIC) of trimethoprim is about 2 μg/ml. At this concentration RNA synthesis and thus protein synthesis is not completely inhibited. Therefore, after adding such amounts of trimethoprim a lot of cells are able to reach the "stationary phase" state. As a consequence the shape of the inactivation curve is similar to that of an ordinary stationary phase culture. Higher amounts of trimethoprim (4 and 8 μg/ml) strongly slow down RNA and protein synthesis, and adaptation of the cells is no longer possible or takes place very slowly. Therefore, cultures incubated for 3 hours with high amounts of the inhibitor are inactivated like exponentially growing cultures.

The more the inhibitor was added, the more the "pressure resistant" fraction of the bacteria disappears. Therefore, we conclude that the biphasic behavior of the inactivation curves reflects two stages of the bacterial development in the cell cycle.

4 Conclusions

Aerobic and anaerobic growth of the same bacterial strain yield specimens with different behavior.

The pH value does not only influence the rate of inactivation but also the characteristics of the kinetic curves; the lag time found for anaerobically grown *E. coli* appears or disappears depending on pH.

E. coli from the exponential phase are more sensitive to pressure than those from the stationary phase. One reason seems to be that stationary phase cells have additional stabilizing proteins.

The biphasic behavior of inactivation curves of a single colony is caused by various stages of the bacterial development in the cell cycle.

Acknowledgments

This work was supported by the EU project AIR 1-CT92-0296.

References

Dabbah, R., Moats, W. A., and Mattic, J. F. (1969). Factors affecting resistance to heat and recovery of heat-injured bacteria, *Journal of Dairy Science*, 52, 608.

Hansen, N. H. and Riemann, H. (1963). Factors affecting heat resistance of nonsporing organisms. *Journal of Applied Bacteriology,* 26, 314.

Moats, W. A. (1971). Kinetics of thermal death of bacteria, *Journal of Bacteriology,* 105, 165.

Mutschler, E. (1991). *Arzneimittelwirkungen*, 6. Auflage, Wissenschaftliche Verlagsgesellschaft, Stuttgart, 605. In German.

Neidhard, F. C. (1987). *Escherichia Coli* and *Salmonella Typhimurium*, In: *Cellular and Molecular Biology*, Vol. 1, American Society for Microbiology, Washington, D.C. p. 3.

Rahn, O. (1929). The non-logarithmic order of death of some bacteria, *The Journal of General Physiology,* 13, 395.

Reeve, C. A., Amy, P. S., and Matin, A. (1984). Role of protein synthesis in the survival of carbon-starved *Escherichia coli* K-12, *Journal of Bacteriology*, 160, 1041.

Rostain, J.C. (1995). Basic and applied high pressure biology, IV. *Medsubhyp Int.*, 5, p. 69

Wallhäußer, K. H. (1995). *Praxis der Sterilisation*, 5 Auflage, Georg Thieme Verlag, Stuttgart. In German.

chapter nineteen

Quality and safety aspects of novel minimal processing technologies

Leon G. M. Gorris and Bernhard Tauscher

Contents

Summary 325
1 Introduction 326
2 The concept of hurdle technology 326
3 Potential hurdles and hurdle-preserved foods 327
4 Minimal-thermal hurdles..328
  4.1 Ohmic heating 328
  4.2 Sous-vide cooking...329
5 Non-thermal hurdles 330
  5.1 Modified atmosphere packaging...330
  5.2 Edible coatings 332
  5.3 Biopreservation...333
  5.4 High-pressure processing (HPP) 334
  5.5 High-voltage impulses 335
6 The future for hurdle technology...336
Acknowledgment...337
References...337

Summary

The concept and applicability of hurdle technology, or combined processes, is reviewed, highlighting their role in providing both high-quality convenience products and microbiologically safe foods. Emphasis is given to the most promising new technologies, namely ohmic heating, sous-vide cooking, modified atmosphere packaging, edible coatings, high-pressure processing and high electric field pulses. The need to adequately control microbial activity in the resulting new habitats for microbial growth is discussed.

1 Introduction

The consumer demand for high-quality foods with "fresh-like" and "natural" characteristics that require only a minimum amount of effort and time for preparation has led to the introduction of ready-to-eat convenience foods preserved by mild technologies. In the fruits and vegetables area, an extensive range of foods are now available that comply with these characteristics, categorized as so-called "IVeme Gamme" preparations (raw or minimally processed vegetables, with or without dressings) and "Veme Gamme" products (sous-vide preparations; cooked vegetable and potato-based dishes). Refrigeration is the main mild preservation technique that these perishable food products rely on after production. Because of the difficulty to maintain sufficiently low temperatures throughout the chain of production, processing, transport, and storage prior to consumption, additional barriers or hurdles are required to control the growth of spoilage and pathogenic microorganisms (Leistner and Gorris, 1995).

Hurdle technology (or combined processes) advocates the deliberate combination of existing and novel preservation techniques in order to establish a series of preservative factors (hurdles) that no microorganism present should be able to overcome. These hurdles may be temperature, water activity (a_W), pH, redox potential, preservatives, etc. It requires a certain amount of effort from a microorganism to overcome each hurdle. The higher a hurdle, the greater this effort is (i.e., the larger the number of organisms needed to overcome it). By combining hurdles, the intensity of the individual preservation techniques can be kept comparatively low, while the overall impact on microbial growth may remain high. Overall, the lower intensities of hurdles utilized should minimize the loss of product quality.

In this chapter, a short introduction is given on the concept of hurdle technology. In addition, several promising novel hurdles are discussed, considering their impact of food quality and safety.

2 The concept of hurdle technology

Hurdle technology has been around for many years as a concept for the production of safe, stable, nutritious, tasty and economical foods. It advocates the intelligent use of combinations of different preservation factors ("hurdles") in order to achieve multitarget, mild preservation effects. The concept is based on the fact that the microbiologic stability and safety of a food product is dependent on the presence of a set of hurdles specific for the particular product in terms of the nature and strength of their effect. Together, these hurdles keep the spoilage or pathogenic microorganisms under control, because these microorganisms cannot overcome all of the hurdles present. Evidently, the selection of hurdles needs to be tailored carefully to the quality attributes of a product and the processing technologies it can be subjected to. In practice, the large variety of foods and microorganisms that are relevant to food preservation make this a considerable effort. The sequence, type, and intensity of hurdles need to be optimized according to the most influencial variables. When only a few specific microorganisms need to be controlled and there is only little variation to be expected, less different hurdles or hurdles of lower intensity may achieve the desired microbiologic stability. The other extreme holds for situations in which usually high numbers of many different target microorganisms occur and the set of hurdles needs to be extended to obtain reliable control.

Under all conditions it should be considered that small variations in the raw material or the processing protocol may lead to reduced control. As for the former, the availability of carbon and energy sources may differ between apparently similar products, and there may also be compounds (e.g., osmoprotectants, essential amino acids, vitamins) present in certain foods that enable or strongly stimulate growth of specific microorganisms (booster effect). Nevertheless, when used with the appropriate care, the hurdle concept is an adequate working hypothesis and can be used to optimize mild food preservation systems.

3   Potential hurdles and hurdle-preserved foods

In a recently concluded survey (Leistner and Gorris, 1994) about 50 different hurdles have been identified in food preservation (Table 1). The most important hurdles commonly used are high temperature, low temperature, low water activity, acidity, low redox potential, and preservatives (e.g., nitrite, sorbate, sulphite). In addition to the use of processing technologies as hurdles, in certain cases "natural" hurdles may be used, i.e., preservative factors that rely on biologic agents that are an integral part of the production process. For instance, with fermented vegetables, the desired product quality and microbial stability are achieved during processing by a sequence of hurdles that arise in different stages of the fermentation process. Vegetables are nonsterile products and when they are stored in water with a high amount of salt added, a selected group of microorganisms start growing and use up the available oxygen. This reduces the redox potential of the product and enhances the Eh hurdle, which reduces the growth of aerobic microorganisms and favors the selection of lactic acid bacteria. They are a competitive flora and cause acidification of the product, thus an increase of the pH hurdle. In effect, a stable product is achieved.

With nonfermented foods, the hurdle technology approach has been established, for instance, in the production of tortellini, an Italian pasta product. In this case, a reduced water activity and a mild heating are the principle hurdles employed during processing, in addition to a modified gas atmosphere or ethanol vapor in the package and a chill temperature during storage and retail. A recent inventory conducted in 10 Latin American countries of foods preserved by traditional hurdle technology identified some

Table 1 Potential Hurdles for the Preservation of Foods (Leistner and Gorris, 1994)

Physical hurdles	High temperature (sterilization, pasteurization, blanching); low temperature (chilling, freezing); ultraviolet radiation; ionizing radiation; electromagnetic energy (microwave energy, radiofrequency energy, oscillating magnetic field pulses, high electric field pulses); photodynamic inactivation; ultra-high pressure; ultrasonication; packaging film (plastic, multilayer, active coatings, edible coatings); modified atmosphere packaging (gaspackaging, vacuum, moderate vacuum, active packaging); aseptic packaging; food microstructure
Physicochemical hurdles	Water activity (a_w), pH, redox potential (Eh), salt (NaCl), nitrite, nitrate, carbon dioxide, oxygen, ozone, organic acids, lactic acid, lactate, acetic acid, acetate, ascorbic acid, sulphite, smoking, phosphates, glucono-∂-lactone, phenols, chelators, surface treatment agents, ethanol, propylene glycol, Maillard reaction products, spices, herbs, lactoperoxidase, lysozyme
Other hurdles	Competitive flora, protective cultures, bacteriocins, antibiotics, free fatty acids, chitosan, chlorine

260 different food items derived from fruit, vegetables, fish, dairy, meat, and cereals, which often have a high water activity (sometimes as high as 0.97) that are stable at ambient temperature (25 to 35°C) for several months (Aguilera Radic et al., 1990).

More examples of traditional and innovative hurdle-preserved foods and details on a range of potential hurdles can be found in several recent publications (Leistner and Gorris, 1994, 1995). In the following text, several promising nonthermal and minimal-thermal hurdles are presented.

4   Minimal-thermal hurdles

4.1   Ohmic heating

The application of ohmic heating to food can be traced back to 1917. Today, a broad spectrum of products is processed in this way in Europe, Asia, and North America (Giese, 1996; Zoltai and Swearingen, 1996). The advantage of the method is optimal, uniform, and fast heating of the medium and of the particles by electric current passing through conductive food. When the electric conductivity of medium and particles is different, different heating rates may result. Therefore, optimal process parameters (voltage, density, heating rate, and holding time) must be determined for each product individually, taking into account the expected variability. Particle sizes may be as large as several cubic centimeters. The shape of the particles is a negligible factor; the method proved successful in the most different shapes. In most ohmic products the concentration of solid components is in the range from 20 to 70%. Particle density should correspond to the density of the medium in order to avoid demixing. Product conductivity is usually between 0.1 and

3.0 s/m, depending on the desirable temperature and on the generator (Larkin and Spinak, 1996; Sastry and Li, 1996).

In less than 90 sec products may be heated up to 140°C and then cooled down again within minutes. Heating and cooling times are much shorter than with conventional heating. Short-time heating preserves valuable components and texture of the food. Particles too are heated uniformly and a larger temperature gradient does not appear. In this way, ohmic heating is comparable to microwave heating, with the advantages over microwaves that ohmic heating allows unlimited penetration depth and that the heating height is determined by the electric conductivity of the product and its residence time in the electric field (Kim et al., 1995, 1996). A disadvantage of the method is that the alternating current of 50 or 60 Hz produces electrolytic effects; metal electrodes, e.g., may be dissolved. To solve the problem, frequencies above 100 kHz are used; then electrodes survive even 3 years of use, nearly uncorroded. With low-frequency current, specific carbon electrodes may also be employed.

Many different food items may be heated by the Ohmic principle, including fruit desserts, yogurt-fruit preparations, ratatouille, pasta, vegetable salads, vegetable curry, tortellini in tomato sauce, etc. In Japan a plant was installed recently to sterilize whole strawberries and other fruit intended to be added to yogurt. In cases where ohmic heating is considered a potential hurdle, it should be checked whether the electrochemical reactions that take place in the food do not affect the desired sensory quality of the product (Reznick, 1996).

4.2 Sous-vide cooking

Sous-vide or vacuum cooking entails packaging a food product in a suitable heat-resistant bag or container, which is evacuated and sealed and subsequently heat-processed under controlled conditions of temperature and time (Bailey, 1995; Schellekens, 1996a, 1996b; Martens, 1995). Typical of the method are lower temperatures and longer cooking times than under conventional conditions. Fish and meat are sous-vide cooked at temperatures below 70°C, vegetables at 95°C. The remaining pressure in the evacuated bag is 10 mbar for solid products and 120 mbar for sensitive produce. After heating, the food is cooled down quickly to 1 to 8°C. Essential prerequisites for successful use of the method are selection of high-quality raw materials, use of packaging materials of proper heat resistance and low gas-permeability, and application of maximal process control. The shelf life of sous-vide cooked food is from 6 to 42 days, depending on the product. An advantage of the method is that water is not required, which avoids the leaching effect, and that hardly any vapors escape, which prevents loss of aroma compounds. Very importantly, the in-bag cooking avoids the problem of postprocess contamination. Economic and quality advantages over conventional cooking are centralized production, uncoupling of production and consumption, better preservation of valuable components, and minimal loss of the original food quality characteristics (Schnellekens, 1996). Guidelines including also sous-vide products were established in Europe and the U.S. (Peck and Stringer, 1996).

The sous-vide processing technique advocated by Gierschner et al. (1996) uses even less liquid than other systems. The containers are closed under vacuum and the product is preheated and sterilized in one and the same container. The product is not leached out and so aroma, taste, and color are perfectly preserved, together with the texture of the raw material. During preheating to 60 to 70°C, cell-wall pectin methylesterase is released and activated, leading to a partial deesterization of the pectin chains. During heating, cell membranes are partly denatured, lose their semipermeability and become permeable also to bivalent ions, which can flow from the inner cell into the region of deesterized pectin chains. The pectin chains are bound to each other through calcium chelates and protected against depolymerization during the subsequent relatively substantial thermal stress. The sensory overall impression of such products comes very close to the impression obtained from fresh produce.

It should be stressed that whereas the heat destroys vegetative microbial cells, it does not destroy bacteria spores. In several food products (vegetables, fish) psychrotrophic strains of the spore-forming pathogen *Clostridium botulinum* can occur on the raw material. When spores are present, they will survive or are only heat-damaged. In case the spores germinate, the anaerobic conditions in the sous-vide package and the nutrients in the food product may sustain their growth to hazardous levels during the potentially long storage period, even when storage is at temperatures as low as 4 to 7°C. To ensure better safety, additional hurdles, e.g., low pH, low a_w organic acids, or bacteriocins may be used (Martens, 1995).

The deliberate use of lysozyme or other lytic enzymes, generally known as potential food preservatives, for the control of *Cl. botulinum* is not advisable due to the recent findings that these may enhance the growth of this pathogen (Lund and Peck, 1994). When spores are only sublethally damaged by the cooking process, their germination system is destroyed but the cell still is potentially viable. Hen egg-white lysozyme, a relatively heat-resistant compound, has been found to bypass the inactivated germination system and, by degrading the spore cortex, allows the cell to be released from the spore. Lysozyme-like lytic activity is found in many foods, including meat, egg, and vegetable products. In several vegetables, the levels of the lytic activity are high enough to increase the apparent heat resistance of sublethally heat-damaged spores of nonproteolytic *C. botulinum* (Stringer and Peck, 1996).

5 Non-thermal hurdles

5.1 Modified atmosphere packaging

Modified atmosphere packaging (MAP) means placing the food inside a material of defined gas-permeability, in which the composition of the atmosphere is not actively controlled (Gorris and Peppelenbos, 1992; Church and Parsons, 1994; Zagory, 1995; Smith et al., 1995; Watada et al., 1996; Phillips, 1996). It is intended to suppress microbial growth, retard respiration, ripening and ageing of fruits and vegetables, and to suppress oxidative reactions requiring free oxygen. The minimally processed produce (either whole, segmented, or cut) is packaged with a semibarrier foil. Immediately before sealing, air or a gas of defined composition is introduced into the package, which is then cold-stored (4 to 7°C). Some produce are less cold-tolerant and need to be stored at somewhat higher temperatures. Due to the respiratory activity of the produce during storage, the product consumes oxygen and releases carbon dioxide, thus producing a so-called product-modified atmosphere.

Modified atmospheres have to be optimized individually for any product as respiration depends on many factors to be taken into account. The respiration rate of apples, for example, is 5 to 10 ml of carbon dioxide per kg and h, of asparagus more than 60 ml. The degree of ripening affects the MAP design as well. In ripe bananas, for example, the respiration rate is 5 times that of green ones. The preparation of the food also has an important influence: sliced carrots produce much more carbon dioxide (6 to 7-fold) than intact ones. In addition, the storage temperature influences the respiration rate directly: generally, it increases 2 to 3-fold per 10°C (Church and Parsons, 1994).

Finally, the permeability of the semibarrier foil to gas and water is decisive. Microperforated films with relatively low barrier properties are required to lower the respiration rate of products with a high respiratory activity, e.g., asparagus or mung bean sprouts. In general, the relative humidity of the package atmosphere must not exceed 90%; decreasing humidity leads to a loss in produce quality whereas increasing humidity affects its microbiologic status (Smith et al., 1995). In practice, with fruits and vegetables, however, the relative humidity is close to saturation.

Typically, the gas composition in MAP for cut vegetable produce is 2 to 5% oxygen and 5 to 10% carbon dioxide. Lower oxygen levels or higher carbon dioxide levels need to be avoided since they may cause undesired physiologic defects. For instance, a reduction of O_2 below 1 to 2% may set off anaerobic respiration (fermentation) of the produce and/or trigger the development of anaerobic, fermentative microorganisms. In both cases, undesirable fermentation products (off-odors, off-flavors) accumulate. Equally detrimental to product quality can be an increase of the CO_2 level over the optimum, since this may cause specific disorders such as brown stains.

In combination with appropriate cooling, storage life can be up to 5 to 7 days. The development of the market for MAP products is promising. In Europe MAP is mainly applied to salads, potatoes, carrots, and cabbage. Great Britain and France supply about 50% and 25%, respectively, of the total European MAP market, which at present handles about 300 million package units of produce, representing a market value of about 1 billion ECU (Day and Gorris, 1993; Brody, 1995). On the U.S. market, a market share of about $8 billion of MAP products has been predicted for the year 2000. This fresh-cut produce market is served mainly with salad, paprika, onions, cabbage, mushrooms, endive, and spinach. One of the most decisive factors for this success of MAP of chilled foods in Europe was its strong promotion by giant food retailers who aimed to sell more high margin fresh or chilled products and also to cut stock losses and in-store waste due to limited shelf life. These retailers have been successful in using their unique position in the marketplace to dictate quality specifications and distribution requirements on their suppliers.

Since the particular gas mixture prevailing in modified atmosphere (MA) packs may selectively promote growth of microorganisms, care should be taken about the safety of the system, especially in practice where the temperature is still not well controlled. Recent research has shown that under optimal MAP conditions for minimally processed produce, especially psychrotrophic foodborne pathogens such as *L. monocytogenes* pose a potential hazard (Berrang et al., 1989; Aytaç and Gorris, 1994; Bennik et al., 1995, 1996). Such studies clearly indicated the need to add additional hurdles to the MAP-cooling system to assure consumer safety.

MA-packed fresh produce has a fair safety record, and their sales and consumption are increasing quickly. However, hazardous situations may arise when products are organoleptically acceptable after prolonged storage, but harbor unacceptable levels of cold-tolerant pathogens. Any food market, including the prepared salad market, is vulnerable to food poisoning scares. In 1988, 60 samples of such salads were tested for *Listeria monocytogenes* and four of them were found to contain *L. monocytogenes* serotype 1/2 (Sizmur and Walter, 1988). The resulting "*Listeria* hysteria" generated by the media cut sales of prepacked salads in the U.K. by 50% overnight. It took 3 to 9 months for sales to recover to their original levels.

5.2 Edible coatings

Biodegradable, even edible, protective coatings may be applied directly to the surface of a food product to act as an additional hurdle for protection of overall food quality, stability against microbial spoilage, and loss of intrinsic product quality. The functional characteristics required for the coating depend on the product matrix (low to high moisture content) and on the deterioration processes the product is subject to. Biodegradable films may also be employed as packaging materials that generate a suitable modified atmosphere around a packaged product and might replace nonbiodegradable plastics used in MAP storage today. Coatings with good resistance to breakage and abrasion properties and with appropriate flexibility have been produced on the basis of wheat gluten and pectin (hydrophilic coatings) and of beeswax and pectin (hydrophobic coatings). The gas and water vapor permeabilities of the various edible films make them suitable for use with minimally processed produce (Guilbert et al., 1996; Gontard et al., 1996).

Currently being investigated is whether food-grade antimicrobials can be enclosed in the coatings in order to allow the use of reduced amounts of additives, because these are fixed at the product surface from which they only slowly migrate. Studies with sorbic acid have proven the potential (Leistner and Gorris, 1994; Guilbert et al., 1996), but the future will see the use of natural antimicrobials (i.e., plant metabolites, bacteriocins) or of coating materials such as chitosan that combine their physical effect with antimicrobial properties.

5.3 Biopreservation

The use of natural antimicrobial systems has been studied because consumers and governments move away from the extensive use of artificial preservatives. Lactic acid bacteria (LAB), for instance, are a well-known example of a natural antimicrobial system and in most instances there is little doubt that they are safe to be exploited in food preservation (Gorris and Bennik, 1994). Their antimicrobial effect resides mainly with the organic acids they form, although some strains produce antimicrobial compounds called bacteriocins additionally. The bacteriocins are small proteins that in most cases only inhibit a certain range of microorganisms, most often Gram(+) bacteria only. Thus, the bacteriocins or their producers can probably not be used as a general safety hurdle, but could still be used to form a specific hurdle toward the growth of *L. monocytogenes*, *Cl. botulinum* and *B. cereus*, since all these pathogens are Gram(+).

On cut vegetables, where *Listeria monocytogenes* is a potential hazard, LAB are present as part of the natural microflora. Although they generally make only about 1% of that microflora, that endogenous source may be used through manipulation of the gas phase composition in a package seeking a selective promotion of LAB growth. Alternatively, producer organisms or their bacteriocins can be added to the product allowing a better control of the preservation system. Using an exogenous source of LAB, the LAB added should be well adapted to the ecosystem in order to grow and produce bacteriocins.

In recently conducted search for suitable LAB, over 900 LAB isolates were obtained from different fresh and MAP stored vegetables (Bennik et al., 1997). Only nine isolates where identified that could adequately control *L. monocytogenes* on artificial growth media. In only one case was sufficient control of the pathogen achieved at 7°C. The isolate was identified as *Enterococcus mundtii* and was able to grow under low O_2/high CO_2 concentrations prevailing in MA-packaged prepared salads and actively produced a bacteriocin at 4 and 8°C. Although this organism is not a LAB, it should be considered as a suitable model to test the potential use of biopreservation as an additional safety hurdle.

Alternative to the use of growing LAB for biopreservation, the application of bacteriocins can also be considered. Bacteriocins such as nisin and pediocin are either on the market or in the process of being approved for a range of food applications. The advantage would be that the control of the preservative effect is much better, since no active growth is needed for sufficient bacteriocin production but appropriate known doses of the bacteriocin can be applied directly. A drawback may be that the bacteriocins are proteinaceous compounds and may be inactivated by enzymes produced by the microorganisms residing in the produce. In addition, adsorption or inactivation by components of the food matrix may reduce the dose of bacteriocin that is actually active. In a recently initiated EU-project (FAIR-CT96-1148), the use of nisin in the vegetable environment is studied, using hurdle technology to overcome possible inactivation mechanisms.

5.4   High-pressure processing (HPP)

One of the most promising novel methods of mild food preservation is the use of high hydrostatic pressure. The natural fresh taste of meat, fish, and of fruit and juices is retained. Under optimal conditions, colors (i.e., chlorophyll, anthocyans, carotenoids) are not or very slightly affected. Novel textures may be produced as phase transitions of biopolymers are enforced by pressure. Proteins and starch exposed to pressure may result in products showing changed textural properties. A number of production plants for application of high hydrostatic pressure on large scale (up to 400-L batch) have become operational in Europe over the last several years. In such plants, pressurization is performed in a high-volume pressure-resistant autoclave. Product pre- and posttreatment handling is comparable to traditional pasteurization processes. Both bulk material and ready-to-use packages may be pasteurized. The demands set upon packages that are exposed to pressure are minimal. They must be flexible enough to compensate for the low compressibility of liquids. Again, the obvious advantage of in-pack treatment in avoiding post-process contamination is an important factor in the success of ultra-high-pressure processing.

Bridgman (1914) was the first to expose foods to very high pressures 100 years ago when he tested the denaturation of hen-egg white at 10000 bar. The method was developed to industrial scale in Japan, where several products have been on the market for many years and where the majority of patents on the systems' design and application originate from. Notably, applications with fruits and vegetables are the subject of twice as many patents as applications to protein containing food, i.e., milk, dairy products, meat and fish.

The first commercially successful applications of high-pressure processing were for low-acid foods, such as fruit juices and marmalades. For example, the production capacity of tangerine juice in Japan has increased up to 1 million L per year at the moment. Another interesting product is pressure pasteurized rice cake, because it seems to have reduced allergenic effect. In the U.S. a HPP pasteurized avocado product was introduced recently, in which browning and other discoloration effects normally seen are much reduced or even completely absent.

Pressure is an important thermodynamic factor, as is temperature. Both pressure and temperature may influence the position of a chemical reaction equilibrium and the reaction rate. For thermal processes, reactions in foods have been thoroughly investigated. Pressure-promoted reactions take place as well and probably as a result of similar reaction mechanisms, but to much lower extent and in some specific cases only (Tauscher, 1995; Butz et al., 1997). On a laboratory scale, experiments are presently conducted in many foods to learn about their behavior under pressure. Among them are egg white and egg yolk, milk and dairy products, fish and meat, pectin and other biopolymers, but also cereals, vegetables, fruits and juices (Butz and Tauscher, 1996). The novel method of pasteurization allows the production of food that has retained its original freshness, but still is "like cooked." High-pressure pasteurization is therefore often referred to as "cold cooking," but might better be called "Bridgmanization" in memory of Bridgman.

Microorganisms, and vegetative cells in particular, are safely inactivated by relatively moderate pressures of some thousand bar, although prominent differences in pressure-sensitivity exist (Leistner and Gorris, 1994). With spores that can resist pressures even over 8000 bar (Butz et al., 1996a), a sequence of pressure treatments can be used. First, a pressure below 1000 bar is used to trigger spores to germinate. Second, the germinated spores are inactivated by high hydrostatic pressure of up to 8000 bar. Alternatively, the combined use of low hydrostatic pressure with certain bacteriocins or high field pulses known to efficiently inactivate vegetative cells of spore-forming bacteria. The effect of high-pressure processing may be enhanced by combined use of nisin or lysozyme (Hauben et al., 1996), maybe allowing the use of moderate pressures to obtain a desired level of microbial inactivation.

Inactivation of enzymes is due to denaturation of the enzyme protein (Heinisch et al., 1995; Butz et al., 1996b, 1996c). Different enzymes are known to respond differently to pressure. To avoid reversible effects and achieve adequate inactivation of all enzymes, pressures over 6000 to 8000 bar are required.

5.5   High-voltage impulses

High-voltage electric fields act on the food in a different way than ohmic heating. They destroy microorganisms essentially without heating the food. The external electric field produces an electric potential in the membranes, resulting in a charge separation in cell membranes. When the membrane potential exceeds a critical value of about 1 volt, pores form in the cell membranes because of repulsion forces between the charge carriers. The critical transmembrane potential and thus the external field strength, too, depend on the type of cell, cell diameter, and cell shape. If the external electric field strength is equal to or little above the critical value, cell-membrane permeability is reversible (Mertens and Knorr, 1992). This phenomenon has been known and utilized for some time in biotechnology. If the field strength is very great, the pores become irreversible and cell membranes are destroyed.

The effect of high-voltage fields on microorganisms has been investigated in good detail. For vegetative microorganisms the critical electric field strength is about 15 kV/cm. The electric impulses last from 1 to 10 msec. Typical field strength is between 20 and 50 kV. These are adequate conditions to treat pumpable food, i.e., foods that flow through a chamber in which two chemically inert electrodes are installed. The duration of exposure to the high electric field impulses is less than 1 sec. Generally, high field strengths are more effective than low field strengths at short exposure times.

When temperatures are kept at 40 to 65°C, they do not affect the food taste (fresh, not cooked) and keep the texture firm. Under practical conditions, unfortunately, the method does not affect endo- and ascospores, which would require much higher field strength. On the other hand, an advantage of high voltage impulses with respect to spores is that it does not trigger spore germination, in contrast to high pressure, temperature, or ultrasound.

In order to fully inactivate even the vegetative microorganisms, it is advised to apply the higher temperatures of up to 65°C for adequate control, since among important psychrotrophic pathogens (e.g., *E. coli* O157:H7, *Listeria monocytogenes, Yersinia enterocolitica*) specific strains or isolates can exhibit quite high heat resistance. Also several enzymes, i.e., alkalic phosphatase, horseradish peroxidase, and pectinesterase, are known to be relatively resistant to high-voltage fields. Obviously, intact enzymatic activity may limit the shelf life of produce pasteurized by high-voltage impulses.

The method was tested in fruit juices and drinks, milk, liquid egg, gravy, purees, syrups, etc. The effectiveness of the processes is preferably tested in milk. No, or scarcely any, significant sensory changes were noted in freshly pressed juices and in other food items pasteurized in this way (Quin et al., 1996). Germany and the U.S. are leading in the development of equipment (Knorr et al., 1994). Much still remains to be done to establish the novel process with its extremely low energy input as a method of pasteurization, which supplies safe products.

6 The future for hurdle technology

With the increasing popularity of ready-to-eat, fresh, and processed foods, which are preserved only by relatively mild techniques, a new habitat for microbial growth may have come into focus. In order to control the growth of food poisoning and spoilage microorganisms in these habitats, while keeping loss of product quality to a minimum, a hurdle technology approach can be chosen to establish a set of mild preservative factors that can adequately ensure product safety. Recent research activities in this field have highlighted that sound information on important factors affecting the survival and growth of such microorganisms under the mild preservation conditions is required. The design and application of hurdle technology in food preservation has experienced a rapid increase over the last years and is expected to be an important concept for further developments. The combination of a number of general or specific antimicrobial hurdles in processing and storage of chilled perishable foods targets primarily in suppressing cold tolerant pathogens that can grow at low oxygen levels such as *L. monocytogenes* and *Cl. botulinum*. In addition to making foods stable and safe, the concept of hurdle technology may also contribute to improve or alter the organoleptic quality or total quality of foods to comply with consumers' expectations, and developments in this respect may be seen in the near future too.

Acknowledgment

Part of the research activities reviewed here were achieved with a financial contribution of the European Commission through contracts AGRF-0037, AIR1-CT92-0125 and AIR1-CT92-0296.

References

Aguilera Radic, J. M., Chirife, J., Tapia de Daza, M. S., Welti Chanes, J., and Parada Arias, E. (1990). *Inventario de alimentos de humedad intermedia tradicionales de Iberoamérica*, Instituto Politécnico Nacional, Unidad Profesional Interdisciplinaria de Biotecnologia, México. In Spanish.

Aytaç, S. A. and Gorris, L. G. M. (1994). Survival of *Aeromonas hydrophila* and *Listeria monocytogenes* on fresh vegetables stored under moderate vacuum, *World Journal of Microbiology Biotechnology*, 10, 670.

Bailey, J.D. (1995). Sous Vide: past, present, and future. In: *Principles of Modified-Atmosphere and Sous Vide Product Packaging*, Farber J.M. and Dodds, K.L. (Eds.), Technomic Publ. Co., Lancaster, U.K., p. 243.

Bennik, M. H. J., Smid, E. J., Rombouts, F. M., and Gorris, L. G. M. (1995). Growth of psychrotrophic foodborne pathogens in a solid surface model system under the influence of carbon dioxide and oxygen, *Food Microbiology*, 12, 509.

Bennik, M. H. J., Peppelenbos, H. W., Nguyen-the, C., Carlin, F., Smid, E. J., and Gorris, L. G. M. (1996). Microbiology of minimally processed, modified-atmosphere packaged chicory endive, *Postharvest Biology and Technology*, 9, 209.

Bennik, M. H. J., Smid, E. J., and Gorris, L. G. M. (1997). Vegetable associated *Pediococcus parvulus* produces pediocin PA-1, *Applied Environmental Microbiology*, 63, 2074.

Berrang, M. E., Brackett, R. E., and Beuchat, L. R. (1989). Growth of *Listeria monocytogenes* on fresh vegetables stored under controlled atmosphere, *Journal of Food Protection*, 52, 702.

Bridgman, P. W. (1914). The coagulation of albumen by pressure, *Journal of Biological Chemistry*, 19, 511.

Brody, A. L. (1995). A Perspective on MAP Products in North America and Western Europe. In: *Principles of Modified-Atmosphere and Sous Vide Product Packaging*, Farber J.M. and Dodds, K.L. (Eds.), Technomic Publ. Co., Lancaster, U.K., p. 13.

Butz, P. and Tauscher, B. (1996). High-pressure treatment of fruit and vegetables: problems and limitations. In: *Proceedings of the European High Pressure Research Group, XXXIVst Annual Meeting*, Leuven, Belgium, September 1–5.

Butz, P., Funtenberger, S., Haberditzl, T., and Tauscher, B. (1996a). High pressure inactivation of *Byssochlamys nivea* ascospores and other heat resistant molds, *Lebensmittel Wissenschaft und Technologie*, 29, 404.

Butz, P., Fister, H., Losch, S., and Tauscher, B. (1996b). Einfluß von hohem hydrostatischem druck auf immobilisierte α-amylase, *Lebensmittelchemie*, 50, 22. In German.

Butz, P., Fister, H., Losch, S., and Tauscher, B. (1996c). Response of immobilized α-amylase to hydrostatic high pressure, *Food Biotechnology*, 10, 93.

Butz, P., Garcia Fernandez, A., Fister, H., and Tauscher, B. (1997). Influence of high hydrostatic pressure on aspartame: instability at neutral pH, *Journal of Agricultural and Food Chemistry*, 45, 302.

Church, I. J. and Parsons, A .L. (1994). Modified atmosphere packaging technology: a review, *Journal Science Food Agriculture*, 67, 143.

Day, B. P. F. and Gorris, L. G. M. (1993). Modified atmosphere packaging of fresh produce on the West-European market, *International Journal of Food Technology, Marketing, Packaging and Analysis*, 44 (1/2), 32.

Gierschner, K., Jahn, W., and Philippos, S. (1996). Neues herstellungsverfahren für naßkonserviertes gemüse hoher qualität sowie über hierbei an den zellwand-hydrokolloiden ablaufende veränderungen, *Die Industrielle Obst- und Gemüseverwertung*, 5, 186. In German

Giese, J. (1996). Commercial development of ohmic heating, *Food Technology*, 114.

Gontard, N., Thibault, R., Cuq, B., and Guilbert, S. (1996). Influence of relative humidity and film composition on O_2 and CO_2 permeability of edible films, *Journal of Agricultural Food Chemistry*, 44, 1064.

Gorris, L. G. M. and Bennik, M. H. J. (1994). Bacteriocins for food preservation, *International Journal of Food Technology, Marketing, Packaging and Analysis*, 45 (11), 65.

Gorris, L. G. M. and Peppelenbos, H. W. (1992). Modified atmosphere and vacuum packaging to extend the shelf-life of respiring food products, *HortTechnology*, 2, 303.

Guilbert, S., Gontard, N., and Gorris, L. G. M. (1996). Prolongation of the shelf life of perishable food products using biodegradable films and coatings, *Lebensmittel Wissenschaft und Technologie*, 29, 10.

Hauben, K. J. A., Wuytack, E. Y., Soontjens, C. C. F., and Michiels, C. W. (1996). High pressure transient sensitization of *Escherichia coli* to lysozyme and nisin by disruption of outer membrane permeability, *Journal of Food Protection*, 59, 350.

Heinisch, O., Kowalski, E., Goossens, K., Frank, J., Heremans, K., Ludwig, H., and Tauscher, B. (1995). Pressure effects on the stability of lipoxygenase: FTIR and enzyme activity studies, *Zeitschrift für Lebensmittel Untersuchung und Forschung*, 201, 562.

Kim, H. J., Choi, Y. M., Yang, A. P. P., Yang, T. C. S, Taub, I. A., Giles, J., Ditusa, C., Chall, S., and Zoltai, P. (1995). Microbiological and chemical investigation of ohmic heating of particulate foods using a 5 kW ohmic system, *Journal of Food Processing and Preservation*, 20, 41.

Kim, H. J., Choi, Y. M., Yang, T. C. S., Taub, I. A., Tempest, P., Skudder, P., Tucker, G., and Parrott, D. L. (1996). Validation of ohmic heating for quality enhancement of food products, *Food Technology*, 253.

Knorr, D., Geulen, M., Grahl, T., and Sitzmann, W. (1994). Food application of high electric field pulses, *Trends in Food Science & Technology*, 5, 71.

Larkin, J. W. and Spinak, S. H. (1996). Safety considerations for ohmically heated, aseptically processed, multiphase low-acid food products, *Food Technology*, 242.

Leistner, L. and Gorris, L. G. M. (1994). *Food Preservation by Combined Processes*, Final report of FLAIR C.A.#7. EUR 15776, European Commission, Brussels, Belgium.

Leistner, L. and Gorris, L. G. M. (1995). Food preservation by hurdle technology, *Trends in Food Science and Technology* 6, 41.

Lund, B. M. and Peck, M. W. (1994). Heat-resistance and recovery of non-proteolytic *Clostridium botulinum* in relation to refrigerated, processed foods with an extended shelf life, *Journal of Applied Bacteriology* Supplement 76, 115S.

Martens, T. (1995). Current status of sous vide in Europe. In: *Principles of Modified-Atmosphere and Sous Vide Product Packaging*, Farber J.M. and Dodds, K.L. (Eds.), Technomic Publ. Co., Lancaster, U.K., p. 37.

Mertens, B. and Knorr, D. (1992). Developments of nonthermal processes for food preservation, *Food Technology*, 124.

Peck, M.W. and Stringer, S.C. (1996). *Clostridium botulinum:* Mild preservation techniques. In: *Proceedings of the Second European Symposium on Sous Vide*, Alma Sous Vide Competence Center, Leuven, Belgium, p. 181.

Phillips, C. A. (1996). Modified atmosphere packaging and its effects on the microbiologic quality and safety of produce, *International Journal of Food Science and Technology*, 31, 463.

Quin, B. L., Pothakamury, U. R., Barbosa-Canovas, G. V., and Swanson, B. G. (1996). Nonthermal pasteurization of liquid foods using high-intensity pulsed electric fields, *Critical Reviews in Food Science and Nutrition*, 36, 603.

Reznick, D. (1996). Ohmic heating of fluid foods, *Food Technology*, 250.

Sastry, S. K. and Li Q. (1996). Modeling the ohmic heating of foods, *Food Technology*, 246.

Schellekens, M. (1996a). New research issues in sous-vide cooking, *Trends in Food Science & Technology*, 7, 256.

Schellekens, M. (1996b). Sous vide cooking: state of the art. In: *Proceedings of the Second European Symposium on Sous Vide*, Alma Sous Vide Competence Center, Leuven, Belgium, p. 9.

Sizmur, K. and Walker, C. W. (1988). *Listeria* in prepacked salads, *Lancet*, May 21, 1167.

Smith, J. P, Abe, Y., and Hoshino, J. (1995). Modified atmosphere packaging — present and future uses of gas absorbents and generators. In: *Principles of Modified-Atmosphere and Sous Vide Product Packaging*, Farber J.M. and Dodds, K.L. (Eds.), Technomic Publ. Co., Lancaster, U.K., p. 287.

Stringer, S. C. and Peck, M. W. (1996). Vegetable juice aids the recovery of heated spores of non-proteolytic *Clostridium botulinum*, *Letters in Applied Microbiology* 23, 407.

Tauscher, B. (1995). Pasteurization of food by hydrostatic pressure: Chemical Aspects, Review, *Zeitschrift für Lebensmittel Untersuchung und Forschung*, 200, 3.

Watada, A. E., Ko, P. N., and Minott, D. A. (1996). Factors affecting quality of fresh-cut horticultural products, *Postharvest Biology and Technology*, 9, 115.

Zagory, D. (1995) Principles and practice of modified atmosphere packaging of horticultural commodities. In: *Principles of Modified-Atmosphere and Sous Vide Product Packaging*, Farber J.M. and Dodds, K.L. (Eds.), Technomic Publ. Co., Lancaster, U.K., p. 175.

Zoltai, P. and Swearingen, P. (1996). Product development considerations for ohmic processing, *Food Technology*, 263.

chapter twenty

Vacuum impregnation: a tool in minimally processing of foods

Amparo Chiralt, Pedro Fito, Ana Andrés, José M. Barat, Javier Martínez-Monzó, and Nuria Martínez-Navarrete

Contents

Summary 341
1   Introduction 342
2   Mathematical model of HDM..342
    2.1   Equilibrium status..343
    2.2   HDM kinetics...345
3   Fruit properties as affected by VI 346
    3.1   Feasibility of VI of fruits..347
    3.2   Changes in mechanical, structural, and other
          physical properties...349
4   Influence of VI on osmotic processes 351
    4.1   Composition changes promoted by VI 352
5   Salting processes by using VI...353
Acknowledgments..353
References..354

Summary

This chapter reviews the advances on the fundamental understanding of the phenomena that preside over the vacuum impregnation of foods and its most recent applications as a pretreatment operation. The benefits of its use are therefore described in terms of its effect on the mechanical, structural, and other physical properties of foods.

1 Introduction

Vacuum pressure was applied in several processes with different aims in the past: minimal processing of fruit to incorporate different additives (Del Rio and Miller, 1979; Santerre et al., 1989), whey and air removal in curd cheddar cheese (Reinbold et al., 1993), osmotic dehydration processes (Zozulevich and D'yachenco, 1969; Hawkes and Flink, 1978). However, no clear reasons explaining the role of vacuum on the different processes had been reported. Fito (1994) explained the action of vacuum pressure in porous products immersed in a liquid phase, by trying to model the faster kinetics of vacuum osmotic dehydration of apples. The low pressure in the system did not justify the observed acceleration of the osmotic process on the basis of the pressure influence on the thermodynamic driving force due to concentration gradients. The pressure gradient in the system when vacuum is imposed, or during the restoring of atmospheric pressure, was another driving force in bulk mass transfer in porous products. Pore internal gas could be expanded or compressed in line with the external solution in-flow while pressure gradients persisted. This phenomenon was called the hydrodynamic mechanism (HDM). The model of this mechanism predicts the entry of an external liquid in a porous product when any overpressure is applied in the system; nevertheless liquid will be released when the product returns to normal pressure. The advantage of the HDM action after a vacuum period in the system lies in the partial gas release throughout this period while the mechanical equilibrium is being achieved; the restoring of atmospheric pressure implies compression only of the residual gas and flow into the pores of the external liquid. Therefore, the product remains filled with the liquid phase at normal pressure. From the HDM model it is possible to predict the amount of liquid that can be introduced in a porous food with different aims: modifying the composition, introducing additives, salting, etc., and it is also possible to evaluate the effectiveness of a vacuum treatment in expelling internal gas or liquid.

The aim of this chapter is to give a general overview of the HDM action during vacuum impregnation (VI) processes: the influence of the mechanical properties of the product, the mechanism kinetics and product equilibrium status, and the physicochemical and structural changes promoted in VI products. Likewise, several applications of the controlled HDM action in VI processes are discussed.

2 Mathematical model of HDM

In a first approach HDM was modeled for rigid products with homogeneous pores of diameter D and length z, which were immersed in a liquid (Fito and Pastor, 1994). The interior of the pore is assumed to be occupied by gas at an initial pressure p_i, whereas in the liquid phase the external pressure is $p_e > p_i$. When the system is at normal pressure, $\Delta p = p_e - p_i$ is due to capillary pressure (p_c), being $\Delta p = p_c$ given by the Young-Laplace equation (Equation (1)) in function of capillary diameter and liquid surface tension (σ).

$$P_c = \frac{4\sigma}{D} \tag{1}$$

2.1 Equlibrium status

The liquid flow into the pores can be estimated from the application of the Hagen-Poiseuille relationship in differential form to the system (Equation (2)). In Equation (2), μ is the liquid viscosity and the liquid penetration was expressed as the pore volume fraction (x_v) occupied by the liquid at each time (t). The x_v value increases with time as internal gas is compressed till Δp equals zero and mechanical equilibrium is reached, then $(dx_v/dt) = 0$. By assuming an isothermal compression and taking into account the relationship between Δp and x_v, Equation (2) can be written at mechanical equilibrium as Equation (3), where p_1 is the initial gas pressure in the pores and p_2 the external system pressure plus the capillary pressure ($p_2 = p_e + p_c$).

$$-\Delta p + \frac{32\mu z^2}{D^2} x_v \frac{dx}{dt} = 0 \tag{2}$$

$$x_v = \frac{p_2 + p_c - p_1}{p_2 + p_c} \tag{3}$$

Equation (3) shows that the volume fraction of a pore penetrated by the external liquid in stiff products depends only on the pressure changes promoted and on capillary effects. If no pressure changes are imposed on the system ($p_1 = p_2$), only capillary effects cause liquid entry, x_v value at equilibrium being given by Equation (4) in this case.

$$x_v = \frac{p_c}{p_2 + p_c} \tag{4}$$

Equation (3) is usually written in a simpler way (Equation (5)), where r is the actual compression ratio (Equation (6)), and **R** the apparent compression ratio (Equation (7)). In most cases, p_c is difficult to estimate because the pore diameter is unknown. Nevertheless, from the mathematical analysis of Equation (5) it can be deduced (Fito, 1994) that the capillary force contribution to liquid penetration can be neglected when the apparent compression ratio is higher than 4 or 6, depending on the pore diameter range.

$$x_v = 1 - \frac{1}{r} = 1 - \frac{1}{R + p_c/p_1} \tag{5}$$

$$r = (p_2 + p_c)/p_1 \tag{6}$$

$$R = p_2/p_1 \tag{7}$$

The sample volume fraction impregnated by the liquid (X) can be obtained by multiplying Equation (5) by the product effective porosity ε_e (Equation (8)). This will be the pore volume fraction available to the HDM action. Equation (8) predicts the amount of sample volume that will be occupied by an external liquid of a determined composition at the mechanical equilibrium status, in terms of product porosity and compression ratio.

$$X = \varepsilon_e\left(1-\frac{1}{r}\right) \approx \varepsilon_e\left(1-\frac{1}{R}\right) \tag{8}$$

In a VI process, a porous product is immersed in an adequate liquid phase and is submitted to a two-step pressure change. In the first step, vacuum pressure is applied to the system for a time t_1, till mechanical equilibrium is achieved, which can be visually detected when no more gas bubbles leave the product. In this moment capillary penetration will be higher than at atmospheric pressure, according to Equation (4). Figure 1 shows the influence of pressure on capillary penetration (x_v (c)) for two cases: for the pore diameter estimated (160 µm) in Granny Smith apples (Fito, 1994) and for a 10 µm diameter. When atmospheric pressure is restored in the system in the second step, compression leads to the HDM product impregnation. Figure 1 shows the x_v values calculated from Equation (5) by considering the actual (r) and apparent (R) compression ratios as a function of the level of vacuum (p_1) applied in the first step. For pore sizes in the order of those found in apples no notable capillary contribution to total impregnation was obtained, even for low p_1 values. Nevertheless, a great contribution will be expected in products with very small pores (~10 µm). In this sense, it may be considered that the small pores are usually associated with a low total porosity, so the overall product impregnation by capillary effect will also be limited (Equation (8)). As deduced from the model, the lower the vacuum pressure, the higher the VI level (e.g., in apples with $\varepsilon_e = 0.20$, a 0.19 volume fraction will be occupied by an external solution by using 50 mbar in the first VI step — Salvatori et al., 1998).

The HDM model was extended for viscoelastic porous products, where pressure changes cause not only gas or liquid flow but also solid matrix deformation-relaxation phenomena (DRP) (Fito and Chiralt, 1996; Fito et al., 1996). During the first VI step the product volume usually swells associated with gas expansion and afterward the solid matrix relaxes; capillary penetration or the expelling of internal liquid also occurs in this period. In the second step, compression causes volume deformation and subsequent relaxation, coupled with the external liquid penetration in the pores. Mechanical properties of the solid matrix and flow properties of the liquid in the pores define the characteristic penetration and deformation-relaxation times, responsible for the final impregnation and deformation of the samples at equilibrium. Equation (9) describes the equilibrium relationship between the compression ratio (r), sample porosity (ε_e), final sample volume fraction impregnated by the external solution (X) and sample volume deformations

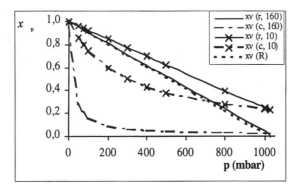

Figure 1 Influence of vacuum pressure on the pore volume fraction impregnated by capillary forces ($x_v(c)$), or by external imposed pressure, considering ($x_v(r)$) and not considering ($x_v(R)$) capillary contribution, for 160 and 10 mm capillary diameters.

at the end of the process (γ) and the vacuum step (γ_1), all these referred to the sample initial volume.

$$\varepsilon_e = \frac{(X-\gamma)\cdot r + \gamma_1}{r-1} \tag{9}$$

2.2 HDM kinetics

Concerning kinetics, a very fast HDM action has been detected, so experimental kinetic analysis is not possible in the real time scale for fruits and low-viscosity solutions. For a stiff porous product, integration of Equation (2) leads to Equation (11), taking into account the function $\Delta p = f(x_v)$ (Equation (10)), and assuming that x_v takes the x_{v0} and x_v values at $t = 0$ and t, respectively. In Equation (11), x_v was expressed in a reduced way (Equation (12)), referred to its value at equilibrium. B and k are, respectively, time dimension and dimensionless constants for specific system and operating conditions. In B, three terms can be distinguished: the liquid viscosity (μ); a product structural number depending on the sample characteristic depth (e), the pore radius (r_p) and tortuosity factor (F_t); and a pressure number depending on the equilibrium pressures in the system before (p_1) and during (p_2) the compression step.

$$-\Delta p = P_2 \left(\frac{1-x_{ve}}{1-x_v} - 1 \right) \tag{10}$$

where x_{ve} was given by Equation (5) in terms of the equilibrium pressures.

$$\frac{t}{B} = -\frac{1}{2}\left(x_r^2 - x_{r_0}^2\right) - k\left[\left(x_r - x_{r_0}\right) + \ln\left(\frac{1-x_r}{1-x_{r_0}}\right)\right] \tag{11}$$

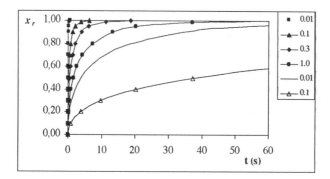

Figure 2 Predicted values of the reduced pore volume fraction impregnated as a function of time, at different external solution viscosities (Pa s), for 160 (closed symbols) and 10 (open symbols) μm pore diameters.

where:

$$x_r = x_v/x_{ve} \tag{12}$$

$$k = 1 + 1/x_{ve} \tag{13}$$

$$B = 8\mu \frac{P_2}{(P_2 - P_1)^2} \left(\frac{eF_1}{r_p}\right)^2 \tag{14}$$

Figure 2 shows the development of x_r as a function of time and liquid viscosity for cylindrical apple samples (2 cm height and diameter, e = 1 cm) considering a tortuosity factor $F_t = 2$ and $r_p = 80$ μm (Fito, 1994). Zero value was taken for x_{v0}. A very short time was required to achieve the x_{ve} value ($x_r = 1$) when viscosity ranges between the usual values of aqueous solution used in VI processes. Even for more viscous solutions (hydrocolloids, concentrated syrups,...) the impregnation time is relatively short, as compared with other process times. When pore radius was reduced considerably to 5 μm the kinetics relents, but for low viscosity the equilibrium time is close to 1 min. This time range is in the order of that required to achieve stationary pressure conditions in the tank. In fact, in different attempts to analyze VI kinetics in a great number of fruits, using sugar isotonic solutions (μ~0.006 to 0.008 Pa s), and varying the length of the compression period from 5 to 30 min, no significant differences between the x_v values at the different times were found (Salvatori et al., 1998a; Sousa et al., 1998).

3   Fruit properties as affected by VI

Most fruits are porous, the pores being the intercellular spaces, which can contain a gas or liquid phase. VI of fruits may offer interesting prospects in fruit processing. VI may promote a fast compositional change of the fruit, this being useful in many cases to assure processed fruit stability (decrease of pH or water activity, introduction of anti-browning agents or microbial preservatives,...) or quality (the improvement of the sweet-sour taste relationship). Nevertheless, VI could imply changes in mechanical, structural, and other physical properties, which will affect the overall fruit quality. Several works (Martínez Monzó et al., 1996, 1998) were developed in order to analyze these changes as well as the feasibility of VI of different fruits commercialized frequently as minimally or deep-processed products (Salvatori et al., 1998a; Sousa et al., 1998).

3.1 Feasibility of VI of fruits

The response of many fruits to VI processes (impregnation and deformation levels at equilibrium in the two process steps) has been characterized experimentally by means of a gravimetric methodology previously described (Fito et al., 1996; Salvatori et al., 1998a). Table 1 gives the VI parameters obtained for a large number of fruits. Positive volume deformations at the end of the vacuum step (γ_1) were obtained in most cases, whereas negative or positive values were achieved at the end of the compression period (γ), depending on the fruit. Concerning the liquid phase flows, losses of native liquid (expelled in line with gas release) occur in almost all the fruits throughout the vacuum step, coupled with the capillary entry ($X_1 < 0$). Because of the viscoelastic properties of the fruits, the level of final impregnation (X) was greatly affected by the penetration-deformation coupling throughout the action of pressure gradients in the system (Salvatori et al., 1998a; Sousa et al., 1998). This fact makes the analysis of the pressure influence on each parameter difficult. However, when the impregnated pore volume fraction, x_v, was calculated taking into account the sample deformation according to Equation (16), the theoretical influence of pressure on the degree of impregnation can be observed. Figure 3 shows the values of the pore volume fraction deformed (γ') and impregnated at different levels of vacuum applied in the first VI step. Both fractions were referred to the pore initial volume and were estimated by Equations (15) and (16) from the X, γ, and ε_e measurements in VI experiments on several fruits. The x_v predictions by Equation (3), without considering capillary contribution, were also plotted (line). The good agreement between experimental and predicted values can be appreciated. On the other hand, the pore deformation at the end of the compression step seems slightly higher when lower vacuum pressures were applied in the first step. Therefore, for practical uses, the vacuum level given by most common industrial vacuum pumps (50 to 100 mbar) will be recommended.

$$\gamma' = \gamma/\varepsilon_{e0} \tag{15}$$

$$x_v = X/(\varepsilon_{e0} + \gamma) \tag{16}$$

where ε_{e0} is the initial sample effective porosity.

Table 1 Mean Values (percentage) of VI Parameters Obtained for Different Fruits, Applying 50 mbar in the Vacuum Step

Fruit, variety	Sample[a]	Brix	X_1	s.e.	γ_1	s.e.	X	s.e.	γ	s.e.	ε_e	s.e.	X_{LN}[b]
Apple, Granny Smith	2×2 cm cylinders	10.4	-4.2	0.3	1.7	0.3	19.0	1.5	-0.6	1.2	21.0	0.9	8.5[c]
Apple, Red Chief	2×2 cm cylinders	13.0	-5.0	0.4	2.1	0.4	17.9	0.7	-2.4	1.0	20.3	0.4	10.6
Apple, Golden	2×2 cm cylinders	15.3	-2.7	0.3	2.8	0.2	11.2	0.8	-6.0	0.5	17.4	0.8	7.5
Mango, Tommy Atkins	1 cm slices	17.8	0.9	0.2	5.4	0.5	14.2	0.5	8.9	0.4	5.9	0.4	0.74
Strawberry, Chandler	Whole fruit	7.2	-2.1	0.2	2.9	0.4	1.9	0.7	-4.0	0.6	6.4	0.3	3.4
Kiwi fruit, Hayward	Fruit quarters	14.3	-0.2	0.2	6.8	0.6	1.09	0.1	0.8	0.5	0.66	0.5	0.34
Peach, Miraflores	2.5 cm side cubes	15.0	-2.29	0.1	2.0	0.3	6.5	0.5	2.1	0.4	4.7	0.3	3.2
Peach, Catherine	2.5 cm side cubes	11.9	-1.4	0.4	0.5	0.3	4.4	0.6	-4.2	0.6	9.1	0.8	2.2[c]
Apricot, Bulida	Fruit halves	13.5	-0.2	0.2	1.5	0.4	2.1	0.3	0.11	0.2	2.2	0.2	0.84[c]
Pineapple, Española Roja	1 cm slices	9.0	-6.5	0.6	1.8	0.4	5.7	0.8	2.3	0.4	3.7	1.3	8.0[c]
Pear, Passa Crassana	2.5 cm side cubes	16.0	-1.3	0.2	2.8	0.2	5.3	0.9	2.2	0.7	3.4	0.5	1.6[c]
Prune, President	Fruit halves	17.6	-1.0	0.1	0.6	0.1	1.0	0.1	-0.8	0.1	2.0	0.2	1.4[c]
Melon, Inodorus	2×2 cm cylinders	10.2	-4.0	0.3	2.0	0.3	5.0	0.2	-0.4	0.2	6.0	0.3	5.1[c]

[a]Size and shape of processed samples.

[b]Volume fraction (percentage) of native liquid lost during the vacuum step.

[c]Mean values obtained from experiments at different vacuum levels.

s.e.: parameter standard error

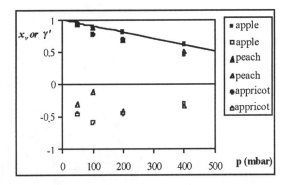

Figure 3 Experimental values of the pore volume fraction impregnated (closed symbols) and deformed (open symbols) as a function of applied vacuum pressure for different fruits. The solid line corresponds to the predicted values for apple.

3.2   Changes in mechanical, structural, and other physical properties.

The effect of VI on mechanical and structural properties of fruits was studied by introducing hypotonic, isotonic, and hypertonic solutions in apple tissue (Martínez Monzo et al., 1996, 1998). Stress relaxation tests (8% constant strain, 200 mm/min deformation rate) were performed to analyze mechanical behavior, and cryo-SEM (Bomben and King, 1982) to observe structural changes.

When isotonic solutions were used (cell turgor unaltered) no significant differences in the initial elasticity modulus between fresh and VI apples were found. Nevertheless, the relaxation rate and the total relaxation level increased in VI samples in line with the impregnation level (X) (Martínez Monzó et al., 1996). On the other hand cryo-SEM observations of isotonic VI samples did not show cellular alterations or debonding, but showed the sample intercellular spaces completely flooded by the external solution (Figure 4a) (Martínez Monzó et al., 1998). These results led to the changes in the viscoelastic behavior of the isotonic VI samples being principally attributed to the exchange of gas (compressible during the mechanical test) for liquid that will flow out from the pores throughout compression (Martínez Monzó et al., 1996).

When hypertonic solutions are used for VI, sample osmotic dehydration simultaneously occurs, thus contributing to changes in chemical and physical properties of the product. Hypertonic liquids promote turgor losses and the cells completely lose their elastic character after plasmolysis (Pitt, 1992). Therefore, the apparent elastic modulus decreases sharply, increasing the viscous character. Nevertheless, by comparing the textural properties of a_w depressed fruits till the same a_w level, a better preserved texture was obtained if VI was used in the fruit processing (Alzamora and Gerschenson, 1997).

VI with hypotonic solution only implied a greater level of stress relaxation, which was explained by flow out of the intracellular liquid in line

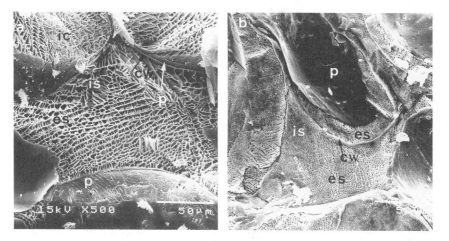

Figure 4 Cryo-SEM micrographs of VI apple tissue with isotonic (a) and 65 Brix hypertonic (b) rectified grape must. **ic**: intracellular content, **is**: intercellular space, **p**: plasmalemma, **cw**: cell wall, **es**: external solution in the intercellular space.

with cell rupture promoted by an excessive turgor (Martínez Monzó et al., 1998).

The most relevant structural change promoted by VI is the filling of the product pores with the external solution. In many cases native liquid was expelled during the vacuum step and was replaced by the external one (products with X_1 negative in Table 1). In VI hypertonic treatments, osmotic dehydration of the tissue promotes plasmolysis but no significant shrinkage of the cellular wall was observed (Figure 4b), contrary to that observed in osmosed tissue at normal pressure (Salvatori et al., 1998b; Barat et al., 1997a, 1997c). On the other hand, the space between plasmalemma and cell wall appeared completely full of solids. Martínez Monzó et al. (1998) suggested that the water loss volume in the cellular cavity was replaced by concentrated solution from the intercellular space due to the pressure gradients produced in the cell cavity.

The reported analysis by TEM and LM of some minimally processed fruits (apple, melon, and kiwi fruit), water activity depressed ($a_w = 0.98–0.93$) (Alzamora and Gerschenson, 1997) are in line with the above cryoSEM conclusions. The final a_w was obtained by osmotic dehydration at normal pressure and by VI in different concentrated humectant solutions (glycerol, glucose, and sucrose). Cell wall observations of the osmosed tissue showed a great alteration and very low electronic density in fruits processed at atmospheric pressure. Nevertheless, the cell wall of VI fruits showed an electronic density close to that of fresh, and the original cell shape and arrangement were retained. These observations were in agreement with an improvement of the mechanical response of these last samples. On the other hand, the higher the sugar molecular weight the better the textural characteristics.

Taking into account the fibrilar structure of the cell wall, these last results suggest some specific interaction between the solution solids and the polymeric fibrils that reinforce cell wall structure in VI processes. This could be promoted by the passing of the external liquid through the cell wall, due to the pressure gradients provoked by cell water loss, in a similar way to a filtration membrane.

Other physical properties such as optical or thermal properties also change due to VI. The gas-liquid exchange in the fruit implies a more homogeneous refraction index throughout the sample and so an increase in the product transparency. Therefore, when color was measured by diffuse reflection a decrease in the reflection coefficients was obtained for VI samples, as compared with fresh samples, thus implying lower color coordinates and small changes in chromatic attributes (Martínez Monzó et al., 1997). Highly porous products such as apple are the most affected in the optical properties by VI.

VI increases thermal conductivity of porous fruits due to gas substitution, although slight modifications were produced in thermal diffusivity because of the simultaneous density increase. Changes are greatly dependent on total porosity and pore distribution in relation with the heat flow sense (Barat et al., 1994).

4   Influence of VI on osmotic processes

Osmotic dehydration (OD) is the usual process to decrease product a_w in minimally processed fruits and vegetables, or in some deep processed fruits such as candies or jam (Shi et al., 1996). The a_w decrease is based on osmotic water flow through cell membranes when the product is immersed in a concentrated solution (osmotic solution, OS) of sugars and/or salt. Vacuum treatment leads to faster osmotic process due to the coupled action of HDM (Shi and Fito, 1993, 1994; Fito, 1994; Shi et al. 1995). Three kinds of osmotic treatments have been defined depending on the pressure applied to the tank: OD (at atmospheric pressure), VOD (at vacuum pressure) and PVOD (pulsed vacuum osmotic dehydration) (Fito et al., 1993; Fito and Chiralt, 1996). In PVOD, VI takes place during the first 5 to 10 min of the process by the action of a vacuum pulse. This implies a fast compositional change in the product that will affect the osmotic driving force and mass transfer kinetics (Fito and Chiralt, 1995, 1998; Barat et al., 1997a). VI also affected the development of osmosed samples to the final equilibrium in terms of mass, volume, density, and structure changes (Barat et al., 1997c; Fito et al., 1998). These aspects are very important in long term osmotic processes.

In OD and VOD processes HDM will only act by capillary forces throughout the process, in greater intensity in VOD according to Equation (4). In VOD, HDM will also act when the system returns to atmospheric pressure if product pores are not collapsed by surface dehydration (Fito and Chiralt, 1995; Fito et al., 1993).

4.1   Composition changes promoted by VI

In minimally processed products, VI with an adequate solution (OS plus the required additives: ascorbic acid, antimicrobials, etc.) permits a fast compositional change with an improvement in texture (Alzamora and Gerschenson, 1997), appearance and microbial and browning stability (Carrol and Tharrington, 1997), as compared with products with similar composition achieved by OD. The mass fraction of any component (water or solutes) reached in a VI product (x_i^{VI}) can be estimated by Equation (17), deduced from a mass balance in the system, in terms of the sample initial composition and the impregnated solution mass fraction (x_{HDM}) and composition. The x_{HDM} value can be obtained (Equation (18)) from the product VI response (impregnated volume fraction: X) and initial product (ρ^O) and solution (ρ^{IS}) densities. From these equations, the required solution concentration of a specific component (water, sugar, acid, additive,...) can be calculated to achieve the desired final level in the product.

$$x_i^{VI} = \frac{x_i^0 + x_{HDM} \cdot y_i}{1 + x_{HDM}} \qquad (17)$$

$$x_{HDM} = X \rho^{IS} / \rho^O \qquad (18)$$

where: x_i = product mass fraction of component i before and after VI
          (0 and VI superscripts).
       y_i = mass fraction of component i in the impregnation solution.

Concentration effects promoted by concentration gradients were not taken into account in Equation (17) due to the short time used in VI. Nevertheless, when driving forces dependent on concentration are high (or for long times), small deviations of the predicted final concentrations can be found. Figure 5 shows the mass fraction of water and soluble solids achieved in VI (5 min at 50 mbar) of Granny Smith apple samples with sucrose syrups (15 min processed) and rectified grape musts (30 min processed) of different concentrations. The values predicted by Equation (17) are also plotted in each case. The lower a_w values of grape musts, and the longer processing time, led to higher solute concentrations than those expected from Equation (17). Nevertheless, osmotic dehydration was not evident for short treatments with sucrose solutions. With 65 Brix sucrose, a lower degree of impregnation than expected was obtained, which could be due to a partial surface pore collapse because of dehydration associated to the high viscosity of the impregnation solution.

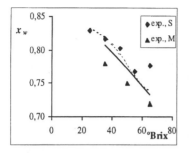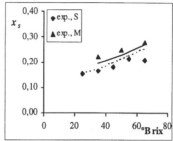

Figure 5 Experimental values of water and soluble solid mass fractions of VI apple samples with sucrose solutions (S) and rectified grape musts (M) of different concentrations. Lines: predicted values for sucrose solutions (*dashed*) and musts (*solid*).

5 Salting processes by using VI

The application of VI to accelerate salting processes has been studied in manchego type cheese, cured ham, and fish (Del Saz et al., 1994; Chiralt and Fito, 1996; Escriche et al., 1996; Barat et al., 1997b). In cheese salting, VI promotes whey and gas release from curd and the capillary entry of brine coupled with the diffusion process. This implies an important reduction in salting time (Chiralt and Fito, 1996) and a flatter salt gradient immediately after salting (Andrés et al., 1997; Gonzalez et al., 1997). Although no relevant influence of VI salting on the development of biochemical ripening indexes was observed, textural properties appeared slightly modified, mainly due to the disappearance of the small gas spots in the cheese matrix due to the vacuum treatment (Guamis et al., 1997). A great influence of factors affecting curd porosity, such as pressing conditions or pore collapse caused by time or temperature effects, was observed in the total salt uptake (Fuentes et al., 1996).

The influence of a VI pretreatment with brine on salting of cod with solid salt was studied in terms of salting kinetics and final textural properties of the product (Escriche et al., 1996). The salt effective diffusivity greatly increased in the VI samples principally during the first salting period. The texture analysis of VI samples showed less hardness, and greater elasticity and cohesiveness, as compared with non-VI samples.

Ham for curing is usually salted with solid salt for 12 to 15 days, depending on ham weight. VI with saturated brine permits a dramatic reduction of salting time (~5 days) and the feasibility to salt frozen pieces directly, thus optimizing the process operation (Barat et al., 1997b).

Acknowledgments

We thank the Comision Interministerial de Ciencia y Tecnología, the European Commission (STD3 programme) and the CYTED Program for financial support.

References

Alzamora, S. M. and Gerschenson, L. N. (1997). Effect of water activity depression on textural characteristics of minimally processed fruits. In: *New Frontiers in Food Engineering: Proceedings of the 5th Conference of Food Engineering,* Barbosa-Cánovas, G. V., Lombardo, S., Narsimhan, G., and Okos, M. R. (Eds.), AIChemE, New York, p. 72.

Andrés, A., Panizzolo, L., Camacho, M. M., Chiralt, A., and Fito, P. (1997). Distribution of salt in Manchego type cheese after brining. In: *Engineering and Food: Proceedings of the ICEF 7,* Jowitt R., Ed., Academic Press, Sheffield, U.K., p.A-133.

Barat, J. M., Martínez-Monzó, J., Alvarruiz, A., Chiralt, A., and Fito, P. (1994). Changes in Thermal Properties due to vacuum impregnation. In: *Proceedings of the IsoPoW Conference,* Poster Session, ISOPOW practicum II, Argaiz, A., López-Malo, A., Palou, E. and Corte, P. (Eds.), Universidad de las Américas, Puebla, Mexico, p. 117.

Barat, J. M., Alvarruiz, A., Chiralt, A., and Fito, P. (1997a). A mass transfer modelling in osmotic dehydration. In: *Engineering and Food: Proceedings of the ICEF 7,* Jowitt R., Ed., Academic Press, Sheffield, U.K., p. G-81.

Barat, J. M., Grau, R., Montero, A., Chiralt, A., and Fito, P. (1997b). Procedimiento de descongelación y salado simultaneo de piezas de carne o pescado, *Spanish Patent,* P9701702. In Spanish.

Barat, J. M., Albors, A., Chiralt, A., and Fito, P. (1997c). Structural changes throughout equilibration of apple in osmotic treatments. In: *Proceedings of the Third Main Meeting,* Project Process Optimization and Minimal Processing of Foods. Vol. 3 — Drying, Oliveira, J.C. and Oliveira, F.A.R. (Eds.), published by Escola Superior de Biotecnologia, Porto, Portugal.

Bomben, J. L. and King, C. J. (1982). Heat and mass transport in the freezing of apple tissue. *Journal of Food Technology,* 17, 615.

Carrol, D. E. and Tharrington, J. B. (1997). Vacuum infiltration process for long term preservation of refrigerated apple slices. Presented at the IFT '97 Annual Meeting, Orlando, June, 14 to 18.

Chiralt, A. and Fito, P. (1996). Salting of *Manchego* Type Cheese by Vacuum Impregnation. In: *Food Engineering 2000,* Fito, P., Ortega-Rodríguez, E., and Barbosa-Cánovas, G. (Eds.), Chapman & Hall, New York, p. 214.

Del Rio, M. A. and Miller, M. V. (1979). Effect of pretreatment on the quality of frozen melon balls, *Bulletin de l'Institut Internationale du Froid,* 59, p. 1172.

Del Saz, A., Mateu, A., Serra, J. A., and Fito, P. (1994). Vacuum Salt Impregnation of Cultivated Trouts (Salmo Gairdneri). In: *Proceedings of the ISOPOW Conference,* Poster Session, ISOPOW practicum II, Argaiz, A., López-Malo, A., Palou E., and Corte, P. (Eds.), Universidad de las Américas, Puebla, Mexico, p. 57.

Escriche, I., Serra, J. A., Fito, P., and Rivero, E. (1996). Estudio de la Influencia de la Deshidratación Osmótica a Vacio en el Salado del Bacalao (*Gadus morhua*). In: *Equipos y procesos para la industria alimentaria,* Vol. II — Análisis cinético, termodinámico y estructural de los cambios producidos durante el procesamiento de alimentos, Ortega, E., Parada, E., Fito, P., Matos-Chamorro, A.R., Sobral, P.J., Chiralt, A., and Alzamora, S.M. (Eds.), Servicio de Publicaciones, Universidad Politécnica de Valencia, p. 142. In Spanish.

Fito, P. (1994). Modelling of vacuum osmotic dehydration of food, *Journal of Food Engineering,* 22, 313.

Fito, P. and Chiralt, A. (1995). An Update on Vacuum Osmotic Dehydration. In: *Food Preservation by Moisture Control: Fundamentals and Applications*, Barbosa-Cánovas, G.V. and Welti-Chaves, J. (Eds.), Technomic Pub. Co. Inc. Lancaster, U.K., p. 351.

Fito, P. and Chiralt, A. (1996). Osmotic dehydration: An approach to the modelling of solid food-liquid operations. In: *Food Engineering 2000*, Fito, P., Ortega-Rodríguez, E., and Barbosa-Cánovas, G. (Eds.), Chapman & Hall, New York, p. 231.

Fito, P. and Pastor, R. (1993). On some non-diffusional mechanism occurring during vacuum osmotic dehydration, *Journal of Food Engineering*, 21, 513.

Fito, P., Andrés, A., Chiralt, A., and Pardo, P. (1996). Coupling of hydrodynamic mechanism and deformation-relaxation phenomena during vacuum treatments in solid porous food-liquid systems, *Journal of Food Engineering*, 27, 229.

Fito, P., Andrés, A., Pastor, R., and Chiralt, A. (1994). Vacuum Osmotic Dehydration of Fruits. In: *Minimal Processing of Foods and Process Optimization: an interface*, Singh, P. and Oliveira, F.A.R. (Eds.), CRC Press, Boca Ratón, FL, p. 107.

Fito, P., Chiralt, A., Barat, J., Salvatori, D., and Andrés, A. (1998). Some advances in osmotic dehydration of fruits, *Food Science and Technology International*, 4, 329.

Fuentes, C., Gonzalez, C., Andrés, A., Chiralt, A., and Fito, P. (1996). Some changes in curd porosity affecting salt uptake in manchego type cheese. In: *Proceedings of the Second Main Meeting*, Project Process Optimization and Minimal Processing of Foods. Vol. 5 — Minimal and Combined Processes, Oliveira, J.C., Oliveira, F.A.R., and Gorris, L.G.M. (Eds.), published by Escola Superior de Biotecnologia, Porto, Portugal.

Gonzalez, C., Fuentes, C., Andrés, A., Chiralt, A., and Fito, P. (1997). Influence of vacuum pressure in the NaCl concentration profiles after cheese salting. In: *Proceedings of the Third Main Meeting*, Project Process Optimization and Minimal Processing of Foods. Vol. 5 — Minimal and Combined Processes, Oliveira, J.C., Oliveira, F.A.R., and Gorris, L.G.M. (Eds.), published by Escola Superior de Biotecnologia, Porto, Portugal.

Guamis, B., Trujillo, J. A., Ferragut, V., Chiralt, A., Andrés, A., and Fito, P. (1997). Ripening control of manchego type ewe's cheese salted by brine vacuum impregnation, *International Dairy Journal*, 7, 185.

Hawkes, J. and Flink, J. M. (1978). Osmotic concentration of fruit slices prior to freeze dehydration, *Journal of Food Processing and Preservation*, 2, 265.

Martínez Monzó, J., Martínez Navarrete, N., Fito, P., and Chiralt, A. (1996). Cambios en las propiedades viscoelásticas de manzana (*Granny Smith*) por tratamientos de impregnación a vacío. In: *Equipos y procesos para la industria alimentaria*, Vol. II — Análisis cinético, termodinámico y estructural de los cambios producidos durante el procesamiento de alimentos, Ortega, E., Parada, E., Fito, P., Matos-Chamorro, A.R., Sobral, P.J., Chiralt, A., and Alzamora, S.M. (Eds.), Servicio de Publicaciones, Universidad Politécnica de Valencia, p. 234. In Spanish.

Martínez Monzó, J., Martínez Navarrete, N., Fito, P., and Chiralt, A. (1997). Effect of vacuum osmotic dehydration on physicochemical properties and texture of apple. In: *Engineering and Food: Proceedings of the ICEF 7*, Jowitt, R., Ed., Academic Press, Sheffield, U.K., p. G-17.

Martínez Monzó, J., Martínez Navarrete, N., Fito, P., and Chiralt, A. (1998). Mechanical and structural changes in apple (var. Granny Smith) due to vacuum impregnation with cryoprotectants, *Journal of Food Science*, 63, 499.

Pitt, R. E. (1992). Viscoelastic properties of fruits and vegetables. In: *Viscoelastic Properties of Foods*, Rao, M.A. and Steffe, J.F. (Eds.), Elsevier Applied Science, New York, p. 49.

Reinbold, R. S., Hansen, C. L., Gale, C. M., and Ernstrom, C.A. (1993). Pressure and temperature during vacuum treatment of 290-kilogram stirred –curd cheddar cheese blocks, *Journal of Dairy Science*, 76, 909.

Salvatori, D., Andrés, A., Chiralt, A., and Fito, P. (1998a). The response of some properties of fruits to vacuum impregnation, *Journal of Food Processing Engineering*, 21, 59.

Salvatori, D., Albors, A., Andrés, A., Chiralt, A., and Fito, P. (1998b). Analysis of the structural and compositional profiles in osmotically dehydrated apple tissue, *Journal of Food Science*. In press.

Santerre, C. R., Cash, J. N., and Vannorman, D. J. (1989). Ascorbic acid/citric combinations in the processing of frozen apple slices, *Journal of Food Science*, 53, 1713.

Shi, X. Q. and Fito, P. (1993). Vacuum osmotic dehydration of fruits, *Drying Technology*, 11, 1429.

Shi, X. Q. and Fito, P. (1994). Mass transfer in vacuum osmotic dehydration of fruits: A mathematical model approach, *Lebensmittel Wissenchaft und Technologie*, 26, 67.

Shi, X. Q., Chiralt, A., Fito, P., Serra, J., Escoín, C., and Gasque, L. (1996). Application of osmotic dehydration technology on jam processing, *Drying Technology*, 14, 841.

Shi, X.Q., Fito, P., and Chiralt, A. (1995). Influence of vacuum treatment on mass transfer during osmotic dehydration of fruits, *Food Research International*, 28, 445.

Sousa, R., Salvatori, D., Andrés, A., and Fito, P. (1998). Analysis of vacuum impregnation of banana (*Musa acuminata* cv. Giant cavendish), *Food Science and Technology International*, 4, 127.

Zozulevich, B. and D'yachenco, E. N. (1969). Osmotic dehydration of fruits, *Konservn. Ovoshchesusch. Prom.*, 7, 32.

chapter twenty-one

Edible and biodegradable polymeric materials for food packaging or coating

Ioannis Arvanitoyannis and Leon G. M. Gorris

Contents

Summary 357
1   Introduction 358
2   Development of biodegradable and edible packaging materials.........359
    2.1   Biodegradable synthetic copolymers and composites..................359
    2.2   Blends of natural and synthetic polymers....................................360
    2.3   Starch-based edible polymer blends...360
3   Development of (active) edible coatings for
    high-moisture foods 362
    3.1   Specific requirements for edible coatings 362
    3.2   Properties of edible coatings...363
    3.3   Types of edible coatings...364
    3.4   Active edible coatings 365
4   Conclusions..366
Acknowledgment...367
References...367

Summary

This chapter reviews recent advances in the production of enviromentally friendly packaging materials, employing synthetic biodegradable polymers, partially degradable polymer blends, or edible materials based on natural polymers. Such materials can be used as a valuable functional hurdle for minimally processed foods, maintaining "natural" and "fresh" characteristics while contributing to increased shelf life and safety. The use of biopolymers in producing edible food coatings for application in high-moisture foods is particularly addressed.

1   Introduction

The increased consumer demand for high-quality, long-shelf life, ready-to-eat foods has initiated the development of mildly preserved products that keep their natural and fresh appearance as long as possible and at the same time are safe to eat (Leistner and Gorris, 1995). Packaging is an important element in many mild preservation concepts. The packaging material provides physical protection and creates the physicochemical conditions around a food product that are essential for obtaining a satisfactory shelf life. The packaging system, based on a proper choice of the packaging material endowed with appropriate gas- and water-barrier properties and of gas atmosphere conditions, prevents product deterioration due to chemical and (micro-)biologic factors and maintains the hygiene status despite frequent handling. Plastic materials are most often used for food packaging to date. It is estimated that, in total, plastics will represent 30% (25 million tons) of the packaging material used in the U.S. in the year 2000. Only ± 2% of all (mostly nondegradable) plastics are yet recycled. The remainder is disposed of in landfills or burned, in both cases putting a serious burden on the environment. The current trend toward environmental protection is also expressed in the current research and development on packaging materials, by designing degradable polymers and composting or recycling the "recalcitrant" polymers.

Edible and biodegradable films offer good alternative packaging options. For one, their advantage over synthetic "recalcitrant" packaging polymers is that they do not contribute to long-term environmental pollution (Arvanitoyannis et al., 1996, Krochta and De Mulder-Johnston, 1997). The potential of biopolymers such as polysaccharides and proteins for the manufacture of edible films, is well appreciated (Guilbert, 1986; Gennadios and Weller, 1990; El Ghaouth et al., 1991a; Wong et al., 1992; Gennadios et al. 1993a; Guilbert and Gontard, 1995; Guilbert et al., 1996), but apart from some very special applications (Kinsella, 1984; Park et al., 1993, 1994; Greener Dohnwe and Fennema, 1994), polysaccharide- and/or protein-based edible films have not yet found extensive applications in the food industry (Kester and Fennema, 1986).

Another interesting development is the application of edible coatings directly on food products or food ingredients, that extend produce shelf life by providing physical protection as well as a semipermeable barrier toward gases and water vapor in analogy to packaging. The aim is to use coating materials made of natural biopolymers that may be eaten together with the food. Functional compounds (antimicrobials, antioxidants, flavors) can be integrated in edible coatings, giving the possibility to accumulate such beneficial compounds at the exact location where their presence is required. Often, lower doses of such compounds can suffice for adequate activity in a coating as compared to dispersed application. Potential applications and properties of edible coatings and films have been reviewed elsewhere (Guilbert, 1986; Kestner and Fenemma, 1986; Baldwin, 1994; Cuq et al., 1994; Koelsch, 1994; Baldwin et al. 1995).

This chapter presents recent advances in the development of more environmentally friendly food packaging materials, employing synthetic biodegradable polymers, partially degradable starch-synthetic polymer blends, or edible materials based on natural polymers. In addition, the use of biopolymers in the construction of edible food coatings for application with high-moisture food products is addressed.

2   Development of biodegradable and edible packaging materials

2.1   Biodegradable synthetic copolymers and composites

The effort to design novel synthetic polymers that are better (bio)degradable than concurrent food packaging materials initially targeted at the synthesis of novel aliphatic polyesters (Vert et al., 1992), which have a higher biodegradability than polyamides and polyanhydrides (Satyanarayana and Chaterji, 1993). Among the aliphatic polyesters, poly(ε-caprolactone), poly(ε-methyl — valerolactone), polylactide, and their copolymers have been suggested as the most promising polymers in terms of potential applications. Low glass transitions and low melting points of most polyesters have oriented several researchers toward exploring other potential avenues such as the direct polymerization of α-aminoacids (Yang et al., 1993) or copolymerization of lactams with lactones (Goodman, 1984) in an attempt to synthesize novel polymers with higher thermal resistance. The biodegradability tests executed with the copolyesteramides were highly promising and in favor of potential applications (Kumar, 1987; Huang, 1985). However, the difficulties encountered in producing high-molecular-weight, biodegradable copolyamides have, application wise, been the restricting factors for the copolyamides (Yang et al., 1993).

The development of novel polymeric materials that degrade slowly is considered a very important research area, especially in view of their various current and potential applications as environmentally degradable materials (Satyanarayana and Chaterji, 1993). Although D- or L- lactide and ε-caprolactone seem to be the most popular monomers, especially in the field of polymeric composite materials mainly related to medical applications, polyamides are another promising class of polymer that appeal to a wider range of applications (Gonsalves et al., 1993). It is anticipated that these novel thermoplastic materials have a lot of potential because of their inherent advantages over the majority of thermoset materials (Arvanitoyannis and Psomiadou, 1994); namely, control of their percentual crystallinity (physicochemical properties), ease of processing, and "friendliness" to the environment (Arvanitoyannis et al., 1995a).

2.2   Blends of natural and synthetic polymers

Although, in the past, synthetic polymers were extensively used for multi-purpose applications because of their satisfactory mechanical and thermal properties, their low biodegradability has complicated and hindered their use (Griffin, 1994; Wool, 1995). On the other hand, natural polymers, though usually endowed with inherent biodegradability, have inferior mechanical properties and present processing problems that are holding them back from widespread use (Greener Dohnwe and Fennema, 1994).

An alternative way to overcome this problem is to resort to blending of natural polymers in order to modify mechanical properties, because it is an easy and straightforward procedure, associated to low costs (Tadmor and Gogos, 1979). The usual target for preparing a novel blend of two or more polymers is not to change drastically the properties of the components, but to capitalize on the maximum possible performance of the blend. On several occasions the initial dispersion of the blend components is further promoted by cross-linking, creation of interpenetrating networks, mechanical inter-locking of components, and use of compatibilizing agents in order to ensure that no demixing will occur at a later stage (Tadmor and Gogos, 1979, Griffin, 1994). Gutta-percha or 1,4 trans-polyisoprene (TPIP) is a natural polymer that has found several applications, among others as an endodontic filling material (Kolokuris et al., 1992a, 1992b). Starch is another natural polymer of great importance because of its inherent biodegradability, its overwhelming abundance, and its annual renewal. Blends of natural–synthetic polymers have been considered promising for preparing polymers with "tailor-made" properties. By now, LDPE/starch blends with favorable functional physical properties and biodegradability, for instance, have already been used commercially for the past 15 to 20 years (Griffin, 1994). LDPE has been thoroughly investigated with regard to its gas permeability (van Krevelen, 1990) because of the extensive use it has in food packaging. LDPE is a semicrystalline polymer with both amorphous and crystalline areas. Apart from disturbing the continuity of the LDPE network and adding to the heterogeneity of the system, the presence of starch particles as fillers within the LDPE matrix enhances the gas permeability of the LDPE/starch composite structure because of their strong hydrophilic character. Water is either strongly bound (0 to 12%) or present as capillary moisture (12 to 30%). The temperature dependence of gas permeability and diffusion of different blends of LDPE and starch is discussed by Arvanitoyannis and Blanshard (1993) and Arvanitoyannis et al. (1994).

2.3   Starch-based edible polymer blends

Edible films or coatings have long been used empirically for food protection and shelf life prolongation (Guilbert and Gontard, 1995). Wax coatings,

casein or lipoprotein-"skin forming" coatings, and polysaccharide film formulations (such as starch and chitosan) are still the most representative
examples of edible film materials (Gennadios and Weller, 1990; El Ghaouth
et al., 1991b; Drake et al., 1991; Wong et al., 1992; Gennadios et al., 1993b;
Lourdin et al., 1995; Guilbert et al., 1996). Edible films have to overcome
several drawbacks, mainly related to their major limitations: their poor mois-
ture-barrier performance (apart from wax), color and appearance, mechanical and rheologic properties (Krochta et al. 1990; Krochta, 1992; Greener
Dohnwe and Fennema, 1994; Koelsch, 1994). As a result, edible films have
not found extensive applications in the food industry yet (Kester and Fennema, 1986; Arvanitoyannis et al., 1996).

Despite these shortcomings, edible films are also endowed with noteworthy properties, such as being environmentally friendly, contribution to
improvement of textural properties, and preservation of volatile components, and coherent structure (Kester and Fennema, 1986). The potential of
polysaccharides and proteins as edible films has already been recognized by
several research groups (Kester and Fennema, 1989a; Gennadios and Weller,
1990) and by the industry as well as being a very promising alternative, but
only for special applications, to the omnipresent synthetic polymer packaging (Greener Donhwe and Fennema, 1994) .

Cellulose, starch, and their constituents are the two most important raw
materials for the preparation of films. Cellulose, the principal structural
component of plants, is the most abundant source of complex carbohydrate
in the world (Engelhardt, 1995). Microcrystalline cellulose (MCC) is a natural
polymer whose chemical name is glycosidic linkage -1,4 and should not be
confused with a -1,4 linkage of starch. Although shorter cellulose fibers
attracted more interest in the past, recently there has been an upsurge in
using lower levels of longer fibers for specialized uses and, mainly, for
enhancing the functional properties of foods (Ang and Miller, 1991;
Yamanaka and Watanabe, 1994). Cellulose can be also "tailored" with appropriate chemical modification to produce cellulose ether/ester derivatives
(Psomiadou et al., 1996). These polymers give films that are flexible and
transparent, have moderate strength and resistance to oil and fat migration,
and act as moderate barriers to moisture and oxygen (Kestner and Fennema,
1986, 1989b; Hagemaier and Shaw, 1990). Methyl cellulose (MC) gives a far
more homogeneous dispersion with starch and has low hydrophilicity, thus
being a promising component for improving the poor moisture barrier performance of starch. In contrast, MCC (Psomiadou et al., 1996) is not well
mixed with starch, since it remains in the form of micro-particles.

Proteins can be good film formers and may be used in coating formulations for fruits and vegetables. Protein films are effective as gas barriers
(O_2 and CO_2), but their water vapor transmission rate is high (Baldwin et al.,
1995). Casein and casein derivatives vary in molecular weight (19000 to
23900) and have been extensively used in the food industry (dairy, meat,
and confectionery) and in medical and pharmaceutical applications as well
(Kinsella, 1984; Southward, 1989). Several publications have already reported

on the film formation from casein/sodium caseinate (Krochta, 1992; Avena-Bustillos and Krochta, 1993; Greener Donhwe and Fennema, 1994) and starches (Ollet et al., 1991; Gennadios et al., 1993c; Arvanitoyannis et al., 1995b).

Collagen and its derivative, gelatin, among other proteins, have been used for sausage casings and as gelling agents because of their abundant supply (Johnston-Banks, 1990). Collagen films extruded in the form of tubular sausage casing may be viewed as a convenient edible packaging material (Hood, 1987). Gelatin coatings, with or without polyols (i.e., glycerol), carrying antioxidants were effective in reducing rancidity when applied to cut-up turkey meat or to smoke-cured chicken by spraying or dipping (Klose et al., 1952; Moorjani et al., 1978). Edible wrappings based on blends of gelatin with farinaceous constituents have been recently marketed (Torres, 1994). Many of the major food-related or industrial uses of gelatin are based on the structure/property relationships of gelatin for which water is an excellent plasticizer of the predominant amorphous regions (Slade and Levine, 1987). Comparatively greater are the number of publications reporting on film formation from starch or its components and their possible applications in the food industry (Ollet et al., 1991, Gennadios et al., 1993d, Arvanitoyannis et al., 1995c).

3   Development of (active) edible coatings for high-moisture foods

3.1   Specific requirements for edible coatings

Coating of high-moisture foods puts specific constraints on the structural integrity and flexibility of the coating material. It should be sufficiently water-resistant to remain intact and should cover all parts of a heterogeneous product adequately when applied via dipping or spraying. With high-moisture foods like fresh and (minimal) processed fruits and vegetables, an extra complicating factor is that the food to be coated consists of living tissues. The metabolic activity of this tissue has to be sustained when coated for the produce to maintain its physiologic quality and resistance toward microbial spoilage. However, respiratory activity of the produce should not fully deplete O_2 or lead to build-up of excessive CO_2 because then physiologic deterioration is initiated, leading to anaerobic respiration and thus off-flavors, abnormal ripening, and spoilage. High storage temperatures increase produce respiration and ripening while low (above freezing) temperatures slow down respiration, delay ripening, and usually retard microbial growth. The O_2 levels that cause anaerobic reactions vary among commodities according to the permeability of the commodity peel, respiration patterns, storage temperature, as well as the type and thickness of coatings applied. As a general rule of thumb, a minimum of 1 to 3% O_2 is required around a commodity to avoid a shift from aerobic to anaerobic respiration.

Fresh and (minimal) processed fruits and vegetables are relatively high-moisture foods that most often are stored under conditions of high relative humidity (90 to 98% RH) in order to minimize water loss via transpiration, that would lead to weight loss and shriveling. Humidity has an important impact on the barrier properties of coatings for gases and water vapor (Gontard et al., 1993, 1994, 1996; Koelsch, 1994; Guilbert et al., 1996). Moisture migration can have adverse effects on the sensory quality and storage quality of many types of food products. With fruits and vegetables, moisture migration within produce is less relevant than moisture transfer between a produce and its environment and/or within different components of mixed food systems. Thus, it is often necessary to retard moisture migration to maximize the storage life and quality.

In addition to the function of coatings for fresh and (minimal) processed fruits and vegetables to retard moisture loss, other functions are to improve appearance/sheen of the surface, maintain structural integrity, improve mechanical handling properties, carry active agents (fungicides, antioxidants, etc.), retain volatile flavor compounds and establish a barrier for gas-exchange between the commodity and the environment. Due to their versatile functionality, edible coatings can be used to improve or substitute several currently employed techniques of preservation of fresh and (minimal) processed produce (Gorris, 1994).

3.2 Properties of edible coatings

Composite edible films consisting of proteins or polysaccharides and lipids, fatty emulsions, or fatty layers can act well as moisture barriers. The fatty compounds reduce water transmission, whereas the proteins or polysaccharides give strength and structural integrity. Among the composites used are methylcellulose or hydroxypropylmethylcellulose and fatty (stearic or palmitic) acids, paraffin or beeswax; carboxymethylcellulose and sucrose fatty acid esters, mono- and diglycerides; casein and acetylated monoglyceride, paraffin, carnauba wax or beeswax; zein and acetylated monoglycerides (see references in Gontard et al., 1993).

Guilbert and co-workers focused on wheat gluten-based materials and provided these with different types and levels of lipids in order to alter water transmission properties (Gontard et al., 1994, 1996; Guilbert et al. 1996). The effects of lipids on the functional properties of gluten-based composite films depended on the lipid characteristics and on the interactions between the lipid and the protein structural matrix. Beeswax, a solid and highly hydrophobic lipid, was the most effective lipid for improving moisture barrier properties, but has inferior structural properties (opaque, brittle). Combining wheat gluten proteins with a diacetyl tartaric ester of monoglycerides reduced water vapor permeability, increased strength, and maintained transparency.

The gaseous barrier properties of edible films and coatings have been studied mainly in relation to temperature (Kester and Fennema, 1989c; Pico-Pena and Torres, 1990; Gennadios et al., 1993e; Brandenburg et al., 1993).

Results from project AIR1-CT92-0125 (Gorris, 1994) have shown that RH has a strong influence on the transmission rates of O_2 and CO_2 for wheat gluten films (Guilbert et al., 1996; Gontard et al., 1996). At high RH values, edible films were found to have a higher permeability for O_2 and CO_2 than synthetic films. The addition of lipid components (beeswax, DATEM) to wheat gluten-based films caused a marked decrease in permeability. This decrease may be related to a reduction in the water content of the film due to the presence of the hydrophobic substances or a strengthening of the protein structural matrix by the lipids whereby the barrier properties are altered. It has been found that the permeability of edible films toward O_2 can be very low, even as to create anaerobic conditions on the food surface (Kester and Fennema, 1989d; Gontard et al., 1996).

The ability of films and coatings to modify gas transport is important for tailoring such materials to specific applications such as fresh fruits and vegetables. Coatings applied to such respiring products should allow for O_2 to penetrate into the package and excessive CO_2 to escape from it. Both wheat gluten and soy protein isolate films have been shown to be very effective O_2 barriers at low RH, whereas they provide poor vapor barriers (Gennadios et al., 1993e; Brandenburg et al., 1993).

3.3 Types of edible coatings

The development of so-called "wax" coatings (which may or may not actually include a wax) emphasized reduction of moisture loss due to hydrophobic components such as waxes, oils, and resins (similar to cutin) and added sheen due to wax or resin components such as shellac or carnauba wax, respectively. Current commercial use of wax preparations is most extensive on citrus, apples, mature green tomatoes, rubatagas, and cucumbers (Baldwin, 1994). There are minor applications with other fruits and some vegetables (Hardenburg, 1967). All coatings used provide for reduced water loss and add sheen, but whether they provide adequate permeability to CO_2 and O_2 is unknown.

Polysaccharide-based coatings are mainly characterized by favorable diffusion properties toward O_2 and CO_2. They may retard ripening and increase shelf life of coated produce, without creating severe anaerobic conditions. Water-soluble coatings, based on carboxymethyl-cellulose (CMC), became commercially available in the mid-1980s (Baldwin, 1994). The first of this type was TAL Pro-long (Courtaulds Group, London). Later followed Pro-long, a composite of sucrose polyesters of fatty acids and the sodium salt of CMC. Semperfresh (United Agriproducts, Greeley, Colorado), a coating of similar composition to TAL Pro-long, claims to be an improved formulation of earlier sucrose polyester products. While reduced O_2 is maintained, Semperfresh and similar coatings are not good barriers to the movement of water.

3.4   Active edible coatings

Loss of color, flavor, or texture and generation of off-flavors are among the deterioration processes that lower product quality. Edible coatings are conducive to the use of antimicrobials and antioxidants that can help to counteract these processes. Also, edible coatings enable fortification of food products with desirable quality characteristics such as color, flavor, spiciness, acidity, sweetness, and saltiness. Enriching coatings with functional additives allows improvement of nutritional and aesthetic quality aspects without destroying the integrity of the food product. In fact, inclusion of these compounds in coatings concentrates them at the produce surface, which is the place where they are needed. This means that possibly only very small amounts of additives are required.

Growth of microbes on the surface of food products is the main cause of spoilage, which may be prevented using antimicrobial agents in coatings. Fungicides are often incorporated into coatings for fruits. Although waxing alone is sometimes reported to control decay, a much better effect is obtained when fungicides are incorporated into the wax.

With regard to the inclusion of active compounds into edible films and coatings, Guilbert (1988) discussed the use of casein or carnauba wax films with sorbic acid to protect papaya and apricot cubes from spoilage by yeasts and fungi, which is a problem at water activities over 0.78. Also, retention of the antioxidant tocopherol in gelatin films was studied for application to margarine. It was found that cross-linking the gelatin was necessary to sufficiently retard migration of the highly fat-soluble tocopherol into the margarine. Other researchers found that edible films composed of pectinate, pectate, or zein that contained citric acid were very useful to prevent rancidity and maintain the desirable texture of nuts.

Considerable research effort has been devoted to the control or prediction of migration of additives from coatings into the food. The migration of antimicrobial additives is a determinative factor, both in terms of activity and impact on the desired food quality. Sorbic acid has been a kind of model additive in migration studies (Giannakopoulus and Guilbert, 1986a, 1986b). It is important to be able to predict and control sorbic acid migration between phases during; (1) food treatments (e.g., absorption of sorbic acid during the processing of dried prunes, or loss during cooking of fabricated foods), (2) storage of composite foods (e.g., dairy products or cakes containing pretreated fruits), (3) storage of foods in contact with wrapping materials or films containing sorbic acid (absorption by dairy products covered with paper saturated with sorbic acid), and (4) storage of foods coated with an external edible layer highly concentrated in sorbic acid.

Guilbert and co-workers studied the migration of potassium sorbate and sorbic acid in composite coatings on model food systems within the framework of the EU project AIR1-CT92-0125. They observed a marked dependence of the migration of the active agents in relation to initial concentration, pH, and temperature. Addition of lipid components such as DATEM and AM, or the more hydrophobic beeswax, to pectin-based coatings resulted is a 50% reduction of the sorbic acid migration, thus causing active retention of sorbic acid. These values are about equivalent to those found in literature for chitosan, cellulose derivatives, and lipids. With lipid compounds only (MA and beeswax), even lower migration may be obtained. With high-moisture foods, such lipid materials would be most advantageously used as a monolayer film or as a bilayer film composed of a hydrophilic base layer coated with a thin layer of lipid-containing sorbic acid.

4　Conclusions

The continuously increasing extent of pollution of the environment has given rise to the design of novel packaging materials of improved (bio-)degradability and of edible coatings that partly substitute for packaging functionalities.

Blending a major synthetic polymer such as LDPE with biodegradable synthetic (poly(ε-caprolactone)) polymers or natural materials (starch) has shown that it is possible to still make good use of the favorable mechanical properties of the less-biodegradable polymer while reducing the burden the packaging material may put on the environment. While LDPE is biodegradable only to some extent, other synthetic polymers such as poly(ε-caprolactone), poly(vinyl alcohol), poly(L-lactide), and poly(hydroxybutyrate) were shown to be fully biodegradable and offer a much better opportunity for the future. The fully biodegradable packaging materials that are currently available have been found inferior to non-(bio)degradable packaging materials due to their rather hydrophilic nature, inadequate mechanical strength, and relatively high costs. However, considerable research effort is now spent to alleviate these weaknesses. For example, poly(hydroxybutyrate), modified starches, chitosan-derivatives, and pullulan appear as potential polymers for single or blended use in food packaging applications. Poly(vinyl alcohol), poly (hydroxybutyrate), and several modified starches (e.g., hydroxy propylated, ethylated) are already being industrially produced by several European and Japanese companies (e.g., Nippon Starch, Japan; Boehringer, The Netherlands). (Active) Edible films and coatings may have very attractive perspectives but are still at an experimental stage. Most importantly, processing, mechanical, and water vapor transmission properties of the natural polymers need to be further improved for practical use. Edible coatings offer unique benefits over conventional packaging materials since they can be consumed along with the foods, provide additional nutrients, enhance sensory characteristics, and can integrate quality-enhancing agents. Coatings with low water barrier properties would be more suitable for foods with a high moisture content and a long shelf life. For (minimal) processed fruits and vegetables, gas barrier properties of coatings should generate proper modified atmosphere conditions and refrigeration temperature and high relative humidity. While at present a wealth of information has been gathered on the physical and chemical properties of the ingredients (building blocks) of edible films and coatings, matching the various properties of edible films and coatings to the specific characteristics of food products is one of the main challenges for the near future.

Acknowledgment

Part of the research discussed in this paper has been conducted with funds from the European Commission under contract AIR1-CT92-0125.

References

Ang, J. F. and Miller, W. B. (1991). Multiple functions of powdered cellulose as food ingredient, *Cereals Food World*, 36 (7), 558.

Arvanitoyannis, I. and Blanshard, J. M. V. (1993). Anionic copolymers of octanelactam with laurolactam (Nylon 8/12 copolymers). VII. Study of diffusion and permeation of gases in undrawn and uniaxially drawn (conditioned at different relative humidities) polyamide films, *Journal of Applied Polymeric Sciences*, 47, 1933.

Arvanitoyannis, I. and Psomiadou, E. (1994). Composition of anionic (co)polyamides (Nylon 6/Nylon 12) with short glass E-fibers — preparation and properties, *Journal of Applied Polymer Science*, 51, 1883.

Arvanitoyannis, I., Kalichevsky, M., Blanshard, J. M. V., and Psomiadou, E. (1994). Study of diffusion and permeation of gases in undrawn and uniaxially drawn films made from potato and rice starch conditioned at different relative humidities, *Carbohydrate Polymers*, 24, 1.

Arvanitoyannis, I., Psomiadou, E., Yamamoto, N., and Blanshard, J. M. V. (1995a). Composites of novel biodegradable copolyamides based on adipic acid, 1,6 hexane diamine and L-proline with short E-glass fibres, I. Preparation and properties, *Polymer*, 36, 493.

Arvanitoyannis, I., Nakayama, A., Kawasaki, N., and Yamamoto, N. (1995b). Synthesis and properties of novel biodegradable oligoesteramides based on sebacic acid, octadecanedioic acid, 1.6-hexane diamine and ε-caprolactone, Part 2, *Polymer*, 36, 857.

Arvanitoyannis, I., Psomiadou, E., Yamamoto, N., and Blanshard, J. M. V. (1995c). Composites of novel biodegradable copolyamides based on adipic acid, 1,6 hexane diamine and L-glycine with short E-glass fibres, 1. Preparation and properties, *Journal of Applied Polymer Science*, 56, 1045.

Arvanitoyannis, I., Psomiadou, E., and Nakayama, A. (1996). Edible films made from sodium caseinate, starches, sugars or glycerol, Part 1, *Carbohydrate Polymers*, 31, 179.

Avena-Bustillos, R. J., and Krochta, J. M. (1993). Water vapor permeability of caseinate-based edible films as affected by pH, calcium crosslinking ligand and lipid content, *Journal of Food Science*, 58, 904.

Baldwin, E. A. (1994). Edible coatings for fresh fruits and vegetables: past, present and future. In: *Edible Coatings and Films to Improve Food Quality*, Krochta J.M., Baldwin E.A., and Nisperos-Carriedo M.O. (Eds.), Technomic Publishing Co, Lancaster, U.K., p. 25.

Baldwin, E. A., Nisperos-Carriedo, M. O., and Baker, R. A. (1995). Use of edible coatings to preserve quality of lightly (and slightly) processed products, *CRC Critical Reviews in Food Science and Nutrition*, 35, 509.

Brandenburg, A. H., Weller, C. L., and Testin, R. F. (1993). Edible films and coatings from soy protein, *Journal of Food Science*, 58, 1086.

Cuq, B., Gontard, N., and Guilbert, S. (1994). Edible films and coatings as active layers. In: *Active Food Packaging*, Rooney, M. (Ed.), Blackie Academic and Professional, London, p. 111.

Drake, S. R., Cavallieri, R., and Kupferman, E. M. (1991). Quality attributes of D'Anjou pears after different wax drying temperatures and refrigerated storage, *Journal of Food Quality*, 14, 455.

El Ghaouth, A., Arul, J., Ponnamplalan, R., and Boulet, M. (1991a). Chitosan coating effect on storability and quality of fresh strawberries, *Journal of Food Science*, 56, 1618.

El Ghaouth, A., Arul, J., and Ponnamplalan, R. (1991b). Use of chitosan coating to reduce water loss and maintain quality of cucumber and bell pepper fruits, *Journal of Food Processing and Preservation*, 15, 359.

Engelhardt, J. (1995). Sources, industrial derivatives and commercial applications of cellulose, *Carbohydrates in Europe*, May 1995, 5.

Gennadios, A. and Weller, C. L. (1990). Edible films and coating from wheat and corn proteins, *Food Technology*, 44, 63.

Gennadios, A., Brandenburg, A. H., Weller, C. L., and Testin, R. F. (1993a). Effect of pH on properties of wheat gluten and soy protein isolate films, *Journal of Agricultural and Food Chemistry*, 41, 1835.

Gennadios, A., Park H. J., and Weller, C. L. (1993b). Relative humidity and temperature effects on tensile strength of edible protein and cellulose ether films, *Transactions ASAE*, 36, 1867.

Gennadios, A., Weller, C. L., and Testin, R. F. (1993c). Property modification of edible wheat gluten films, *Transactions ASAE*, 36, 465.

Gennadios, A., Weller, C. L., and Testin, R. F. (1993d). Modification of physical and barrier properties of edible wheat gluten-based films, *Cereal Chemistry*, 70, 426.

Gennadios, A., Weller, C. L., and Testin, R. F. (1993e). Temperature effect on oxygen permeability of highly permeable, hydrophilic edible films. *Journal of Food Science*, 58, 212.

Giannakopoulos, A. and Guilbert, S. (1986a). Determination of sorbic acid diffusivity in model food gels, *Journal of Food Technology*, 21, 339.

Giannakopoulos, A. and Guilbert, S. (1986b). Sorbic acid diffusivity in relation to the composition of high and intermediate moisture model gels and foods, *Journal of Food Technology*, 21, 477.

Gonsalves, K. E., Chen, X., and Cameron, J. A. (1993). Degradation of nonalternating poly (esteramides), *Macromolecules*, 25, 3309.

Gontard, N., Duchez, C., Cuq, J. L., and Guilbert, S. (1994). Edible composite films of wheat gluten and lipids: water vapor permeability and other functional properties, *International Journal of Food Science and Technology*, 29, 39.

Gontard, N., Guilbert, S., and Cuq, J. L. (1993). Water and glycerol as plasticizers affect mechanical and water vapor barrier properties of an edible wheat gluten film, *Journal of Food Science*, 58, 206.

Gontard, N., Thibault, R., Cuq, J. L., and Guilbert, S (1996). Influence of relative humidity and film composition on oxygen and carbon dioxide permeabilities of edible films, *Journal of Agriculture and Food Chemistry*, 44, 1064.

Goodman, I. (1984). Copolyesteramides-IV: Anionic copolymers of -caprolactam with -caprolactone, *European Polymer Journal*, 20, 548.

Gorris, L. G. M. (1994). Novel mild preservation techniques, *Minimal Processing of Foods*, Ahvenainen R., Mattila-Sandholm, T., and Ohlsson, T. (Eds.), VTT Symposium series number 142, 37.

Greener Donhwe, I. and Fennema, O. (1994). Edible films and coatings: characteristics, formation, definitions and testing methods. In: *Edible Coatings and Films to Improve Food Quality*, Krochta J.M., Baldwin E.A., and Nisperos-Carriedo, M.O. (Eds.), Technomic Publishing Co, Lancaster, U.K., p. 1.

Griffin, G. J. L. (1994). *Chemistry and Technology of Biodegradable Polymers*, Blackie Academic & Professional, London U.K, p. 135.

Guilbert, S. (1986). Technology and application of edible protective film. In: *Food Packaging and Preservation*, Matathlouthi, M. (Ed.), Elsevier Applied Science Publishers, London, UK, p. 371.

Guilbert, S. (1988). Use of superficial edible layer to protect intermediate moisture foods: application to the protection of tropical fruit dehydrated by osmosis. In: *Food Preservation by Moisture Control*, Seow, C. C. (Ed.), Elsevier Applied Science Publishers, London, UK, p. 199.

Guilbert, S. and Gontard, N. (1995). Edible and biodegradable food packaging. In: *Foods and Packaging Materials*, Royal Society of Chemistry, Oxford, U.K., p. 159.

Guilbert, S., Gontard, N., and Gorris, L. G. M. (1996). Prolongation of the shelf life of perishable food products using biodegradable films and coatings. *Lebensmittel-Wissenschaft und Technologie*, 29, 10.

Hagemaier, R. D. and Shaw, P. E. (1990). Moisture permeability of edible films made with fatty acids and (hydroxyl)methylcellulose, *Journal of Agriculture and Food Chemistry*, 38, 1799.

Hardenburg, R.E. (1967). Wax and related coatings for horticultural products — a bibliography, USDA/ARS Publication #51, p. 15.

Hood, L. L. (1987). Collagen in sausage casings, *Advances in Meat Research*, 4, 109.

Huang, S. J. (1985). Biodegradable polymers. In: *Encyclopedia of Polymer Science & Engineering*, Vol. 2, Klingsberg, A., Muldoon, J., and Salvadore, A. (Eds.), John Wiley & Sons, New York, p. 220.

Johnston-Banks, F. A. (1990). Gelatin. In: *Food Gels*, Harris, P. (Ed.), Elsevier Applied Science Publishers Ltd, London, p. 233.

Kester, J. J. and Fennema, O. R. (1986). Edible films and coatings: a review, *Food Technology*, 40 (12), 47.

Kester, J. J. and Fennema, O. (1989a). An edible film of lipids and cellulose ethers: barrier properties to moisture vapor transmission and structural evaluation, *Journal of Food Science*, 54, 1383.

Kester, J. J. and Fennema, O. (1989b). An edible film of lipids and cellulose ethers: performance in a model frozen-food system, *Journal of Food Science*, 54, 1390.

Kester, J. J. and Fennema, O. R. (1989c). Temperature influence on oxygen and water vapor transmission through stearyl alcohol film, *Journal of the American Oil Chemists Society*, 66 (8), 1154.

Kester, J. J. and Fennema, O. (1989d). The influence of polymorphic form on oxygen and water vapor transmission through lipid films, *Journal of the American Oil Chemists Society*, 66 (8), 1147.

Kinsella J. E. (1984). Milk Proteins: physicochemical and functional properties, *CRC Critical Reviews in Food Science and Nutrition*, 21, 197–262.

Klose, A. A., Macchi, E. P., and Hanson, H. L. (1952). Use of antioxidants in the frozen storage of turkeys, *Food Technology*, 6, 308.

Koelsch, C. (1994). Edible water vapor barriers: properties and promise, *Trends in Food Science & Technology*, 5 (3), 76.

Kolokuris, I., Arvanitoyannis, I., Blanchard, J. M. V., and Robinson, C. (1992a). Thermal analysis of commercial gutta-percha using differential scanning calorimeter and dynamic mechanical thermal analysis, *Journal of Endodontics*, 18, 4.

Kolokuris, I., Arvanitoyannis, I., Robinson, C., and Blanchard, J. M. V. (1992b). Effect of moisture and aging on gutta-percha, *Journal of Endodontics*, 18, 583.

Krochta, J. M. (1992). Control of mass transfer in foods with coatings and films. In: *Advances in Food Engineering*, Singh, R.P. and Wirakartakysumah, M.A. (Eds.), CRC Press Inc., Boca Raton, FL, p. 517.

Krochta, J. M. and De Mulder-Johnston, C. (1997). Edible and biodegradable polymer films: Challenges and opportunities, *Food Technology*, 51, 61.

Krochta, J. M., Pavlath, A.E., and Goodman N. (1990). Edible films from casein-lipid emulsions for lightly-processed fruits and vegetables. In: *Engineering and Food*, Vol. 2 — Preservation Process and Related Techniques, Spiess, W.E. and Schubert, H. (Eds.), Elsevier Science Publishers, New York, p. 329.

Kumar, G. S. (1987). *Biodegradable Polymers, Prospects & Progress*, Marcel Dekker, New York.

Leistner, L. and Gorris, L. G. M. (1995). Food preservation by hurdle technology, *Trends in Food Science & Technology*, 6, 41.

Lourdin, D., Della Valle, G., and Colonna, P. (1995). Influence of amylose content on starch films and foams, *Carbohydrate Polymers*, 27, 261.

Moorjani, M. N., Raja, K. C. M., Puttarajapa, P., Khabade, N. S., Mahendrakar, V. S., and Mahadevaswamy, M. (1978). Studies on curing and smoking poultry meat, *Indian Journal of Poultry Science*, 13 (1), 52.

Ollet, A. L., Parker, R., and Smith, A. C. (1991). Deformation and fracture behaviour of wheat starch plasticized with glucose and water, *Journal of Material Science*, 26, 1351.

Park, H. J., Weller, C. L., Vergano, P. J., and Testin, R. F. (1993). Permeability and mechanical properties of cellulose-based edible films, *Journal of Food Science*, 58, 1361.

Park, H. J., Chiman, M. S., and Shewfelt, R. L. (1994). Edible coating effects on storage life and quality of tomatoes, *Journal of Food Science*, 59, 568.

Pico-Pena, D. C. and Torres J. A. (1990). Oxygen transmission of an edible methylcellulose-palmitic acid film, *Journal of Food Process Engineering*, 13, 125.

Psomiadou, E., Arvanitoyannis, I., and Yamamoto, N. (1996). Edible films made from natural resources; Microcrystalline Cellulose (MCC), Methylcellulose (MC) and corn starch and polyols; Part 2, *Carbohydrate Polymers*, 31, 193.

Satyanarayana, D. and Chaterji, P. R. (1993). Biodegradable Polymers: Challenges and strategies, *Journal of Macromolecular Sciences — Reviews in Macromolecular Chemistry and Physics*, 33, 349.

Slade, L. and Levine, H. (1987). Polymer-chemical properties of gelatin in foods. In: *Advances in Meat Research*, Vol. 4 — Collagen as Food, Pearson, A.M., Dutson, T.R., and Bailey, A.J. (Eds.), Van Nostrand Reinhold, New York, p. 251.

Southward, C. R. (1989). Uses of casein and caseinates. In: *Developments in Dairy Chemistry* — Vol. 4, Fox, P.F. (Ed.), Elsevier Applied Science Publishers Ltd, London, p. 173.

Tadmor, Z. and Gogos, G. G. (1979). *Principles of Polymer Processing*, Wiley-Interscience, New York.

Torres, J. A. (1994). Edible films and coatings from proteins. In: *Protein Functionality in Food Systems*, Hettiarachchy, N.S. and Ziegler, G. R. (Eds.), IFT Basic Symposium Series, Marcel Dekker, New York, p. 467.

Van Krevelen, D.W. (1990). *Properties of Polymers*, 3rd Edition, Elsevier, Amsterdam, Netherlands, p. 189.

Vert, M., Li, S. M., Spenlehauer, G., and Guerin, P. (1992). Bioresorbability and biocompatibility of aliphatic polyesters, *Journal of Material Sciences: Materials for Medicine*, 3, 432.

Wong, D. W. S., Gastineau, F. A., Gregorski, K. S., Tillin, S. T., and Pavlath, A. E. (1992). Chitosan-lipid films: microstructure and surface energy. *Journal of Agriculture and Food Chemistry*, 40, 540.

Wool, R. P. (1995). The science and engineering of polymer composite degradation. In: *Degradable Polymers*, Scott, G. and Gilead, D. (Eds.), Chapman & Hall, London, p. 207.

Yang, J. Z., Wang, M., and Otterbrite, R. L. (1993). Synthesis of novel copolyamides based on -aminoacids, *Journal of Material Sciences — Pure Applied Chemistry (A)*, 30, 503.

Yamanaka, S. and Watanabe, K. (1994). Applications of bacterial cellulose. In:*Cellulosic Polymers, Blends and Composites*, Guilbert, R.D. (Ed.), Hanser Publishers, Münich, Germany, p. 207.

chapter twenty-two

The role of ionizing radiation in minimal processing of precut vegetables with particular reference to the control of Listeria monocytogenes

József Farkas, László Mészáros, Csilla Mohácsi-Farkas, Tamás Sáray, and Éva Andrássy

Contents

Summary 374
1 Introduction 374
    1.1 Occurrence and behavior of *Listeria monocytogenes* on
          prepacked vegetables 375
    1.2 Low-dose irradiation as potential intervention to control
          non-spore-forming pathogenic bacteria on packaged
          vegetables 376
    1.3 Objectives 377
2 Materials and methods 378
    2.1 Source and preparation of vegetables 378
    2.2 *Listeria monocytogenes* inoculum: inoculation and
          packaging procedures 378
    2.3 Irradiation treatment and postirradiation storage 378
    2.4 Microbiologic and vitamin C analyses..................................378
    2.5 pH measurements...379
    2.6 Estimation of the sensorial shelf life.....................................379

3   Results and discussion 379
   3.1   Microbiologic changes and sensorial shelf life 379
      3.1.1   Shredded cabbage 379
      3.1.2   Sliced radish 381
   3.2   pH of vegetable tissues 382
   3.3   Vitamin C content 382
4   Conclusions 382
Acknowledgments 385
References 385

Summary

Minimally processed fresh-cut vegetables and salads are prone to pathogen contamination. The incidence of *Lysteria monocytogenes* is reviewed in this chapter. Low-dose irradiation may be used as part of a policy to prevent foodborne hazards. This chapter describes the results of a study concerning cabbages and radishes inoculated with *L. monocytogenes* using gamma irradation of 1 kGy dose at a 0.5 kGy/h rate. Samples were then stored at 5, 10, and 15°C for 10 days. The treatment proved effective in drastically reducing the number of pathogen cells, but a good control of the temperature during storage is still required.

1   Introduction

There is an increasing consumer trend for fresh-like, or at least less severely processed, still convenient, or ready-to-eat foods with reduced levels of chemical preservatives. The catering industry is also strongly interested in such minimally processed nonfrozen meals and meal components. Among these foods are the minimally processed and chilled vegetables usually carrying Pseudomonads, Enterobacteria, and lactic acid bacteria as natural microflora. The high moisture and numerous cut surfaces of minimally processed vegetables provide excellent conditions for the growth of microorganisms.

Modified atmosphere packaging (MAP) of vegetables by producing elevated concentrations of CO_2 or N_2 can suppress the growth of the spoilage bacteria, especially *Pseudomonas* species. However, these extended-shelf-life chilled produce potentially involve a microbiologic hazard due to the growth of psychrotrophic pathogenic bacteria (such as *Listeria monocytogenes, Aeromonas hydrophila,* and *Yersinia enterocolitica*), particularly, because there is always a risk of temperature abuse (Beuchat, 1995). At abuse temperatures, even some mesophilic pathogens can proliferate (Nguyen-the and Carlin, 1994). Food poisoning outbreaks related to fresh produce have been increasingly reported in the last two decades (Schlech et al., 1983; Archer, 1996).

1.1   Occurrence and behavior of Listeria monocytogenes on prepacked vegetables

Considering that *L. monocytogenes* is the most tolerant among psychrotrophic non-spore-forming pathogens to environmental stress factors, a brief survey on its occurrence and behavior on prepackaged vegetables seems to be worthwhile.

L. monocytogenes is an ubiquitous environmental contaminant. It is widely distributed on plant vegetation, including raw vegetables. Its presence on plant materials is likely due to contamination from decaying vegetation, animal feces, soil, river and canal waters, or effluents from sewage treatment operations. The organism is known to survive on plant materials for 10 to 12 years (Beuchat, 1996). According to Beuchat (1996), plants and plant parts used as salad vegetables play — directly or indirectly — a key role in the spread of the pathogen *L. monocytogenes* from natural habitats to the human food supply. Some observations suggest, however, that the majority of strains indigeneous to decaying plant vegetation are incapable of causing illness in animals (Welsheimer, 1968).

Sizmur and Walker (1988) detected *L. monocytogenes* in 4 of 60 prepackaged ready-to-eat salads in the U.K. Vegetables that are particularly susceptible to *Listeria* contamination are endive (*Cichorium endivia*), radish, cucumber, lettuce, cabbage, and sprouts (Heisick et al., 1989; Zollinger, 1990; Nguyen-the and Carlin, 1994). A survey of 1000 samples of 10 types of fresh produce at the retail level in the U.S. has revealed the presence of *L. monocytogenes* on cabbage, cucumbers, potatoes, and radishes (Heisick et al., 1989). Beckers et al. (1989) detected *L. monocytogenes* in 11 of 25 samples of fresh cut vegetable in the Netherlands and Harvey and Gilmour (1993) reported that 7 of 66 samples of salad vegetables and prepared salads produced in Northern Ireland contained the pathogen. Lack et al. (1996) in Germany detected *L. monocytogenes* in 3 of 10 different commercially prepackaged salad mixtures studied. Farber et al. (1989), on the other hand, did not detect *Listeria* species in 110 samples of vegetables, including lettuce, celery, tomatoes, and radishes analyzed in Canada. Likewise, Petran et al. (1988) failed to detect *L. monocytogens* in market samples of fresh vegetables in the United States. A higher rate of contamination of prepared salads can be attributed to cross contamination of the pathogen during chopping, mixing, and packaging (Velani and Roberts, 1991). In an outbreak of listeriosis, the source of the pathogen was coleslaw (Schlech et al., 1983; Heisick et al., 1989) made from contaminated cabbage.

The ability of *Listeria* to grow on raw vegetables and prepackaged mixed salads has been proven by several authors (Heisick et al., 1989; Zollinger, 1990; Carlin and Nguyen-the, 1994). The growth and survival of *L. monocytogenes* in raw shredded cabbage was studied by Beuchat et al. (1986). Survival and growth characteristics of *L. monocytogenes* on lettuce stored at 5 and 25°C increased by several logs during 14 days of storage, while in some trials the organism was not detected at the end of the storage period. Carlin and Nguen-the (1994) reported also that *L. monocytogenes* grows on lettuce and endive held at 10°C. Growth of *L. monocytogenes* on chicory endive at 6.5°C has been also reported (Aytac and Gorris, 1994). The behavior of *L. monocytogenes* on raw broccoli, cauliflower, and asparagus, stored under air and modified atmoshperic gas packaging conditions at 4 to 15°C has been studied (Berrang et al., 1989). Growth at 15°C resulted in populations as high as 6.0×10^7 CFU/g of broccoli florets, 3.7×10^7 CFU/g of cauliflower florets, and 1.1×10^7 CFU/g of asparagus, before these vegetables were considered inedible by subjective evaluation.

The effects of shredding, chlorine treatment, and modified atmosphere packaging on survival and growth of *L. monocytogenes* on lettuce stored at 5 and 10°C were studied by Beuchat and Brackett (1990a). With the exception of shredded lettuce that had not been chlorine treated, no significant changes in populations of *L. monocytogenes* were detected during the first eight days of incubation at 5°C. However, significant increases occurred within 3 days when lettuce was stored at 10°C; after 10 days, populations reached 10^8 to 10^9 CFU/g of product. Chlorine treatment, modified atmosphere (3% O_2, 97% N_2) and shredding did not influence growth of *L. monocytogenes*.

The efficacy of chlorine for killing *L. monocytogenes* on inoculated Brussels sprouts has been evaluated by Brackett (1987). The viable population was reduced by about 2 $\log_{10}$ CFU/g from an initial population of about 6 $\log_{10}$ CFU/g when the Brussels sprouts were dipped for 10 s in water containing 200 µg/ml chlorine. Brackett (1987) concluded that hypochlorite was ineffective in removing *L. monocytogenes* from contaminated vegetables.

In spite of some conflicting reports, the above survey shows that in order to avoid potential hazards from pathogens in practice, adequate measures should be applied to prevent contamination during cultivation and processing of salad vegetables. In addition, extra safety factors may be required for the safe use of minimally processed vegetables.

1.2   Low-dose irradiation as potential intervention to control non-spore-forming pathogenic bacteria on packaged vegetables

From the survey it is evident that sanitizing with chemical means, such as chlorine treatment referred to above, is not very reliable. Modified atmosphere packaging, although suppressing the aerobic spoilage flora, may even favor the growth of microaerophilic or anaerobic pathogens. Obviously, there is a need for a procedure that can reduce the incidence of pathogenic microorganisms associated with fresh produce, without changing their minimally processed character. Ionizing radiation offers a physical means for pasteurization without changing the fresh state of these commodities.

In his pioneering studies in the Netherlands, Langerak (1978) demonstrated that gamma radiation treatment with a dose of 1 kGy extended the shelf life of prepackaged leafy vegetables stored at 10°C by reducing microbial populations. Irradiation of endive in unperforated polyethylene bags assured good quality (8 on a 10-point scale) even after 5 days, while the quality of the unirradiated control fell below that level after 2 days. The decreased oxygen and elevated carbon dioxide atmosphere, caused by the respiration activity of the vegetable, slowed down discoloration, diminished loss of vitamin C, and helped to retard senescence. More recently, Howard et al. (1995) reported promising results on the effects of gamma processing (1 kGy) for extending the shelf life of refrigerated "Pico De Gallo," a Mexican style cold salad prepared by chopping and mixing fresh tomatoes, onions, and jalapeno peppers.

1.3   Objectives

Because non-spore-forming pathogenic bacteria (e.g., enterotoxigenic *E. coli*, *Salmonella*, *Shigella*, *L. monocytogenes*, *Aeromonas*, etc.) are relatively radiation sensitive, one can assume that ionizing radiation at relatively low doses may be able to help to minimize the above hazards from prepackaged vegetables, without changing their sensorial or freshness properties. Thus, the aim of our studies was to investigate the effect of low-dose gamma radiation (1 kGy) of selected vegetables inoculated with *L. monocytogenes*, which has the highest radiation resistance and environmental stress-tolerance among the non-spore-forming pathogens mentioned above (Monk et al., 1995). Our results with gamma irradiated precut bell pepper and carrot cubes stored in polyethylene pouches at various temperatures within the range of 1 to 16°C have been reported in detail elsewhere (Farkas et al., 1997). In the case of yellow bell pepper, the microbiologic shelf life defined by storage times elapsed until the aerobic plate count reached the 10^8 CFU/g level was approximately doubled by the radiation treatment: however, in some cases the time period of marketable quality was extended less due to physiologic deterioration of the produce. *L. monocytogenes* grew readily on this chilled product. Low-dose irradiation reduced the viable cell counts of the pathogen by at least four log cycles. The microbiologic shelf life of unirradiated carrot cubes was shorter at each storage temperature tested than the shelf life judged by appearance. The low-dose irradiation drastically reduced the load of viable spoilage bacteria, thereby improving considerably the microbiologic shelf life and extending the sensorial keeping quality. Literature information on anti-listerial effect of fresh carrot tissue (Beuchat and Bracket, 1990b; Nguyen-the and Lund, 1991) was confirmed in our studies, and the low-dose irradiation complemented efficiently this specific anti-listerial effect. In the present chapter our further studies on the effect of gamma irradiation on the fate of the native bacterial flora and a high inoculum level of *L. monocytogenes* in shredded white cabbage and sliced radish are reported. On the basis of preliminary sensory panels, a dose of 1 kGy has been selected again for radiation treatment, which did not change the sensory properties of the test materials.

2 Materials and methods

2.1 Source and preparation of vegetables

White summer-cabbage and white autumn-radish were purchased from the local market. Outer leaves of cabbages were removed and discarded and inner leaves were mechanically shredded into narrow strips. The radishes were diced into approx. 5-mm thick slices.

2.2 Listeria monocytogenes inoculum: inoculation and packaging procedures

The *Listeria monocytogenes* strain 4ab No.10 was the same test organism used in our former studies with bell pepper and carrots (Farkas et al., 1997). The stock culture of this strain was activated by culturing in brain heart infusion broth (BHI, OXOID CM 225) at 30°C for 18 h and this culture was diluted with tap water to yield a suspension of approx. 10^5 CFU/ml. Two batches of six kg each of precut cabbage or radish were mixed in 20-liter volumes of suspensions of *L. monocytogenes* for 1 min, allowing an initial contamination level of 10^4 to 10^5 CFU/g, then thoroughly drained. Approx. 100-g portions of precut inoculated vegetables were placed in low-density polyethylene pouches of approx. 15×10 cm each and sealed under air. The oxygen transmission rate at 25°C of this type of 25 μm thick foil is approx. 7800 cm³m⁻² per 24 h, the nitrogen transmission rate at 25°C is approx. 2800 cm³m⁻² per 24 h, the CO_2 transmisson rate at 25°C is approx. 42,000 cm³m⁻² per 24 h, while the water vapor transmission rate at 38°C is 18 g.m⁻² per 24 h at 90% relative humidity.

2.3 Irradiaton treatment and postirradiation storage

Half of the inoculated samples were irradiated with a dose of 1 kGy at a dose rate of 0.5 kGy/h at room temperature by a panoramic ⁶⁰Co irradiator. Equal amounts of unirradiated control and irradiated batches were stored at 5, 10, and 15°C for 10 days. Directly after irradiation and periodically during the storage, duplicate samples from both untreated and irradiated batches were analyzed.

2.4 Microbiological and vitamin C analyses

Besides total aerobic plate counts (APC), selective estimations of *L. monocytogenes*, counts of presumptive lactic acid bacteria, coliform counts, and HPLC-analyisis of the ascorbic acid and dehydro-ascorbic acid content of the vegetable samples were performed as described previously (Farkas et al., 1997).

2.5   pH measurements

pH values were determined by mixing a portion of the vegetables with approximately equal portions of distilled water and by measuring with an electrical pH-meter.

2.6   Estimation of the sensorial shelf life

Changes in the purchase acceptability of the prepackaged vegetable samples were followed subjectively as a function of radiation treatment and storage on the basis of scores on appearance, discoloration, and softness to touch, ranging from score 5 as excellent to score 1 as nonmarketable. The time period elapsed until the scores declined to the level of 2 ("still acceptable") was considered as a limit of the sensorial keeping quality.

3   Results and discussion

3.1   Microbiological changes and sensorial shelf life

3.1.1   Shredded cabbage

The aerobic plate count and the viable cell count of *L. monocytogenes* of samples of shredded cabbage are shown in Figure 1 as a function of the radiation treatment, storage temperature, and time. The sensorial shelf life values are also indicated on the graphs. It can be seen that the radiation treatment reduced the total aerobic plate count by almost three log-cycles whereas the viable cell count of *L. monocytogenes* was diminished by the radiation treatment by more than 4 log-cycles. The *Listeria* grew readily on the heavily inoculated unirradiated shredded cabbage even at 5°C. At the end of the respective sensorial shelf life periods, the total aerobic plate counts of the untreated samples were around 10^8 CFU/g whereas the total APC of irradiated samples were only around 10^5 CFU/g. In the irradiated samples, no recovery of *Listeria* above the detection level was observed in samples stored at 5°C. Growth of *Listeria* began at 10°C only when the samples were already of unacceptable sensorial quality. However, at 15°C recovery of the surviving *Listeria* was noted within two days in the irradiated samples and a four log-cycle growth occurred until the end of the shelf life.

The fate of coliforms and presumptive lactic acid bacteria is shown in Figures 2 and 3. Compared to the total aerobic plate counts, coliforms were not the dominant spoilage bacteria on the aerobically packaged shredded cabbage. The limiting factor of the sensorial shelf life was mainly a darkening discoloration. Its intensity was strongly enhanced by the increased storage temperature whereas discoloration was remarkably retarded in the irradiated samples. The low-dose irradiation extended the sensorial shelf life by 1 to 3 days, depending on the storage temperature (Figure 2). The coliforms were radiation sensitive. While the coliforms grew quite readily even at the

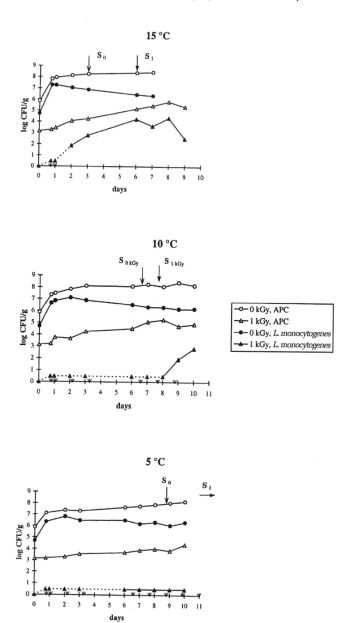

Figure 1 Comparison of the total aerobic plate counts (APC) and the viable counts of *Listeria monocytogenes* on untreated and 1 kGy-irradiated shredded packaged cabbage at various storage temperatures. Sensorial shelf life periods (S) are indicated with arrows over the growth curves.

lowest storage temperature in the untreated samples, no recovery of coliforms were noted in the irradiated batch during the entire storage period at 5°C (Figure 2).

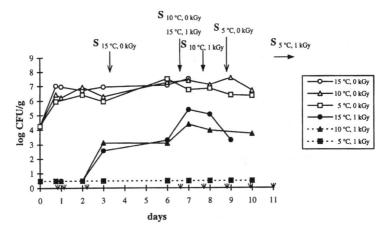

Figure 2   Effects of the storage temperature and time on the growth of coliform counts of unirradiated and 1 kGy-irradiated shredded cabbage. Sensorial shelf life periods (S) are indicated with arrows over the growth curves.

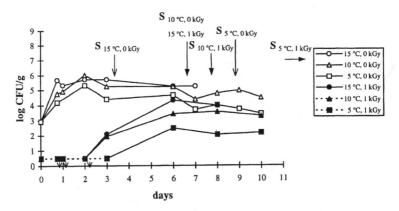

Figure 3   Effects of the storage temperature and time on the growth of presumptive lactic acid bacteria on unirradiated and 1 kGy-irradiated shredded cabbage. Sensorial shelf life periods (S) are indicated with arrows over the growth curves.

   Lactic acid bacteria were only a small fraction of the native microflora of shredded cabbage. They were a little bit less sensitive to radiation treatment than the coliforms because they recovered at all storage temperatures.

3.1.2   Sliced radish

The fate of microflora on untreated and irradiated radish samples is illustrated by Figures 4 to 6 in the same way as above. Reduction of viable cell counts as an effect of irradiation was similar to what has been observed with shredded cabbage. The results of the microbiologic analyses during storage also showed similar patterns to those described previously. However, recovery of *Listeria* was slightly better in the irradiated radish samples than in the cabbage samples and the *Listeria* counts fluctuated more among the irradiated samples than on the untreated ones. In general, *Listeria monocytogenes* growth was less pronounced in the unirradiated sliced radish than in the shredded cabbage. At 5°C, the initial *Listeria* population remained practically unchanged during storage in the untreated radish samples. Coliforms and lactic acid bacteria played even less of a role in the spoilage of precut radish than in the case of shredded cabbage.

3.2   pH of vegetable tissues

The initial pH values were 5.54 and 5.81 for untreated and irradiated shredded cabbage, respectively. pH increases of 0.4 to 0.7 were noted at the end of the respective shelf life intervals. Only slight decreases of 0.04 to 0.3 pH values were observed in the irradiated samples at various storage temperatures. In sliced white radish, the initial pH of untreated and irradiated samples were 6.29 and 6.14, respectively. pH decreases of 0.06 at 5°C, 0.2 at 10°C, and 0.4 at 15°C were measured in the untreated samples during the course of shelf life. Insignificant pH decreases (0.01 to 0.16) were found in the irradiated samples.

3.3   Vitamin C content

The ascorbic and dehydroascorbic acid contents of both vegetables were determined at the beginning and at the end of their shelf life. The sum of both is shown in Table 1 as vitamin C content. It can be seen that the vitamin C content of cabbage immediately after treatment decreased as an effect of the 1 kGy radiation dose by a maximum of 13% in shredded cabbage, and by about 30% in the sliced radish. Until the end of the sensorial shelf life, no significant further decrease of the vitamin C content was observed neither in the unirradiated or irradiated cabbage samples, nor in the irradiated radish samples. Vitamin C content of the unirradiated radish samples diminished by 13 to 30% during their respective shelf life at various storage temperatures.

4   Conclusions

These results together with those described in a previous report communication (Farkas et al., 1997) show that low-dose irradiation (1 kGy of gamma rays) is able to improve the microbiologic safety and stability of specific precut, prepackaged vegetables. The studies also underline the importance of good temperature control in the chill chain. The radiation treatment diminished remarkably the population of *L. monocytogenes* inoculated onto the prepared vegetables. The high inoculum level used should be viewed as a "worst case" scenario (Kallender et al., 1991). The same test strain applied in these experiments showed a D_{10} value of 0.4 kGy in neutral pH phosphate

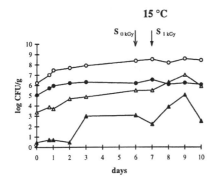

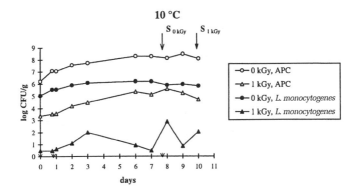

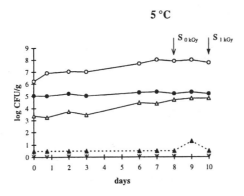

Figure 4 Comparison of the total aerobic plate counts (APC) and the viable cell counts of *Listeria monocytogenes* on packaged sliced radish. Shelf life periods (S) are indicated with arrows over the growth curves.

buffer in our previous studies (Farkas et al., 1995) and proved to be more radiation sensitive in contact with the vegetable tissue. In agreement with our previous model studies (Farkas et al., 1995), and due to its radiation

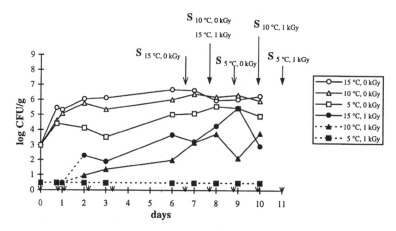

Figure 5 Effects of storage temperature and time on the coliform counts of unirradiated and 1 kGy-irradiated packaged sliced radish. Shelf life periods (S) are indicated with arrows over the growth curves.

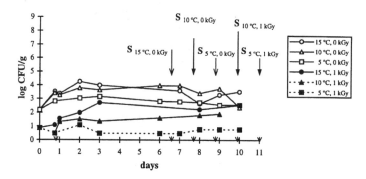

Figure 6 Effects of storage temperature and time on the growth of presumptive lactic acid bacteria on unirradiated and 1 kGy-irradiated packaged sliced radish. Shelf life periods (S) are indicated with arrows over the growth curves.

damage, the surviving population of the *Listeria* inoculum in the irradiated samples seems to be also more susceptible to low temperature than its unirradiated population. Therefore, considering that low contamination levels of *L. monocytogenes* are frequently found in fresh vegetable produce (Nguyen-the and Lund, 1991) and the extent of decrease of its viable cell count by 1 kGy irradiation, this low-dose treatment may practically eliminate the notorious pathogenic bacteria from the noninoculated prepared vegetables under conditions of good manufacturing practices.

Loss of the vitamin C content as a direct effect of low-dose irradiation does not seem to be higher than the loss that may occur in unirradiated samples over shelf life. Considering that the gram-negative, common spoilage bacteria of fresh produce are among the most radiation sensitive organisms

Table 1   Effect of Irradiation and Storage on the Vitamin C Content*
of Prepared Packaged Vegetables

Vegetable	Storage temp. (°C)	Storage time (days)	Ascorbic acid + dehydroascorbic acid (µg/g)	
			Unirradiated	1 kGy — irradiated
Shredded		0	298.4 (26.1)	260.5 (8.6)
cabbage	15	3	279.4 (25.9)	
		7		281.4 (9.6)
	10	6	290.3 (21.0)	
		8		290.5 (5.7)
	5	9	303.2 (10.4)	
		10		274.0 (8.0)
Sliced		0	255.6 (48.4)	175.9 (15.5)
radish	15	3	220.6 (15.5)	
		7		184.9 (18.8)
	10	6	176.1 (20.7)	
		8		176.6 (4.3)
	5	9	178.3 (11.0)	
		10		196.4 (27.5)

* Averages and standard deviations (in parentheses) of two samples analyzed in two replicates
each.

and that the *Listeria monocytogenes* test strain used here proved to be remarkably sensitive to the low-dose treatment under the prevailing experimental conditions, this physical technology shows promise for use in minimally processed vegetable products. Its use is not only restricted to extending produce shelf life but may also minimize the microbiologic safety hazard regarding certain psychrotrophic non-spore-forming pathogens.

Acknowledgments

The authors thank Ms. Gabriella Kiskó and Ms. Mercedes Pálmai for their skilful technical assistance and Dr. Daood Hussein for HPLC-analyses. Partial support of this work by the International Atomic Energy Agency under Research Contract No. 9603/R0 is gratefully acknowledged.

References

Archer, D.L. (1996). *Listeria monocytogenes.*The science and policy, *Food Control*, 7, 181.
Aytac, S.A. and Gorris, L.G.M. (1994). Survival of *Aeromonas hydrophila* and *Listeria monocytogenes* on fresh vegetables stored under moderate vacuum, *World Journal of Microbiology and Biotechnology*, 10, 670.
Beckers, H. J., Huis in't Veld, P., Sventoro, P.S.S., and Delfgou-van-Asch, E.H.M. (1989). The occurrence of *Listeria* in food. In: *Proceedings of the Symposium on Foodborne Listeriosis*, Sept. 7, Wiesbaden, Germany, 85.
Berrang, M.E., Brackett, R.E., and Beuchat, L.R. (1989). Growth of *Listeria monocytogenes* on fresh vegetables stored under a controlled atmosphere, *Journal of Food Protection*, 52, 702.

Brackett, R.E. (1987). Antimicrobial effect of chlorine on *Listeria monocytogenes, Journal of Food Protection*, 50, 999.

Beuchat, L.R. (1995). Pathogenic microorganisms associated with fresh produce, *Journal of Food Science*, 59, 204.

Beuchat, L.R. (1996). *Listeria monocytogenes*: incidence on vegetables, *Food Control*, 7, 223.

Beuchat, L.R. and Brackett, R.E. (1990a). Survival and growth of *Listeria monocytogenes* on lettuce as influenced by shredding, chlorine treatment, modified atmosphere packaging and temperature, *Journal of Food Science*, 55, 755.

Beuchat, L.R. and Brackett, R.E. (1990b). Inhibitory effects of raw carrots on *Listeria monocytogenes, Applied. Environmental Microbiology*, 56, 1734.

Beuchat, L.R., Brackett, R.E., Hao, D.Y.-Y., and Conner, D.E. (1986). Growth and thermal inactivation of *Listeria monocytogenes* in cabbage and cabbage juice, *Canadian Journal of Microbiology*, 32, 791.

Carlin, F. and Nguyen-the, C. (1994). Fate of *Listeria monocytogenes* on four types of minimally processed green salad, *Letters of Applied Microbiology*, 18, 222.

Farber, J.M., Sanders, G.W., and Johnston, M.A. (1989). A survey of various foods for the presence of *Listeria* species, *Journal of Food Protection,*52, 456.

Farkas, J., Andrássy, É., Mészáros, L., and Bánáti, D. (1995). Growth of untreated and radiation-damaged *Listeria* as affected by environmental factors, *Acta Microbiology and Immunology of Hungary*, 42, 19.

Farkas, J., Sáray, T., Mohácsi-Farkas, Cs., Horti, K., and Andrássy, É. (1997). Effect of low dose gamma radiation on shelf life and microbiologic safety of precut/prepared vegetables, *Advances in Food Science (CMTL)*, 19, (3/4), 111.

Harvey, J. and Gilmour, A. (1993). Occurrence and characteristics of *Listeria* in foods produced in Northern Ireland, *International Journal of Food Microbiology*, 19, 193.

Heisick, J.E., Wagner, D.E., Niermand, M.L., and Peeler, J.T. (1989). *Listeria* spp. found on fresh market produce, *Applied Environmental Microbiology*, 55, 1925.

Howard, L.R., Miller, G.H., Jr., and Wagner, A.B. (1995). Microbiological, chemical and sensory changes in irradiated Pico De Gallo, *Journal of Food Science*, 60, 461.

Kallender, K.D., Hitchins, A.D., Lancette, G.A., Schmieg, J.A., Garcia, G.R., Solomon, H.M., and Sofos, J.N. (1991). Fate of *Listeria monocytogenes* in shredded cabbage stored at 5 and 25°C under a modified atmosphere, *Journal of Food Protection*, 54, 302.

Lack, W.K., Becker, B., and Holzapfel, W.H. (1996). Hygienischer Status frischer vorverpackter Mischsalate im Jahr 1995, *Arch. f. Lebensmittelhyg.*, 47, 129. In German.

Langerak, D.I. (1978). The influence of irradiation and packaging on the quality of prepackaged vegetables, *Ann. Nutr. Alim.*, 32, 569.

Monk, J.D., Beuchat, L.R., and Doyle, M.P. (1995). Irradiation inactivation of foodborne microorganisms, *Journal of Food Protection*, 58, 197.

Nguyen-the, C. and Carlin, F. (1994). The microbiology of minimally processed fresh fruits and vegetables, *Critical Reviews in Food Science and Nutrition*, 34, 371.

Nguyen-the, C. and Lund, B.M. (1991). The lethal effect of carrot on *Listeria* species, *Journal of Applied Bacteriology*, 70, 479.

Petran, R.L., Zottola, E.A., and Gravini, R.B. (1988). Incidence of *Listeria monocytogenes* in market samples of fresh and frozen vegetable, *Journal of Food Science*, 53, 1238.

Schlech, W.F., Lavigne, P.M., Bortolussi, R.A., Allen, A.C., Haldane, E.V., Wort, A.J., Hightower, A.W., Johnson, S.E., King, S.H., Nicholls, E.S., and Broome, C.V. (1983). Epidemic listeriosis — Evidence for transmission by food, *New England Journal of Medicine*, 308, 203.

Sizmur, K. and Walker, C. W. (1988). *Listeria* in prepacked salads, *Lancet*, i, 1167.

Velani, S. and Roberts, D. (1991). *Listeria monocytogenes* and other *Listeria* spp. in prepacked salad mixes and individual salad ingredients, *PHLS Microbiology Digest (U.K.)*,8, 21.

Welsheimer, H. J. (1968). Isolation of *Listeria monocytogenes* from vegetation, *Journal of Bacteriology*, 80, 316.

Zollinger, W. (1990). Mikrobiologie — Spezielle Aspekte bei Früchten und Gemüsen, *Lebensmittel Technologie*, 23, 262. In German.

chapter twenty-three

Development of perforation-mediated modified atmosphere packaging for fresh-cut vegetables

Susanna C. Fonseca, Fernanda A. R. Oliveira, Jeffrey K. Brecht, and Khe V. Chau

Contents

Summary 389
1 Fresh-cut vegetables 390
2 Issues in the postharvest chain for extending shelf life.........................391
3 Modified atmosphere packaging...392
4 A case study...396
5 Conclusions...402
Nomenclature 402
Acknowledgment...402
References..403

Summary

Fresh-cut vegetables are a recent and increasing market niche. This chapter presents an overview of key issues in the manufacture and handling of these products: sanitation, temperature management, and packaging. Modified atmosphere packaging (MAP) is an important technique for produce preservation, but often fresh-cut products require CO_2 levels higher than those that can be reached in conventional MAP. Perforation-mediated MAP may be a suitable alternative for these situations. The application of these systems to shredded Galega kale is presented and discussed.

0-8493-7905-9/99/$0.00+$.50
© 1999 by CRC Press LLC

1 Fresh-cut vegetables

Consumer demand for freshness and convenience has led to the evolution and increased production of fresh-cut fruits and vegetables. In Europe this market grew explosively in the early 1990s (Ahvenainen, 1996). In the United States the sales of fresh-cut products are expected to increase more than three times from 1994 to 1999 (Hodge, 1995).

These products are submitted to a minimal process or combination of minimal processes, taking advantage of the hurdle concept. These processes include operations of handling or preparation and preservation. Preparation includes: operations of separation, such as peeling, coring, trimming, selecting, sorting, and grading, as well as operations to reduce the size, such as chopping, slicing, dicing, granulating, and shredding. This results in convenient, fresh-like products that can be prepared and consumed in less time. Other advantages of fresh-cut products are: (1) prepackaging allows for more efficient control, (2) labor costs are reduced, (3) solid waste disposal problems for retail consumers are reduced, (4) demands on limited refrigerated space decrease, (5) inventory of raw products is minimized and (6) a supply of excellent uniformity and quality is available (Schlimme, 1995).

Consumption of fresh-cut vegetables had a higher increase in the market than fresh-cut fruits, because the preparation of vegetables is more time-consuming. Some examples of commonly marketed fresh-cut vegetables are: carrots in slices, in sticks and in shreds; peeled onions; tomatoes in slices, in dices, and in wedges; cleaned and trimmed spinach; chopped lettuce; peeled beets; broccoli florets; sliced cucumber; and cut vegetables mixes to prepare soups (Schlimme, 1995). Other common denominations for this type of products are: minimally, lightly, fresh or partially processed, fresh-prepared, pre-cut, preprepared, cut-prepared or "ready-to-eat" vegetables. These products are characterized by a shelf life shorter than the unprocessed raw materials (Bolin and Huxsoll, 1991), in contrast to most processed foods, and refrigeration (Bolin et al., 1977; McDonald et al., 1990) and packaging are essential to their preservation.

The increase in convenience for the consumer has however a detrimental effect on the product quality. Physiologically, the operations of preparation will damage the integrity of the cells, promoting contact between enzymes and substrates, the entry of microorganisms, and creating stress conditions. The consequences of wounding are: (1) increase in the respiration rate, (2) production of ethylene, (3) oxidative browning, (4) water loss, and (5) degradation of the membrane lipids (Brecht, 1995). These alterations will increase the susceptibility of the produce to decay and therefore are accompanied by an accumulation of metabolites (ethanol, lactic acid and ethyl acetate) and enzymatic oxidation of polyphenols. The microflora of vegetables include *Pseudomonas* spp., *Xanthomonas* spp., *Enterobacter* spp., *Chromobacterium* spp., yeasts and lactic acid bacteria.

Thus, attention must be focused on extending shelf life by maintaining quality and assuring food safety throughout the postharvest chain — harvesting, handling, packaging, storage, and distribution.

2 Issues in the postharvest chain for extending shelf-life

The most important tools to extend the shelf life in the postharvest chain of fresh-cut vegetables are: (1) proper sanitation, (2) refrigeration, and (3) adequate packaging.

In order to extend the shelf life by minimizing the microbial contamination level it is important to control (1) the initial microbial load of the product and (2) the field sanitation (fertilizers, water, personal hygiene, cleanable containers or bins) and (3) to implement good manufacturing practices in the processing plant (handler sanitary training, positive air pressure, sanitized and well-designed equipment). The good quality of raw products also depends on the cultivar or variety. Varieties that give the most juicy product are not appropriate for shelf lives of several days (Ahvenainen, 1996). In ready-to-eat-salads, it is of special interest to include vegetables that have antimicrobial activity (e.g., red chicory) (Guerzoni et al., 1996). The cutting method (cut, tear, or shred) influences the product respiration rate (Bolin and Huxsoll, 1991). Garg et al. (1990) concluded that shredders are a major source of contamination. Sharpness of blade, cut method (chop or tear), thickness, along with centrifugation, washing, initial microbial load, chlorine and sulphur dioxide were shown to have positive influence on storage life of shredded lettuce (Bolin et al., 1977).

Chemical and physical treatments may be used to assure food safety. Chlorination is a common practice in fresh-cut products and as a disinfectant controls postharvest decay. The removal of the excess water after chlorination, to avoid microbial contamination, may be done by centrifugation or air drying. Water removal may also be achieved by passing over vibrating screens, blotting or air blast drying (Bolin and Huxsoll, 1991). But care has to be taken when applying these treatments, because they also have negative effects on the product quality (loss of cellular fluids and desiccation). Guerzoni et al. (1996) concluded that the effectiveness of the washing operation (quality of the water used to wash, the delay between the washing and cutting phases and the level of active chlorine in the water), plays an important role in the reduction of the microbial spoilage species. Warm water has a negative influence on the storage life of shredded lettuce (Bolin et al., 1977). Restrictions to the use of chlorine are nowadays being considered due to the formation of trihalomethanes, which are carcinogenic compounds. Ozone and irradiation are under review as alternatives to the use of chlorine (Hurst, 1995).

Temperature management is doubtless one of the most important aspects of maintaining the quality and assuring the safety of fresh-cut products. Low temperatures decrease respiration rate and enzymatic and microbial activity.

Another effect of low temperature is the reduction of water loss from the product, which is particularly important in fresh-cut products, as these products do not retain their natural transpiration barriers. Temperature abuse in the retail display is one concern that can be overcome by the use of temperature-sensitive labels (Cameron et al., 1995).

Optimal storage temperature varies with the product. Temperatures around zero are the most usual. However, some vegetables, such as beans, cucumbers, okra, pepper, and tomatoes are sensitive to low temperatures, and their optimum temperature is around 10°C (Saltveit, 1997).

Packaging is the other key issue in the storage and distribution chain. Packing the product has two different main functions: product quality and marketing issues. Packaging is essential in the protection of the product against the outside environment, avoiding mechanical damage and chemical or biologic contamination from the outside. It allows the control of water loss, which is an important symptom of loss of quality. Many flexible films with perforations are currently available in the market and used to prevent moisture loss without development of anaerobic conditions. On the other hand, the marketing issues include the convenience of the package, the identification of the product, the brand identity, and product information.

A very important function of packaging is the build-up and/or maintenance of an atmosphere that extends the product shelf life and this function brings us to the modified atmosphere packaging (MAP) concept.

3   Modified atmosphere packaging

MAP relies on modification of the atmosphere inside a package, achieved by the natural interplay between two processes: (1) the respiration of the product and (2) the transfer of gases through the packaging. The low levels of O_2 and the high levels of CO_2 in MAP reduce the produce respiration rate, with the benefit of delaying senescence, thus extending the storage life of the fresh produce (Kader et al., 1989). The benefits of MAP in fresh-cut vegetables were reported by several authors, such as Ballantine et al. (1988), Barriga et al. (1991), Barth et al. (1993), Kaji et al. (1993), Lee et al. (1996), and Watada et al. (1996).

The atmosphere concentrations recommended for preservation depend on the product. Recommended concentrations of O_2 and CO_2 for some intact and fresh-cut vegetables are presented in Figure 1 [data collected from Kaji et al. (1993), Saltveit (1993), Gorny, (1997) and Saltveit (1997)]. Data for fresh-cut vegetables are scarce in the literature. In general, fresh-cut products are more tolerant to higher CO_2 concentrations than intact products, because the resistance to diffusion is smaller (Kader et al., 1989). For example, lettuce is not tolerant to CO_2 but shredded lettuce can tolerate concentrations from 10 to 15% (Figure 1). Recommended concentrations also depend on the quality parameter under evaluation, the cultivar, the maturity stage at harvest, temperature and duration of storage. However, the extension of product shelf life by use of low O_2 and high CO_2 is not convenient for all products. Some

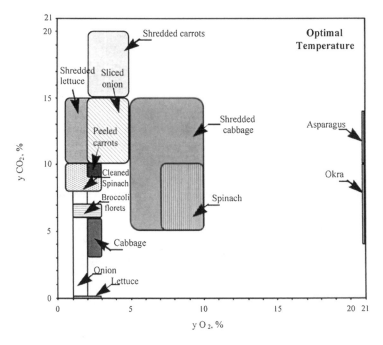

Figure 1 Recommended O_2 and CO_2 concentrations for preservation of intact and fresh-cut vegetables at optimal storage temperature ($0<T<5°C$, except okra: $7<T<12°C$).

vegetables, such as asparagus and okra, do not benefit from low O_2 levels (Figure 1). Other technologies currently being investigated are the use of atmospheres rich in O_2 that may also extend storage life of the produce (Amanatidou et al., 1998).

Polymeric films are the most usual packaging material. Due to the increase in the consumption of fresh-cut products, which are products with higher respiration and transpiration rates and higher tolerance to CO_2, alternative materials are being investigated. The perforation-mediated package is one such alternative, where the regulation of the gas exchange is achieved by single or multiple perforations or tubes that perforate an otherwise impermeable covering (Emond and Chau, 1990; Emond et al., 1991). The simultaneous use of polymeric film patches and perforations in the package is another alternative that can be explored.

Some potential advantages of a perforation-mediated package are (Fonseca et al., 1997):

1. The high values of mass-transfer coefficients, implying that a reduced size and number of perforations for gas exchange are required, thus high-respiring produce can be packed in this system.
2. MAP using perforations can be adapted easily to any impermeable container, including large bulk packages. Polymeric films are not

strong enough for packs much larger than those used for retail but perforations can be applied to retail packages as well as shipping boxes, because rigid materials can be used. Rigid packages also prevent mechanical damage of the product.

3. A flexible system is obtained due to the ability to change the gas transfer coefficients by selecting the adequate size and shape of the perforations.

4. Commodities requiring high CO_2 concentrations and relatively high O_2 concentrations can be packed with this system. Thus perforations have special interest in fresh-cut products.

But this type of package also shows some limitations (Fonseca et al., 1997):

1. Although it may be applied to products that could not be packed in conventional MA packages, the range of products is not very wide, as the CO_2/O_2 transfer coefficients ratio averages 0.8, the ratio of the diffusion values of CO_2 and O_2 in air. This may eventually be overcome by the use of perforations packed with materials with different affinities for O_2 and CO_2.

2. Water loss in the product may become a problem, but packed perforations may solve this limitation.

3. Nonuniformity of concentrations inside the package due to gas stratification may also be a problem in large containers.

MA packages should be carefully designed, as a system incorrectly designed may be ineffective or even shorten the storage life of the product. A good MA package should also minimize the time required to achieve the desired atmosphere. If the desired atmosphere is not established rapidly, the package has no benefit, and if O_2 and/or CO_2 levels are not within the recommended ranges of O_2 and CO_2 concentrations, the product may experience serious alterations, and its storage life is shortened. This technology may inadvertently lead to a product potentially hazardous to consumers. An outbreak of botulism in 1987 was suspected to be caused by coleslaw in MAP (Hurst, 1995). The use of predictive mathematical models is an efficient way to properly design a MA package. The important design variables may be divided in packaging factors, environmental factors and commodity factors. The commodity factors include: (1) the characteristics of the produce (e.g., bulk porosity, moisture content), (2) its mass, (3) the recommended atmosphere composition, and (4) the respiration rate and its dependence on temperature and atmosphere composition. The packaging factors include: (1) the package dimensions and shape, (2) the permeability of the package material to O_2, CO_2, and water vapor, and (3) the gas exchange mechanism. The environmental factors are (1) the storage temperature and (2) relative humidity, and (3) the composition of the atmosphere surrounding the package.

Modeling the respiration rate is important to predict the CO_2 production rate and the O_2 consumption rate as a function of the influencing variables. Because of the stress caused in the tissues, a peak in the respiration rate after wounding is expected. Thus, there should be an influence of time on the respiration rate. Additionally, history effects may be relevant. The other influencing variables are: CO_2 concentration, O_2 concentration and temperature. The respiratory quotient, the ratio between produced CO_2 to consumed O_2, is another important parameter to design MA packages. The models presented in literature relating respiration rate to O_2 and CO_2 are usually based on the Michaelis-Menten equation with different inhibitions for CO_2 (uncompetitive, noncompetitive, competitive, and both competitive and uncompetitive). The dependence on temperature is usually described by an Arrhenius type equation.

The gas exchange through a polymeric film can be modeled according to the Fick's law of mass transfer. The important variables are: (1) the permeability coefficient, (2) the surface area, and (3) the thickness of the film. The permeability coefficient varies with the temperature. The β ratio, the ratio between the CO_2 to O_2 permeability, is the other important factor and normally ranges between 4 and 6 in commercial polymeric films (Mannapperuma et al., 1989). The minimum β ratio for polymeric films found in literature was 2.2 for a hydrochloride film (Exama et al., 1993).

The gas exchange through a single perforation or multiple perforations can also be modeled according to Fick's law of mass transfer. Considering that the perforation is a tube with diameter (D) and length (L), the total flow of gas i (N_i) is proportional to the number of perforations (n), the difference in concentrations between the outside and the inside of the package ($y_i^e - y_i$), and the mass transfer coefficient (K_i), which depends on the tube diameter and length, and on the temperature:

$$N_i = nK_i\left(y_i^e - y_i\right) \tag{1}$$

It was earlier reported (Fonseca et al., 1997) that the dependence of the O_2 mass transfer coefficient on perforation diameter and length could be described by the following equation:

$$K_{O_2} = (9.12 \pm 3.53) * 10^{-6} * D^{(1.47 \pm 0.08)} * L^{(-0.55 \pm 0.03)} \tag{2}$$

$$9\,mm < D < 17\,mm;\, 6\,mm < L < 30\,mm$$

Temperature had no significant effect in the range 5 to 20°C and the β ratio was independent of the perforation dimensions and averaged 0.82±0.04.

Figure 2   Galega kale.

4   A case study

Shredded Galega kale is a fresh-cut vegetable of special interest in the Portuguese market. Galega kale *(Brassica oleracea var. acephala)* is a primitive cultivar that grows year round and is well adapted to impoverished soils and adverse climatic conditions (Figure 2). Galega kale represents an important contribution to the total production and consumption of vegetables in Portugal, representing 22% of total Brassica production. Cruciferous vegetables in Portugal represent nearly one quarter of the total Portuguese vegetable production with a per capita consumption of 65 kg annually (Portas and Costa, 1977), one of the highest in Europe. Galega kales were found to have higher levels of protein, calcium, and magnesium than other Brassica crops (Rosa and Almeida, 1996). The shredded Galega is used in a soup called "Caldo Verde," where the leaves are cut very thin (0.5 to 1.25 mm wide). The shredded leaves are sold in traditional stores and supermarkets with no special care with regard to sanitation, refrigeration, and packaging. There is almost no temperature management in the Portuguese distribution chain, thus the product is exposed to temperatures that can reach values of 20°C or higher. Shredded Galega kale is usually packed in trays covered by a flexible film or polymeric bags perforated or nonperforated (Beaulieu et al., 1997a). Condensation was visible in the commercial packs of shredded Galega kale and critical levels of O_2, which would lead to anaerobic respiration, were detected (Beaulieu et al., 1997b). Thus, these packages do not maintain quality and, even worse, are not health safe.

A case study will be presented to illustrate the use of a perforation-mediated MAP for preservation of shredded Galega kale.

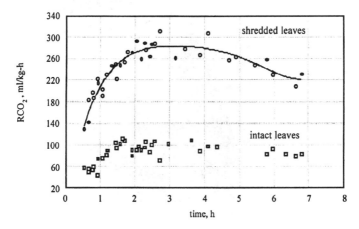

Figure 3 Production rate of CO_2 along time after shredding at 20°C.

In this case study, the evaluation of recommended concentrations, the modeling of the respiration rate for this fresh-cut product, and a simple methodology to define the adequate perforated package for shredded kale will be presented.

The evaluation of O_2 and CO_2 concentrations that extend this product's shelf life was not found in literature, because it is a product grown in a very specific region. Samples of shredded kale stored at different atmospheres were compared for sensorial attributes as perceived by the consumer, as well as color, chlorophyll content, and respiration rate. Preliminary experiments were conducted to evaluate the tolerance to low O_2 levels, maintaining the level of CO_2 constant. Samples in atmospheres with 1, 2, and 3% of O_2 or air with 0% CO_2 at 20°C were compared every day for up to 6 days. No injury and no anaerobic respiration were noticed even at the 1% O_2 level. The 1% O_2 level was also the one that maintained the quality for the longest time. A second set of experiments were then conducted to evaluate the tolerance to high CO_2 while maintaining the level of O_2 constant (21% O_2). The levels tested were: 10, 15, and 20% CO_2, and air was used as a control. The 20% CO_2 level gave the best results for preservation of the shredded kale. Finally, combinations of low O_2 and high CO_2 were assessed. The values tested were combinations of 1 and 2% O_2 and 15 and 20% CO_2 and no significant differences were found. From the results, we concluded that an atmosphere with 1 to 2% O_2 and 15 to 20% CO_2 at 20°C would be adequate for the storage of shredded kale, and the shelf life of the product increased from 2½ days at atmospheric air to 4½ days at this atmosphere at 20°C.

The respiration rate after cutting was also studied. The wounding of the cells is expected to affect the product metabolism, thus increasing the rate of the process that supplies energy to the cells: the respiration rate. A peak of CO_2 production after approximately 3 hours is clear in Figure 3. In this study we also observed that the respiration rate of shredded kale is approximately 2½-fold that of the intact leaves.

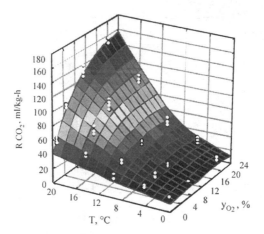

Figure 4 Fit of the Michaelis-Menten equation with noncompetitive inhibition of CO_2 to the experimental data and exponential dependence on temperature ($1<T<20$ °C and $0\%CO_2$).

Different models were tested to describe the respiration rate as a function of O_2 and CO_2 concentration and temperature, and it was found that the best fit was obtained with the enzymatic kinetics of Michaelis-Menten with non-competitive inhibition of CO_2 and an exponential dependence of the model parameters with temperature:

$$R_{CO_2} = \frac{5.036 * 10^{-6} \exp(0.116T) * y_{O_2}}{\left[0.00342\exp(0.118T) + y_{O_2}\right] * \left[1 + \dfrac{y_{CO_2}}{0.18943\exp(0.043T)}\right]} \quad (3)$$

Figure 4 relates CO_2 production rate with O_2 concentration for different temperatures and for a constant value of CO_2.

A methodology to design perforation-mediated modified atmosphere packaging was then applied based on a simplified model, assuming a constant free volume and no gas stratification inside the package, constant temperature and external atmosphere composition, no interactive effects between multiple perforations, and a constant respiration rate.

The integration of the differential equations of mass balance for O_2 and CO_2 in a perforation-mediated package at steady state leads to:

$$y_{O_2}^{eq} = y_{O_2}^{e} - \frac{R_{O_2}M}{nK_{O_2}} \quad (4)$$

$$y_{CO_2}^{eq} = y_{CO_2}^{e} + \frac{R_{CO_2}M}{nK_{CO_2}} \quad (5)$$

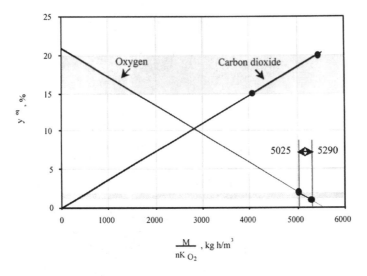

Figure 5 Plot for selecting the O_2 mass-transfer coefficient required for packing shredded Galega kale (the lines represent Equations (4) and (5); the recommended O_2 and CO_2 concentration range is indicated by the shadowed areas).

The concentrations at equilibrium are thus linear functions of $M/(nK_i)$, with a slope equal to R_i (Figure 5).

An example of a calculation can be given to clarify how these equations can be easily used by any producer as a package design methodology. Taking the case of a 2 kg pack of shredded Galega kale, we start by determining the optimum range of values of M/nKO_2 from Figure 5. The optimum range of values for the M/nKO_2 ratio, for simultaneously obtaining the recommended O_2 and CO_2 concentrations for shredded Galega kale, is controlled by the narrow range for O_2 and corresponds to 5025 to 5290 kg h/m³. From this range and from the equation relating the mass-transfer coefficients to the diameter and length of perforation (Equation (2)) it is easy to determine the minimum and maximum value of the tube length for obtaining the optimal concentrations for preservation, once the produce mass is specified. This can be done graphically from Figure 6, where for a perforation diameter of 10 mm, the minimum and maximum lengths can be easily determined for any value of the product mass up to 5 kg. For 2 kg, tubes from 13.8 to 15.1 mm long can be used. If the length chosen is above the lines defined, the levels of O_2 will decrease and the levels of CO_2 will increase more than the recommended levels, with risks of anaerobiose.

It is possible to use more than one tube. The length of n tubes that give the same effect as a single tube, can be determined from Figure 7. For example, when the length of a single tube ranges from 13.8 to 15.1 mm, the length of two tubes ranges from 49 to 53 mm (note that we are assuming that there are no interactions resulting from the use of multiple tubes).

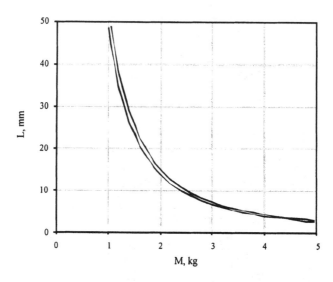

Figure 6 Plot for selecting perforation (tube) length (D = 10 mm) for packing 2 kg of shredded Galega kale (the lines result from the combination of Equations (2), (4), and (5)).

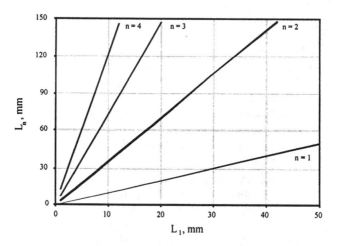

Figure 7 Plot for selecting perforation (tube) length for packing 2 kg of shredded Galega kale with multiple perforations.

It is also important to determine the free volume inside the package, so that steady state is reached within a reasonable time. The required free volume of the package can be calculated with the equations of the transient period:

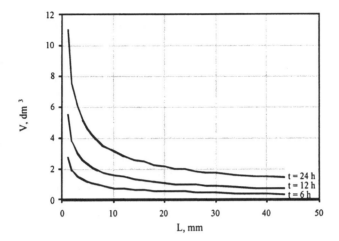

Figure 8 Effect of the perforation (tube) length of a single 10 mm diameter perforation (tube) on the allowable free volume inside the package, to guarantee that concentrations close to equilibrium (dimensionless concentration = 0.95) are reached in the indicated time period (the lines represent Equations (6) and (7)).

$$\frac{y_{O_2} - \left(y^e_{O_2} - \dfrac{R_{O_2}M}{nK_{O_2}}\right)}{y^o_{O_2} - \left(y^e_{O_2} - \dfrac{R_{O_2}M}{nK_{O_2}}\right)} = e^{-\frac{nK_{O_2}t}{V}} \tag{6}$$

$$\frac{y_{CO_2} - \left(y^e_{CO_2} + \dfrac{R_{CO_2}M}{nK_{CO_2}}\right)}{y^o_{CO_2} - \left(y^e_{CO_2} + \dfrac{R_{CO_2}M}{nK_{CO_2}}\right)} = e^{-\frac{nK_{CO_2}t}{V}} \tag{7}$$

For any specified equilibration time, the free volume can be calculated from the length of a single tube that relates easily with the mass transfer coefficients, considering a dimensionless concentration near equilibrium (and not equilibrium because the transient equation does not have a real solution for a dimensionless concentration of 0). The free volume inside the package can be determined from Figure 8 using the range of lengths of a single tube. For the example of a perforation diameter of 10 mm, the free volume ranges between 2.5 and 2.6 dm³ for achieving these concentrations in 24 hours (Figure 8).

Shredded kale is however a material with a very large bulk porosity. Without pressing the kale, its bulk porosity is 80% or even greater. This means that 2 kg of shredded kale imply a free volume of 8 dm³ simply due to its bulk porosity. For packages with smaller free volumes, the kale would have to be significantly compressed, which would have many detrimental effects on its quality. Figure 8 shows that for a free volume of 8 dm³ an equilibration time much greater than 24 h is needed and this would be undesirable.

Although a perforation-mediated package is suitable for storing shredded Galega kale under MAP, it would be advantageous to reduce the time required to reach the equilibrium atmosphere composition. It is therefore advisable to flush the package with the required gas concentration when packing.

5   Conclusions

As demand for fresh-cut vegetables continues to increase, application of HACCP plans, maintenance of low temperatures, and improved technology in packaging throughout the storage and distribution chain must be a constant concern of all intervenients in the chain. Perforation-mediated MAP presents an interesting potential as these products require high CO_2 concentrations, but more research is needed in this field.

Nomenclature

D   = tube diameter, m
K_i = gas i mass transfer coefficient, m³/s
L   = tube length, m
M   = mass of produce, kg
N_i = flow of gas i, m³/s
n   = number of tubes
R_i = consumption/production rate, m³/(kg s)
RQ = respiratory quotient, dimensionless
t   = time, s
T   = temperature, °C
V   = free volume inside package, m³
y_i^e = outside volumetric gas i concentration, dimensionless
y_i = volumetric gas i concentration, dimensionless
y_i^0 = initial volumetric gas i concentration, dimensionless
y_i^{eq} = volumetric gas i concentration at equilibrium, dimensionless

Subscripts

i   = O_2 or CO_2

Acknowledgments

The first author acknowledges financial support from PRAXIS XXI program, Portugal.

References

Ahvenainen, R. (1996). New approaches in improving the shelf life of minimally processed fruit and vegetables. *Trends in Food Science and Technology*, 7, 179–187.

Amanatidou, A., Smid, E.J., and Gorris, L.G.M. (1998). High O_2 modified atmosphere packaging, a novel approach with minimally processed vegetables. In: *Proceedings of the Third Main Meeting*, Project Process Optimization and Minimal Processing of Foods. Vol. 5 — Minimal and Combined Processes, Oliveira, J.C., Oliveira, F.A.R., and Gorris, L.G.M. (Eds.), published by Escola Superior de Biotecnologia, Porto, Portugal.

Ballantine, A., Stark, R., and Selman, J.D. (1988). Modified atmosphere packaging of shredded lettuce. *International Journal of Food Science and Technology*, 23, 267–274.

Barriga, M.I, Trachy, G., Willemot, C., and Simard, R.E. (1991). Microbial changes in shredded iceberg lettuce stored under controlled atmospheres. *Journal of Food Science*, 56 (6), 1586–1588.

Barth, M.M., Kerbel, E.L., Broussard, S., and Schmidt, S.J. (1993). Modified atmosphere packaging protects market quality in broccoli spears under ambient temperature storage. *Journal of Food Science*, 58, 1070–1072.

Beaulieu, J.C., Oliveira, F.A.R., Fernandes-Delgado,T., Fonseca, S.C., and Brecht, J.K. (1997a). Fresh-cut kale: quality assessment of Portuguese store-supplied product for development of a MAP system. In: *Proceedings of the 7th International Controlled Atmosphere Research Conference*, Vol. 5, Gorny, J.R. (Ed.), University of California Davis, U.S., p. 145–151.

Beaulieu, J.C., Oliveira, F.A.R., Fonseca, S.C., Fernandes-Delgado, T., and Poças, M.F. (1997b). A tecnologia pós-colheita como factor essencial no aproveitamento do potencial dos produtos hortofrutícolas nacionais. In: *Livro de actas do 3º Encontro de Química de Alimentos*, Faro, Portugal. In Portuguese. p. 17–20.

Bolin, H.R. and Huxsoll, C.C. (1991). Effect of preparation procedures and storage parameters on quality retention of salad-cut lettuce. *Journal of Food Science*, 56 (1), 60–67.

Bolin, H.R., Stafford, A.E., King, Jr. A.D., and Huxsoll, C.C. (1977). Factors affecting the storage stability of shredded lettuce. *Journal of Food Science*, 5, 1319–1321.

Brecht, J.K. (1995). Physiology of lightly processed fruits and vegetables. *HortScience*, 30, 18–21.

Cameron, A.C., Talasila, P.C., and Joles, D.W. (1995). Predicting film permeability needs for modified atmosphere packaging of lightly processed fruits and vegetables. *HortScience*, 30, 25–34.

Emond, J.-P. and Chau, K.V. (1990). Use of perforations in modified atmosphere packaging. *American Society of Agricultural Engineers*, Paper No. 90, 6512.

Emond, J.-P., Castaigne, F., Toupin, C.J., and Desilets, D. (1991). Mathematical modelling of gas exchange in modified atmosphere packaging. *American Society of Agricultural Engineers*, 34 (1), 239.

Exama, A., Arul, J., Lencki, R.W., Lee, L.Z., and Toupin, C. (1993) Suitability of plastic films for modified atmosphere packaging of fruits and vegetables. *Journal of Food Science*, 58 (6), 1365–1370.

Fonseca, S.C, Oliveira, F.A.R., Chau, K.V., and Brecht, J.K. (1997). Modelling the effects of perforation dimensions and bed porosity on gas exchange in perforation-mediated modified atmosphere packaging. In: *Proceedings of the 7th International Controlled Atmosphere Research Conference*, Vol. 5, Gorny, J.R. (Ed.), University of California Davis, U.S., p. 77–82.

Garg, N., Churey, J.J., and Splittstoesser, D.F. (1990). Effect of processing conditions on the microflora of fresh-cut vegetables. *Journal of Food Protection*, 53, 701–703.

Gorny, J.R. (1997). A summary of CA and MA requirements and recommendations for fresh-cut (minimally processed) fruits and vegetables. In: *Proceedings of the 7th International Controlled Atmosphere Research Conference*, Vol. 5, Gorny, J.R. (Ed.), University of California Davis, U.S., p. 30–66.

Guerzoni, M.E., Gianotti, A., Corbo, M.R., and Sinigaglia, M. (1996). Shelf-life modelling for fresh-cut vegetables. *Postharvest Biology and Technology*, 9, 195–207.

Hodge, K. (1995). Fresh-cut and the Perfect Meal. *Fresh-Cut*, 3 (8), 12–20.

Hurst, W.C. (1995). Sanitation of lightly processed fruits and vegetables. *HortScience*, 30, 22–24.

Kader, A.A., Zagory, D., and Kerbel, E.L. (1989). Modified atmosphere packaging of fruits and vegetables. *CRC Critical Reviews in Food Science and Nutrition*, 28, 1–30.

Kaji, H., Ueno, M., and Osajima, Y. (1993). Storage of shredded cabbage under a dynamically controlled atmosphere of high O_2 and high CO_2. *Bioscience, Biotechnology and Biochemistry*, 57, 1049–1052.

Lee, K.S., Park, I.S., and Lee, D.S. (1996). Modified atmosphere packaging of a mixed prepared vegetable salad dish. *International Journal of Food Science and Technology*, 31, 7–13.

Mannapperuma, J.D., Zagory, D., Singh, R.P., and Kader, A.A. (1989). Design of polymeric packages for modified atmosphere storage of fresh produce. In: *Proceedings of the 5th International Controlled Atmosphere Research Conference*, Vol. 2, Wenatchee, WA, U.S., p. 225–233.

McDonald, R.E., Risse, L.A., and Barmore, C.R. (1990). Bagging chopped lettuce in selected permeability films. *HortScience*, 25, 671–673.

Portas, C.M. and Costa, P.C. (1977). Produção, comercialização e consumo hortícolas em Portugal Continental. In: *Actas do I Colóquio Nacional de Horticultura e Floricultura*, Lisboa, Portugal. p. 9–30. In Portuguese.

Rosa, E. and Almeida, D. (1996). The influence of growing season on protein and mineral content of several Brassica crops. *Acta Horticulturae*, 407, 261–268.

Saltveit, Jr. M. E. (1993). A summary of CA and MA requirements and recommendations for the storage of harvested vegetables. In: *Proceedings of the 6th International Controlled Atmosphere Research Conference*, Ithaca, NY.

Saltveit, Jr. M. E. (1997). A summary of CA and MA requirements and recommendations for harvested vegetables. In: *Proceedings of the 7th International Controlled Atmosphere Research Conference*, Vol. 4, Saltveit, M.E. (Ed.), University of California Davis, U.S., p. 98–117.

Schlimme, D.V. (1995). Marketing lightly processed fruits and vegetables. *HortScience*, 30, 15–17.

Watada, A.E., Ko, N.P., and Minott, D.A. (1996). Factors affecting quality of fresh-cut horticultural products. *Postharvest Biology and Technology*, 9, 115–125.

Index

A

Acidity, 29, 36
Actinidia chinensis, 303
Activation energy, 87
Aerobic growth, 315, 316
Aerobic plate counts (APC), 378
Aggregation, polymolecular, 24
Air velocity, 156, 231, 236, 237
Air-blast freezing, 9, 145, 146
Amino acids, 327, 359
a-Amylase, 17, 20, 23, 273
Amylopectin, 116
Antimicrobials, 250, 358
Antioxidants, 358, 363
APC, see Aerobic plate counts
Apples, recent advances in drying of,
    229–248
    drying kinetics, 232–241
        drying under constant conditions,
        232–234
        drying under variable conditions,
        234–241
    properties of dried material, 241–243
Apricot, 291
Argentinus silus, 165
Aroma, 2, 98
Arrhenius model, 73
Ascorbic acid, 168, 255, 382
Aseptic packaging, 83
Aseptic processing, 77
Asparagus, 92, 393
Aspartame, 291
Aspergillus flavus, 284, 285

B

Baby foods, 68
Bacillus
    amyloliquefaciens, 19, 25, 274, 274
    cereus, 284, 333
    licheniformis, 19
    species, 17, 29, 35
    stearothermophilus, 84, 86
    subtilis, 19, 260, 273, 274, 282

Bacteria, 68
    common spoilage, 384
    from exponential growth phase, 313
    Gram(+), 285, 333
    grown with, 317
    lactic acid, 285, 286, 327
    spore-forming, 6
Bacteriocins, 333
Beans, 138, 207, 392
Beeswax, 363, 364
Bioconversion rates, 253
Biomaterials science, application of to quality
        optimization of frozen foods,
        107–130
    application to frozen foods, 114–122
        complexity of food matrixes, 116–117
        deteriorative phenomena, 117–122
        glass transition in frozen food
            products, 114–115
    basic concepts, 109–114
        experimental measurement of T_g,
        112–113
        factors affecting T_g, 111–112
        glass transition below freezing point,
        113–114
        glass transition and T_g, 109–111
    novel developments in analysis of T_g in
            frozen foods, 122–126
        application to determination of T_g in
            frozen sugar solutions, 125–126
        dynamic mechanical thermal analysis,
        122–123
        measurement of T_g in DMTA tests,
        123–125
Biopolymers, 7, 250
Biotechnology, 127
Blackberries, 251
Blanching, 139, 208, 255
Blueberry, 291
Boundedness, 54
Browning
    nonenzymatic, 83, 120
    oxidative, 390
Brussels sprouts, 158, 159

C

Cabbage, 205, , 379
Candying, 190
Capillary effects, 343, 344
Capillary forces, 187
Carbonylhemoglobin, 272
Carboxymethyl cellulose, 168
Carnauba wax, 363
Carrots, 138, 205, 211
Case hardening, 211, 221
Casein, 362
Caseinates, 164
Cauliflower, 205
Celery, 206
    prisms, 142
    raw, 139
    roots, 138, 140
Cell
    cultures, 128, 262
    wall damage, 142
Cellulose, 361
Cereal products, 101
CFD, see Computational fluid dynamics, for
        thermal food process optimization
Cheese(s), see also High-pressure treatment,
        of fruit, meat, and cheese products
    Gouda, 290
    microorganisms determined in, 286
    ripened, 282, 285, 288
Chenopodium rubrum, 262
Cherry, 291
Chicken, 103, 159
    product, breaded, 158
    smoke-cured, 362
Chloramphenicol, 314, 320, 321
a-Chymotrypsin, 276
Chromobacterium, 390
Cichorium endivia, 375
Citric acid, 168
Citrus aurantium, 303
Citrus fruits, 302
Clostridium
    botulinum, 19, 29, 73, 330
    sporogenes, 84, 85, 86, 88
Coalescence, 118
Coliforms, 287
Color, 2, 7, 309
Commercial sterilization, 68
Computational fluid dynamics (CFD), for
        thermal food process
        optimization, 41–66
    examples in thermal food processing,
        59–61
        calculation of local surface heat
        transfer coefficients, 60–61

        forced convection oven, 59–60
    numerical solution procedures, 53–58
        discretization, 53–56
        solution of discretized equations,
            56–58
    physical background, 44–53
        additional features of CFD, 46–52
        additional information, 44–45
        conservation equations, 44
        initial and boundary conditions,
            52–53
Computer simulation techniques, 101
Conservativeness, 54
Consumer
    demands, 2, 15, 194
    panels, 117
Consumption requirements, 136
Convection, 55, 59
Cooking
    chamber, 59
    cold, 335
    sous-vide, 325, 329
Cooling, 33, 93
    fast, 110, 113
    forced convection, 49
    media, 149
Copernicus Project, 4
Corn syrup solids (CSS), 178, 181
Cowpeas, 208
Creeping flow, 48
Cryogenic freezing, 117, 145
Cryoprotectants, 6, 117, 163, 167, see also
        Europe, developments in fish
        freezing in, with emphasis on
        cryoprotectants
Crystal melting, 125
Crystal nucleation, 126
Crystallization, 118, 126
CSS, see Corn syrup solids
Cucumbers, 392

D

DAC, see Diamond anvil cell
Dairy ingredients, 164, 168
Data analysis, 21, 26
DDM, see Dynamic dispersion medium
DE, see Dextrose equivalent
Defrosting, 153
Denaturation, 271
Density, 194
Design freedom, 104
Dextrose equivalent (DE), 178, 181
Diamond anvil cell (DAC), 273
Dielectric heating, 100

Differential scanning calorimetry (DSC), 110, 123, 125
Diffusivity, 119
Dihydrofolate reductase, 321
Dipole heating, 100
Disaccharides, 167
Discretization, 53, 56, 58
DMTA, see Dynamic mechanical thermal analysis
Drying, 3, 177
    conditions, sudden changes in, 229
    parameters, 230
    processes, new, 202
    rates, 191
    techniques, 210
    times, 241
DSC, see Differential scanning calorimetry
Dynamic dispersion medium (DDM), 145, 149, 160
Dynamic mechanical thermal analysis (DMTA), 113, 122, 124

E

Eddy viscosity, 50
Edible coatings, 325, 332–333
Electric conductivity, 99
Electric heating methods, minimal processing of foods with, 97–105
    comparing frequencies, 103–104
    electric resistance/ohmic heating, 99–100
        applications, 100
        equipment, 99
        fundamentals, 99
    electric volume heating method for foods, 98–99
    high-frequency heating, 100–101
        applications, 101
        equipment, 101
        fundamentals, 100
    microwave heating, 101–103
        applications, 103
        equipment, 102
        fundamentals, 101–102
    safety aspects, 104
    thermal heating approach to minimal processing, 98
Electron spin resonance (ESR), 112
Energy equation, 44
Enterobacter, 390
Enterococcus mundtii, 333
Enthalpy formulation, 44
Enzyme
    activation, 303

activity, 17, 302
    inactivation, 9, 22
Enzymic time–temperature integrators, 13–40
    application of enzymic time–temperature integrators for thermal process evaluation, 29–35
    determination of coldest zone in retort, 32–35
    materials and methods, 30–32
    development of enzymic isolated extrinsic time-temperature integrators, 19–29
    evaluation of integrating properties under variable temperature conditions, 25–29
    isothermal calibration of enzymic time–temperature integrators, 19–25
    feasibility of thermostable enzymes for TTI development, 17
    general aspects on time–temperature integrators, 15–17
    classification of time–temperature integrators, 16–17
    criteria for time–temperature integrator, 15–16
    definition of time–temperature integrator, 15
    methods for process impact assessment, 14–15
    research objective, 18–19
    pasteurization process, 18
    sterilization process, 18–19
Equilibrium
    concentration, 72
    constant, 90
    stages, 186
    status, 343
    time, 401
Escherichia coli, influence of culturing conditions on pressure sensitivity of, 313–324
    materials and methods, 314–315
    materials, 314
    methods, 314–315
    results and discussion, 315–323
    E. coli strain ATCC 11303, 315–319
    E. coli strain ATCC 39403, 319–323
ESR, see Electron spin resonance
Ethylene, 390
EU, see European Union
Europe, developments in fish freezing in, with emphasis on cryoprotectants, 163–174

cryoprotectants, 167–172
  dairy ingredients as cryoprotectants,
    168–169
  inclusion level and freezing rate,
    169–170
  sensory properties, 170
  tests in underutilized species, 170–172
freezing conditions/equipment, 165–166
fresh fish history, 164–165
multiple freezing, 167
storage conditions, 166
thawing conditions, 167
European Union (EU), 1

F

FDA, see Food and Drug Administration
Fermentation, 286
Fick's law, 395
Fish, 103
  fresh, 164
  species, enzymatic ripening of pelagic, 7
Flair-Flow Europe dissemination project, 2
Flavor, 6, 74, 245, 358
FLP, see Food liquid phase
Fluidization, 152
Food and Drug Administration (FDA), 89
Food
  engineering, 219
  liquid phase (FLP), 184
  poisonings, 288
  preservation, 14, 253
  science, research programs in, 294
  solid matrix (FSM), 184
Food packaging or coating, edible and
        biodegradable polymeric
        materials for, 357–371
  development of biodegradable and edible
        packaging materials, 359–362
    biodegradable synthetic copolymers
        and composites, 359–360
    blends of natural and synthetic
        polymers, 360
    starch-based edible polymer blends,
        360–362
  development of edible coatings for high-
        moisture foods, 362–366
    active edible coatings, 365–366
    properties of edible coatings, 363–364
    specific requirements for edible
        coatings, 362–363
    types of edible coatings, 364
Fourier transform infrared (FTIR)
        spectroscopy, 270, 277

Fragaria ananassa, 303
Free-volume theory, 110
Freezing, 3
  air-blast, 145, 146
  in bulk, 139
  cryogenic, 117, 145
  damage, 133
  equipment, FGP rotary, 166
  multiple, 163, 167
  point, 113
  rapid, 137
  rates, 9, 169
  time, prediction of, 146, 154
Freezing, influence of frozen storage time
        and, on changes in plant tissue,
        131–143
  changes of texture and other alterations
        in fruits, 133–136
    other fruits, 136
    raspberries, 133–134
    strawberries, 134–136
  effects of freezing and frozen storage on
        texture and other properties of
        vegetables, 137–142
    other vegetables, 142
    umbellifers, 138–142
    vegetable legumes, 138
Freezing, intensification of, 145–162
  comparison of sensory quality of range of
        foods frozen by air-blast and
        cryogenic processes, 157–160
  intensification of freezing using dynamic
        dispersion medium, 146–154
    development and evaluation of air-
        blast freezing, 146–149
    dynamic dispersion medium, 149–154
  modeling approaches to prediction of
        freezing times, 154–157
    factors affecting validation of
        predictive models for freezing,
        155–157
    results, 157
Fresh-cut vegetables, see Modified
        atmosphere packaging (MAP),
        development of for fresh-cut
        vegetables
Fructose, 127
Fruit(s), see also, High-
        pressure/temperature treatments,
        for quality improvements of fruit-
        derived products; High-pressure
        treatment, of fruit, meat, and
        cheese products
  citrus, 302
  osmotic dehydration of, 7

properties, 346
purees, hot filled pasteurized, 77
storage of, 7
FSM, see Food solid matrix
FTIR spectroscopy, see Fourier transform
  infrared spectroscopy
Fungicides, 363

G

Gamma irradiation, 7
Gas stratification, 398
Gauss's theorem, 53
Gel
  compression values, 172
  formation, 273
Gelatinization, 116
Glass transition, 110, 113, 123, 272
Glassy state, 110
Glucose, 127
Glycerol, 111, 167, 350
Gram(+) bacteria, 285, 333
Gram(–) bacteria, 285
Green beans, 138, 207
Green peppers, 207
Grid generators, 63

H

HACCP, see Hazard analysis and critical
  control point
Hazard analysis and critical control point
  (HACCP), 84
HDM, see Hydrodynamic mechanism
Heat
  distribution, 6, 76
  exchangers, tubular, 6
  -transfer, 69
    coefficient, 43, 146, 149
    theory, 154
Heating
  dielectric, 100
  dipole, 100
  electric, 98, 99
  high-frequency, 97, 100
  microwave, 101
  ohmic, 97, 328
  vacuum, 98
High-frequency heating, 100
High-pressure processing (HPP), process
  assessment of, 249–267
  advantages of high hydrostatic pressure
    application, 250–252
  challenges for high pressure R&D in food

science and technology, 259–264
  interactions between food
    components and high pressure,
    263–264
  microbial morphology, 261–262
  plant cell culture model systems,
    262–263
  technical/engineering challenges, 264
  opportunities of high hydrostatic
    pressure, 252–259
  membrane permeabilization, 257–259
  modifications of foods and related
    substances, 257
  opportunities for high-pressure
    processing of foods, 253–256
  phase transition in food systems, 257
  preservation of food and related
    substances, 252–253
High-pressure/temperature treatments, for
    quality improvements of fruit-
    derived products, 301–312
  materials and methods, 303–305
    experimental design, 304–305
    high-pressure equipment and
      treatments, 303–304
  effects of combined high
    pressure/temperature treatments
    on enzyme activities, 305–307
  effects of combined high-
    pressure/temperature treatments
    on microbiologic quality, 308–309
  quality and stability of pressurized fruit-
    derived products, 309–310
High-pressure treatment, of fruit, meat, and
    cheese products, 281–300
  comparison of influence of high-pressure
    treatment on survival of *Listeria
    monocytogenes* in minced meat,
    sliced, cured ham, and ripened
    sliced cheeses, 288–290
  conclusions, 290
  experimental details, 289
  results, 289–290
  effect of high pressure on selected strains
    of lactic acid bacteria, raw cow's
    milk, and ripening cheeses,
    285–288
  conclusions, 287–288
  experimental details, 285–286
  results, 286–287
  effect of ultrahigh pressure on vegetative
    microorganisms and spores of
    chosen bacteria and molds,
    283–285
  conclusions, 285

experimental details, 283
results, 284
high-pressure experimental techniques, 294–298
pressure medium, 297
pressure packaging, 298
pressure and temperature measurements, 297–298
pressure vessel, 297
principle of operation, 296
quality studies of high-pressure processed fruit products, 290–294
conclusions, 294
experimental details, 291–292
results, 292–293
High temperature short time (HTST), 69, 98
conditions, 84
kinetic studies, 92
principle, 76, 103
processes, 83
sterilization, 89
Hilum, 202
HPP, see High-pressure processing, process assessment of,
HTST, see High temperature short time
Hurdle(s)
minimal-thermal, 328
technology, 326–327, 336
Hydrodynamic mechanism (HDM), 183, 185

I

Ice
formation, 275
recrystallization, 115, 117
Immersion media, 218
Inlet boundaries, 52
In-pack pasteurization, 68
In-pack sterilization, 75
Ionizing radiation, role of in minimal processing of precut vegetables, 373–387
low-dose irradiation as potential intervention to control non-spore-forming pathogenic bacteria on packaged vegetables, 376–377
materials and methods, 378–379
estimation of sensorial shelf life, 379
irradiation treatment and postirradiation storage, 378
Listeria monocytogenes inoculum, 378
microbiological and vitamin C analyses, 378
pH measurements, 379

source and preparation of vegetables, 378
objectives, 377
occurrence and behavior of *Listeria monocytogenes* on prepacked vegetables, 375–376
results and discussion, 379–382
microbiological changes and sensorial shelf life, 379–382
pH of vegetable tissues, 382
vitamin C content, 382
Irradiation
gamma, 7
low-dose, 376, 379
treatment, 378

J

Jams, 255
Juice production, 7

K

Kale
Galega, 389, 396
shredded, 397, 402
Kinetic parameters, 22
Kinetic studies, data analysis of, 21
Kinetics
convective dehydration, 191
of drying process, 230
of enzyme systems in UHP, 7
first-order, 91, 221
HDM, 345
heat inactivation, 27
modeling, 78, 220
reaction, 72
thermal inactivation, 90
water sorption, 193
Kiwi fruit, 301, 306

L

LAB, see Lactic acid bacteria
Labor costs, 390
Lactic acid bacteria (LAB), 285, 327, 333, 390
Lactose, 127
Leafscale gulper shark, 171
Lethality distribution, 32, 35
Lettuce, chopped, 390
Listeria monocytogenes, 10, 29, 282, 288, see also Ionizing radiation, role of in minimal processing of precut

vegetables
Low Reynolds number (LRN), 50
LRN, see Low Reynolds number

M

MA, see Modified atmosphere
MAP, see Modified atmosphere packaging,
        development of for fresh-cut
        vegetables
Marmalades, 334
Mathematical modeling, 74
MCC, see Microcrystalline cellulose
Meat, see also High-pressure treatment, of
        fruit, meat, and cheese products
   lean beef, 289
   products, 7, 103
   raw, 290
Melon, 350
Metmyoglobin, 270, 275
MIC, see Minimal inhibitory concentration
Microbial assessment, quality assessment
        and, in thermal processing, 83–96
   microbiological assessment, 84–88
   quality assessment, 89–93
Microbial growth, 115, 118
Microbial morphology, 261
Microbial safety, 114
Microcrystalline cellulose (MCC)
Microflora, 333
Microorganisms
   fermentative, 331
   inactivation of, 259
Micropyle, 202
Microscopy, 132
Microwave
   energy, transfer of, 102
   heating, 77, 97, 101
   sterilization, 6
Milk
   contamination of, 287
   pasteurized, 288
   protein isolate, 168
   raw cow's, 285
   skimmed, 286
Minimal inhibitory concentration (MIC), 323
Minimal processing technologies, quality
        and safety aspects of novel,
        325–339
   concept of hurdle technology, 326–327
   future for hurdle technology, 336–337
   minimal-thermal hurdles, 328–330
      ohmic heating, 328–329
      sous-vide cooking, 329–330

non-thermal hurdles, 330–336
   biopreservation, 333–334
   edible coatings, 332–333
   high-pressure processing, 334–335
   high-voltage impulses, 335–336
   modified atmosphere packaging,
      330–332
potential hurdles and hurdle-preserved
      foods, 327–328
Model
   first-order, 90
   multifraction, 90
   Plank's, 155
Modeling
   kinetics, 78
   mathematical, 74
   process, 183
   research work on, 75
   structural changes, 222
Modified atmosphere (MA), see 332
Modified atmosphere packaging (MAP),
        development of for fresh-cut
        vegetables, 389–404
   case study, 396–402
   fresh-cut vegetables, 390–391
   issues in postharvest chain for extending
      shelf-life, 391–392
   modified atmosphere packaging, 392–395
Molds, 283
Momentum transfer, 71
Monosaccharides, 167
Mung bean sprouts, 331
Mushrooms, 207

N

NMR, see Nuclear magnetic resonance
Nonlinear regression, 22, 26
Nuclear magnetic resonance (NMR), 112, 122
Nutrient content, 73
Nutrition, 132

O

OD, see Osmotic dehydration, advances in
Off-odors, 331
Ohmic heating, 77, 97, 328
Okra, 392, 393
Onions, 207, 211, 220
OP, see Osmotic preconcentration
OS, see Osmotic solution
Osmotic dehydration (OD), advances in,
        175–199
   osmotic preconcentration as pretreatment

in convective drying, 190–194
general considerations, 190–191
impact of osmotic dehydration on
convective dehydration kinetics,
191–192
impact of osmotic dehydration on
final product properties, 192–193
impact of osmotic dehydration on
water sorption kinetics, 193–194
process analysis and modeling, 183–190
equilibrium stages and controlling
mechanisms, 186–188
scope and objectives of process
modeling, 183–184
system description, 184–186
volume and compositional changes,
188–190
process variables, 178–183
osmotic solution, 180–181
phase contacting, 181–182
process duration, 183
process pressure, 183
process temperature, 182
product identity, 178–179
product pretreatment, 179–180
research and development needs,
194–195
Osmotic preconcentration (OP), 189
Osmotic solution (OS), 189
Ovalbumin, 272
Oven
forced convection, 59
multimode cavity, 102
Oxidation, 116, 118
Oxygen, 315, 317, 361

P

Package design, 264, 399
Packaging, 298
aseptic, 83
materials, 2
perforated-mediated, 10
Pasteurization, 103
continuous, 75
in-pack, 68
of tropical fruit juice, 6
Pathogen(s), 375
cold-tolerant, 332, 337
non-spore-forming, 385
PDM, see Pseudo-diffusion mechanisms
Peaches, 251
Pears, 251
Peas, 159, 251

Pectin(s)
assay of, 134
content, of celery prisms, 142
methylesterase (PME), 138, 307, 310
Peppers, 207, 211
Peroxidase (POD), 302, 305, 310
Persil, 207
pH, 68, 85, 112
reduction, 253
of vegetable tissues, 382
Picard iteration, 57
Plank's model, 155
Plasticizer, 25, 111
Plum, 136
PME, see Pectin methylesterase
POD, see Peroxidase
Polymer
blends, 357, 359
synthetic, 360
Polymorphism, 112, 117
Polyols, 111
Polyphenoloxidase, 302
Polyphenol oxidase activity (PPO), 305, 310
Polysaccharides, 112, 216, 256
Porosity, 194, 212
Postharvest treatment, 252
Postirradiation storage, 378
Potato(es), 207
cell cultures, 263
high-pressure blanching of, 255
rapid cooking of, 100
tissues, 178
PPO, see Polyphenol oxidase activity
Prandtl number, 50
Preservation, 326
Pressure
boundary, 52
inactivation, 317, 318
medium, 297
packaging, 298
Process
design, 75
duration, 183
temperature, 182
Process optimization, minimal processing
and, 1–11
activities, 10–11
dissemination, 11
program of short stays, 11
workshops, 11
participants and management, 3–5
plugged-in research projects, 6–10
overview, 6–8
topical issues in project, 8–10
Product(s)

formulation, 210
pretreatment, 179
properties, analysis of, 245
shelf-stable, 252
stability, 194
Protease inactivation, 287
Protective coatings, 332
Protein(s), 116
coagulation, 98
extractability, 165
gelling behavior of, 257
stability diagrams of, 271
synthesis, 323
Pseudo-diffusion mechanisms (PDM), 185
Pseudomonas species, 374, 390
Pulsed vacuum osmotic dehydration
(PVOD), 184
Purees, 255
PVOD, see Pulsed vacuum osmotic
dehydration

Q

Quality
factors, 89, 90, 91
losses, 254
Quenching, 109

R

Radish, 377, 381
Rancidity, reducing, 362
RANS equations, see Reynolds averaged
Navier-Stokes
Raspberries, 133, 215, 251
Raw sake, 255
Reaction
kinetics, 72
rate, 119
Recalcitrant packaging polymers, 358
Recrystallization, 115, 126
Rehydration, 202, 229
capacity, 205–208, 214–215
effect of air constant velocities on, 242
factors affecting, 204
rate, 219
ratio, 213
Rehydration, of dried plant tissues, 201–227
factors affecting rehydration, 204–218
extrinsic factors, 217–218
intrinsic factors, 204–217
modelling rehydration processes,
218–223
kinetics modeling, 220–222

mechanisms of rehydration, 218–219
structural changes modeling, 222–223
Relaxation phenomena, 217
Renormalization group (RNG), 51
Residence time distribution (RTD), 71, 72
Retort(s)
coldest zone in, 32
rotary, 77
temperature, 33
water-cascading, 34
Reynolds averaged Navier-Stokes (RANS),
49, 51
Ribonuclease, 24
RNG, see Renormalization group
RTD, see Residence time distribution

S

Safety
criteria, 7
microbial, 7, 114
Salmonella enteritidis, 283
Salt, 168, 180, 353
Sanitation, 264
Scanning electron microscopy (SEM), 140,
159
Sea urchin paste, 255
SEM, see Scanning electron microscopy
Sensory
analysis, 74
evaluation, 159, 292
properties, 73, 170, 243
quality, 9, 114, 157
Shape retention, 243, 245
Shelf life, 391
evaluation of, 6
sensorial, 379
Shrinkage, 194, 211, 231
Solvent, role of, 272
Sorbic acid, 365
Sorbitol, 168
Sous-vide cooking, 325, 329
Spinach, 207, 211, 390
Spores, heat resistance of, 85
Staphylococcus aureus, 283
Starch, 116
denaturation in, 271
description of, 109
gelatinization, 71
properties of, 6
swelling, 98
synthesis, 137
syrup, 191, 193
Sterilization, 18, 68

continuous, 75
heat, 84
HTST, 89
in-pack, 75
Strawberry(ies), 136, 158, 159, 215, 301
desserts, physicochemical parameters of,
293
osmotic pretreatment of, 191
pasteurization of, 258
purees, 256
Streptococcus faecalis, 29
Stress relaxation tests, 349
Sucrose, 127, 193
Sugars, 112, 117
crystallization of amorphous, 115
forms of, 116, 193

T

TBARS, see Thiobarbituric acid substances
TDT, see Thermal Death Time
Temperature effects, comparison between
pressure effects and on food
constituents, 269–280
case studies, 274–277
pressure-assisted cold denaturation of
metmyoglobin, 275–276
pressure effects on emulsions and
inverted micelles, 276–277
pressure-induced gelation of starch,
277
stability diagrams of amylases,
274–275
in situ observation of protein
denaturation, aggregation, and
gel formation, 274
pressure compared with temperature
effects, 271
stability diagrams of proteins, 271–273
kinetics of denaturation, 272
role of solvent, 272–273
thermodynamics of denaturation,
271–272
Texture, 2, 74, 91
changes of, 92, 133
characteristics, 133
destruction, 93
preservation of, 132
softening, 98
Thawing conditions, 163, 167
Thermal Death Time (TDT), 21, 73
Thermal diffusivity, 351
Thermal processing conditions,
methodologies to optimize, 67–82

process design, 75–76
process modelling, 69–75
reaction kinetics, 72–74
thermal processing applications,
74–75
transport phenomena, 69–72
process optimization, 76–78
experimental validation, 77
process assessment, 77–78
Thermobacteriology, 6
Thermomechanical analysis (TMA), 113
Thermoresistance, 90
Thiobarbituric acid substances (TBARS), 165
Time–temperature integrators (TTIs), 14, 78
development, 17
enzymic, 29
major advantage of, 15
as monitoring devices, 30
TMA, see Thermomechanical analysis
TMAO, see Trimethylamine oxide
Tomato, 207, 211, 301
Transportiveness, 54
Transport phenomena, 69
Trehalose, 24
Trimethoprim, 321
Trimethylamine oxide (TMAO), 166
TTIs, see Time–temperature integrators
Turbulent flow, 49

U

UHP, see Ultrahigh pressure
Ultrahigh pressure (UHP), 283
Ultrasounds, 6

V

Vacuum
cooking, 329
heating, 98
Vacuum impregnation (VI), 188, 341–356
fruit properties as affected by VI, 346–351
changes in mechanical, structural, and
other physical properties, 349–351
feasibility of VI of fruits, 347–349
influence of VI on osmotic processes,
351–352
mathematical model of HDM, 342–346
equilibrium status, 343–345
HDM kinetics, 345–346
salting processes by using VI, 353
Valerolactone, 359
Vegetable(s), 327
fresh-cut, 390

frozen storage of, 131
MAP stored, 333
product, ready-to-use, 7
storage of, 7
VI, see Vacuum impregnation
Viscosity, 50, 119
brine, 31
of glassy matrix, 113
Vitamins, 327, 382
Voronoi diagrams, 222, 223

W

Water
activity, 68, 87, 108
-holding capacity (WHC), 112, 164
removal, 203
retention ability, 137
Wax coatings, 364
WHC, see Water-holding capacity
Whey protein concentrates, 164
Wigner-Seitz cells, 223
WLF equation, 119

X

Xanthomas, 390

Y

Yersinia enterocolitica, 336
Yogurt, 329

Z

z-value, 16, 18, 27, 36, 73